SEROTONIN AND THE CARDIOVASCULAR SYSTEM

Serotonin and the Cardiovascular System

Editor

Paul M. Vanhoutte, M.D.
Professor of Physiology and Pharmacology
Department of Physiology and Biophysics
Mayo Clinic
Mayo Foundation
Rochester, Minnesota

Raven Press ■ New York

Raven Press, 1140 Avenue of the Americas, New York, New York 10036

Made in the United States of America

Library of Congress Cataloging in Publication Data
Main entry under title:

Serotonin and the cardiovascular system.

Bibliography: p.
Includes index.
1. Cardiovascular system—Diseases—Etiology.
2. Serotonin—Physiological effect. I. Vanhoutte, Paul M.
[DNLM: 1. Cardiovascular Diseases—drug therapy.
2. Serotonin—pharmacodynamics. 3. Serotonin—therapeutic use. QV 126 S4848]
RC669.S47 1985 616.1′07′1 84-27688
ISBN 0-88167-101-0

The material contained in this volume was submitted as previously unpublished material, except in the instances in which credit has been given to the source from which some of the illustrative material was derived.

Great care has been taken to maintain the accuracy of the information contained in the volume. However, Raven Press cannot be held responsible for errors or for any consequences arising from the use of the information contained herein.

Preface

Ever since serotonin was discovered by Dr. Irvine Page and his colleagues (see Introduction), its role in health and disease has remained elusive. Indeed it can be characterized as the "parent pauvre" of the catecholamines and other neurohumoral mediators that have key functions in the regulation of the cardiovascular system. Since serotonin is present in the platelets and is released when these aggregate, it has the potential to affect blood vessels in any part of the body, either directly or by augmenting the response to other vasoactive agents. Certainly its actions on the components of the blood vessel wall are complex and the observed responses are the result of many interactions between cells. Interest in the potential role of serotonin was increased by the demonstration that in several cardiovascular diseases, the platelet turnover is exaggerated and that the aging and diseased blood vessel wall becomes hyperresponsive to serotonin. However, it was the advent of new potent serotonergic antagonists and the discovery that one of these, ketanserin, acutely lowers the arterial blood pressure in elderly hypertensive patients that gave credence to the concept of the involvement of serotonin in cardiovascular disease. This discovery provided the momentum for further studies on the subtypes of serotonergic binding sites and the multiple sites of action of serotonin in the cardiovascular system.

In view of the enhanced interest in serotonin, both in health and disease, it is timely to summarize current knowledge and to speculate on its potential role in cardiovascular function. Obviously, it is too early to reach final conclusions, as the points of view on many of the issues covered still diverge, sometimes considerably.

The first chapters of this monograph deal with the biochemistry, physiology, and pharmacology of serotonin and explain its effects on the different target cells that ultimately can alter cardiovascular function. A second series of chapters discusses the possible involvement of serotonin in hypertension and the potential therapeutic use of S_2-serotonergic antagonists as antihypertensive agents. The final part of the book deals with other types of cardiovascular disease in which serotonin could partly contribute to the symptoms.

As always with multiauthored texts, the responsibility for the scientific content rests with the individual authors. Hence, all the statements made are not necessarily endorsed by the editor. His task has been mainly to select the authors, to streamline their texts, to achieve as much as possible uniformity of presentation, and to avoid overt overlap and repetition.

Serotonin and the Cardiovascular System will be of interest not only to physiologists and pharmacologists puzzled by the complexity of the actions of serotonin and the difficulty of comprehending the consequences of serotonergic blockade, but also to clinical researchers and to the physicians who treat patients with cardiovascular diseases. Indeed, serotonergic inhibitors may well turn out to be a useful way of treating hypertension and vasospastic disorders.

Paul M. Vanhoutte

Acknowledgment

This monograph would not have been possible without the total dedication of Mr. Robert R. Lorenz who took the responsibility of the illustrations and of compiling the reference list. The invaluable administrative help of Mrs. Janet Beckman was instrumental in keeping the project within reasonable deadlines. The editor would like to thank most sincerely Dr. Paul A. Janssen and Dr. Jan Symoens, of the Janssen Research Foundation, who supported the endeavor to try to make this monograph into a reasonably uniform text, despite the many contributors. He also would like to thank the authors most sincerely for their collaboration and understanding when faced with extensive editing of their manuscript. Last, but not least, Ms. Rita Scheman and the staff of Raven Press should be complimented for a most efficient handling of the manuscripts.

Contents

Contributors

Håken Ahlman
Department of Surgery I
Institute of Neurobiology
University of Goteborg
S-413 45 Goteborg, Sweden

Richard A. Cohen
Robert Dawson Evans Department of Clinical Research
Boston University School of Medicine
Boston, Massachusetts 02215

Fred De Clerck
Laboratory of Haematology
Janssen Pharmaceutica Research Laboratories
B-2340 Beerse, Belgium

John A. Dormandy
St. James' and St. George's Hospitals
Sarsfeld Road
London SW12, England

Michael D. Gershon
Department of Anatomy and Cell Biology
Columbia University
College of Physicians and Surgeons
New York, New York 10032

C. Norman Gillis
Departments of Pharmacology and Anesthesiology
Yale University School of Medicine
333 Cedar Street
New Haven, Connecticut 06510

Ulf Haglund
Department of Surgery
University of Lund
Malmö General Hospital
S-214 01 Malmö, Sweden

Herbert B. Hechtman
Department of Surgery
Harvard Medical School
Brigham and Women's Hospital
75 Francis Street
Boston, Massachusetts 02115

Norman K. Hollenberg
Departments of Radiology and Medicine
Harvard Medical School and
Brigham and Women's Hospital
50 Binney Street
Boston, Massachusetts 02115

Holm Holmsen
Department of Biochemistry
University of Bergen
Arstadveien 19
N-5000 Bergen, Norway

W. J. Janssens
Department of Pharmacodynamics
Janssen Pharmaceutica
B-2340 Beerse, Belgium

Donald M. Kuhn
Section on Biochemical Pharmacology
National Institutes of Health
National Heart, Lung, and Blood Institute
Bethesda, Maryland 20205

J. E. Leysen
Department of Biochemical Pharmacology
Janssen Pharmaceutica
B-2340 Beerse, Belgium

Walter Lovenberg
Section on Biochemical Pharmacology
National Institutes of Health
National Heart, Lung, and Blood Institute
Bethesda, Maryland 20205

L. H. Opie
Ischaemic Heart Disease Research Unit
The Medical Research Council of South Africa
University of Cape Town
Observatory 7925
Cape Town, South Africa

Irvine H. Page
Box 516
Hyannis Port, Massachusetts 02647

Robert S. Reneman
Department of Physiology
University of Limburg
Maastricht, The Netherlands

S. Saman
Ischaemic Heart Disease Research Unit
The Medical Research Council of South Africa
University of Cape Town
Observatory 7925
Cape Town, South Africa

M. A. D. H. Schalekamp
Department of Internal Medicine I
Erasmus University
3015 Rotterdam, The Netherlands

James R. Seibold
Division of Rheumatology
University of Medicine and Dentistry of New Jersey
Academic Health Sciences Center
New Brunswick, New Jersey 08903

J. F. Stoltz
Centre Régional de Transfusion et d'Hématologie
Brabois, Avenue de Bourgogne
54500 Vandoeuvre-lès-Nancy, France

Hadassah Tamir
Department of Psychiatry
New York State Psychiatric Institute
New York, New York 10032

F. T. Thandroyen
Ischaemic Heart Disease Research Unit
The Medical Research Council of South Africa
University of Cape Town
Observatory 7925
Cape Town, South Africa

Gertrude M. Tyce
Department of Physiology
Mayo Clinic
Rochester, Minnesota 55905

Paul M. Vanhoutte
Department of Physiology and Biophysics
Mayo Clinic
Mayo Foundation
Rochester, Minnesota 55905

J. M. Van Nueten
Department of Pharmacodynamics
Janssen Pharmaceutica
B-2340 Beerse, Belgium

M. Verstraete
Center for Thrombosis and Vascular Research
Department of Medical Research
University of Leuven
Leuven, Belgium

R. Clinton Webb
Department of Physiology
University of Michigan
Ann Arbor, Michigan 48109

Carl P. Weiner
Department of Obstetrics and Gynecology
Division of Maternal Fetal Medicine
University of Iowa Hospitals and Clinics
Iowa City, Iowa 52242

William S. Wolf
Section on Biochemical Pharmacology
National Institutes of Health
National Heart, Lung, and Blood Institute
Bethesda, Maryland 20205

Introduction

The Neonatology of Serotonin

We are all part of history, but seldom know it until most of the action has subsided. We speak of "looking back" on history for lessons to guide the future. And so it has been with the isolation and neonatology of serotonin. Thirty-six years ago I participated along with Maurice Rapport and Arda Green in the discovery of serotonin (874). For a few years thereafter I knew everything there was to know about it. Still later I began to know progressively less and less about it. Dr. Green died and Dr. Rapport sought other pastures.

But the subject grew into adolescence at what could only be considered an alarming rate as reflected in an avalanche of papers I was unable to follow. Betty Twarog and I had found serotonin in brain and that simple observation did it! Psychiatrists, neurologists, neurophysiologists soon were in full cry, and we knew we had met our match, and must either give full attention to the field or be merely observers. I confess to regret at having to make the latter decision, since the "chemistry of the brain" was my first major love from my 1928–1931 period in Munich, Germany. History shows that when you are sidelined in research, you are forgotten unless you transform yourself into a disease, like Prinzmetal, Addison, Cushing, or Lou Gehrig—so the name, at least, is remembered.

But how did I become involved with serotonin in the first place? In the early 1930s I was searching for a humoral agent responsible for hypertension. The woods were full of vague suggestions and undefined physiological effects of extracts of plasma and organs. The most immediately disturbing was one that resulted when blood clotted. Plasma carefully prepared had no vasoconstrictor properties, but serum had. This clearly posed a problem for me. Any vasoconstrictor I isolated from blood of hypertensives would always be suspect, because one could never be sure some unseen coagulation had not occurred. Platelets and their secretions are not easily controllable. Before we could progress in the search for a pressor agent in hypertension it was imperative to control this elusive vasoconstrictor resulting from coagulation.

Very briefly, the substance was isolated and its structure identified (Fig. 1). Synthesis by the Upjohn Company followed later. We were made aware only then of the work of Erspamer, a gifted biochemist who had made extracts of stomach and intestine and found that they contracted uterus. Although no purification was achieved, the active substance was labeled "enteramine" and thought to be a "phenolic compound" on the basis of little evidence. I mention this because Erspamer,

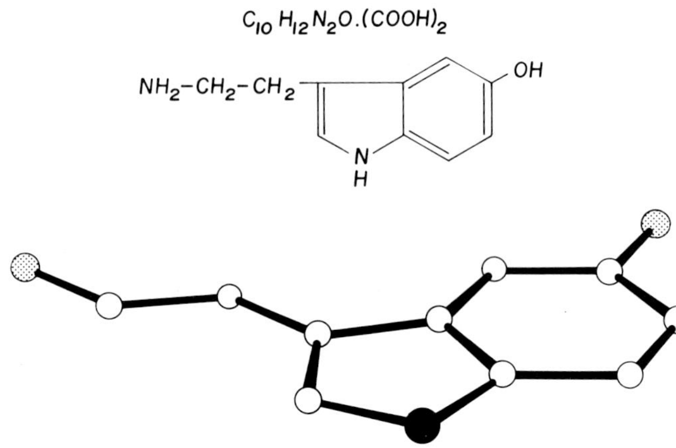

FIG. 1. Chemical structure of 5-hydroxytryptamine (serotonin). The perspective drawing is derived from X-ray crystallographic data (modified from ref. 1046).

unbeknown to us, helped to identify the problem, although he did not solve it. His later work in invertebrate biochemistry proved him to be a man of accomplishment, although he did nothing further with his enteramine except to show in 1952 that it was indeed 5-hydroxytryptamine (319).

Having launched serotonin, it became apparent to me within a year or so that I could not keep up my work in clinical and experimental hypertension and study serotonin as well. After writing reviews covering 153 papers in 1954 (816) and abstracting 530 more papers in 1958 (819), I summarized what I knew in 1968 in a book titled, as you might suspect, *Serotonin* (820).

The serotonin episode has taken about 35 years to develop. What I learned from it was that substances in the body that are nuisances to one person give tenure to others. Clearly, serotonin has been shown to be involved in such varied functions as neural transmission, the symptoms of carcinoid tumors, and endomyocardial fibrosis. Without doubt one of the happiest outcomes for older people like me is the result of experiments in rats showing that on either low or high doses of alcohol, memory becomes poor. Zimelidine, an inhibitor of serotonin reuptake, reversed this alcohol-induced impairment. So it was not Alzheimer's disease after all! If the serotonin neurotransmitter system mediates learning and memory in human beings and simultaneously determines the effects of alcohol, I must say I am impressed.

Serotonin, in short, has taken on a life of its own and we are no longer living together. I have no regrets, because a long life has taught me that the natural history of one's active participation in a discovery is about 5 to 10 years. Then the subject grows complicated, new very bright young faces appear with their better methods and they take over. If they are aware that anything preceded their own work, they give no indication of it—which is nature's way of preventing constipation of the mind.

Serotonin has been slow to grow up and I suspect it is about to enter adolescence and weed-like growth. The important thing now is to keep investigations moving ahead along creative lines. It is easy to fall heir to Brownian movement and perform like the vehicles described in an Arizona newspaper: "The motorized rail car provides service semi-bi-hourly and will run east and west simultaneously."

Irvine H. Page

— 1 —

Biochemistry of Serotonin

Gertrude M. Tyce

Department of Physiology, Mayo Clinic, Rochester, Minnesota 55905

DISTRIBUTION OF SEROTONIN

Serotonin has a widespread occurrence. In plants, it is present in edible fruits and vegetables but has also been associated with poisonous species (380). In the animal kingdom, serotonin can be detected both in vertebrates and in invertebrates. The important depots for serotonin in mammals are the enterochromaffin cells in gastrointestinal mucosa, the brain (where it is thought to be a neurotransmitter in certain neurons), the pineal gland, where it serves as a precursor of melatonin, and platelets. In addition, serotonergic neurons have been identified in the peripheral nervous system in the gastrointestinal tract in mice, rats, and guinea pigs (392). Significant amounts of serotonin have also been detected in other tissues of mammals, notably in heart, kidney, spleen, and thyroid, although the significance of the presence of serotonin in these tissues is not yet understood. A thorough review of the distribution of serotonin in tissues and body fluids of several mammalian species has been published by Essman (322).

Serotonin is synthesized in most of the tissues in which it is stored and has a rapid turnover in most of these. Half-lives as rapid as 10 to 17 hr have been reported for serotonin in the gastrointestinal tract (320,1061). However, serotonin is not synthesized in platelets, and the serotonin stored in platelets has its origin in other tissues, notably the gastrointestinal tract (320). The turnover of serotonin in platelets is relatively slow, the reported half-life being between 33 and 48 hr (1061).

Consistent with this rapid turnover of serotonin in most tissues, there are large amounts of 5-hydroxyindoleacetic acid, the major metabolite of serotonin, excreted in urine. Up to 9 mg of 5-hydroxyindoleacetic acid is excreted daily by normal individuals (1058). It is of interest that the amounts of 5-hydroxyindoleacetic acid that are excreted in urine are rather similar to the amounts of vanillylmandelic acid, the major metabolite of norepinephrine (1058). Only small amounts of serotonin (between 10 and 120 μg) are excreted in urine by normal individuals (588).

METHODS OF ASSAY OF SEROTONIN

Overview

Early methods for measuring serotonin were almost exclusively bioassay procedures. These methods were very sensitive but tedious, and they lacked specificity. Presently the major methods for measuring the amine are fluorometric, radioenzymatic, and high performance liquid chromatographic procedures. A radioimmunoassay procedure has been described (847).

Since serotonin is metabolized very rapidly, in many experiments more information can be obtained by measuring 5-hydroxyindoleacetic acid than by measuring serotonin itself. In addition, experimental procedures sometimes involve the administration of 5-hydroxytryptophan (the precursor of serotonin) in an attempt to increase the concentrations of serotonin in tissues. Since under these circumstances it is often desirable to measure all three of the 5-hydroxyindoles, methods for their measurement are briefly mentioned.

Fluorometric Procedures

Fluorometric methods for measuring serotonin are relatively nonspecific and must be preceded by separation of serotonin from other 5-hydroxyindoles and other interfering compounds. Each of the three common methods used for separation (solvent extraction, ion-exchange chromatography, and gel filtration) will be briefly described. This will be followed by a description of major fluorometric methods used for quantitation of serotonin (and of other 5-hydroxyindoles) after separation has been achieved.

Separation

The original method for extracting serotonin from tissues was by solvent extraction as described by Bogdanski, Pletscher, Brodie, and Udenfriend (106), and this method, or some modification thereof, is still in use. Tissues are homogenized in 0.1 N hydrochloric acid and are then centrifuged. Aliquots of supernatant solution are adjusted to pH 10, saturated with sodium chloride, and serotonin is extracted into *n*-butanol. Aliquots of the *n*-butanol extract are transferred to tubes containing heptane and the serotonin is extracted back into acid. The serotonin in this fraction is assayed by fluorometric techniques. A modification of this extraction procedure is used to measure serotonin concentrations in whole blood (221).

If 5-hydroxytryptophan has been administered in the course of an experiment, it is essential that the *n*-butanol fraction, containing the serotonin, be washed three times with a pH 10 buffer to remove this amino acid. However, these washes can be omitted in determinations of serotonin in tissues from untreated animals since these do not contain sufficient 5-hydroxytryptophan to interfere in the assay.

When ion-exchange chromatography is to be used for separation of serotonin and metabolites, it is necessary that the tissue extract be protein-free prior to separation. Tissues are usually homogenized in 0.4 M perchloric acid and proteins are precipitated by centrifugation. Since 5-hydroxyindoles are not stable in perchloric acid (602), the perchlorates are removed rapidly by adjustment to pH 7 with 1 M potassium hydroxide, followed by centrifugation. Usually cysteine and disodium ethylenediaminetetraacetate are added to the perchloric acid to protect serotonin from oxidation. Alternatively, tissues can be homogenized in a mixture of acetone/1 N formic acid (85:15, vol/vol). The acetone extract is then washed with a mixture of heptane/chloroform (8:1, vol/vol) to remove lipids. The organic layer is removed and the aqueous layer that is formed is evaporated to dryness. The residue is dissolved in water and is ready to be applied to the ion-exchange resin (981).

A weak cation-exchange resin (BioRex 70 or Amberlite CG 50) is used most frequently for the separation of serotonin (981). When the protein-free extract is applied to the resin, serotonin is adsorbed and can subsequently be eluted with acid. 5-Hydroxytryptophan and 5-hydroxyindoleacetic acid are not adsorbed on the resin and are present in the effluent from the column. Catecholamines are adsorbed on the resin and can be eluted together with serotonin, or alternatively, the catecholamines can be first eluted from the resin with 0.67 M boric acid and then serotonin can be eluted with 0.5 N acetic acid (448).

A strong cation-exchange resin (Dowex 50) can also be used for the separation of serotonin and other 5-hydroxyindoles (46). A protein- and perchlorate-free extract of tissues is adjusted to pH 2 and applied to small columns of Dowex 50. All three major 5-hydroxyindoles are adsorbed on the resin at this pH. 5-Hydroxyindoleacetic acid can be eluted first with 60% methanol, and then 5-hydroxytryptophan can be eluted with a neutral buffer. Finally serotonin can be eluted with a mixture of ethanol/1 N hydrochloric acid (1:1, vol/vol).

Separation of 5-hydroxyindoles can also be achieved using gel filtration (296). Protein- and perchlorate-free extracts of tissues are adjusted to pH 2 to 3 and are applied to columns of Sephadex G10. 5-Hydroxytryptophan is found in the effluent from the column. Serotonin is first eluted with a mixture of ethanol/0.01 M HCl (10:80, vol/vol), and then 5-hydroxyindoleacetic acid is eluted with dilute sodium carbonate.

Quantitation

Once the 5-hydroxyindoles have been separated, either by solvent extraction or by column chromatography, they can be quantitated in their respective fractions using fluorescence. All 5-hydroxyindoles fluoresce at neutral pH at A 300 nm F 340 nm. However, the extracts are usually made acid by adding concentrated hydrochloric acid and their fluorescence then measured at A 300 nm F 540 nm. The sensitivity of this assay is reported to be 200 ng for serotonin or for other 5-hydroxyindoles (602).

Two fluorescence procedures exist which increase the sensitivity for the measurement of serotonin. These involve reaction with *O*-phthaldialdehyde (45) or with ninhydrin (1066). The sensitivities of both procedures are reported to be 10 ng for serotonin. The ninhydrin reaction will not proceed in acid solution and it is necessary to adjust extracts to neutrality before proceeding with this reaction. Although all 5-hydroxyindoles react with *O*-phthaldialdehyde the optimum conditions for the reaction vary (1017).

Choice of Methods

The use of column chromatography is gradually supplanting solvent partition methods for the extraction of serotonin. This is partly because the health hazards associated with handling organic solvents can be avoided. In addition, column chromatography is easier to perform and is generally more economical if only small amounts of serotonin are to be measured. However, as with all analytical procedures, it is essential to monitor recoveries of added authentic standards, and in addition, volumes of washes and of eluents should be checked with each batch of resin that is used.

If column chromatography is to be used, the Dowex 50 separation is a good choice when it is desired to measure all three major 5-hydroxyindoles, since all can be separated using a single column. The Amberlite CG 50 is useful if it is necessary to measure catecholamines as well as serotonin. It is also the resin of choice if low concentrations of serotonin are to be measured, since the monoamine can be eluted from the CG 50 resin with volatile acids, and the eluate can subsequently be lyophilized (448). However, the Amberlite CG 50 is unusually sensitive to high salt concentrations, and if solutions that are applied to this resin have high ionic strength, serotonin will be poorly adsorbed.

In all fluorescence procedures it is necessary to check for "quenching" of the analyte and also to prepare suitable "blank" solutions. For a discussion of the precautions to be taken when using fluorescence, the reader is referred to the classic monograph of Udenfriend (1060).

Radioenzymatic Method

Saavedra, Brownstein, and Axelrod (929) described a method for measurement of serotonin based on the transformation of the amine into melatonin by successive enzymatic reactions as represented by the following equations:

(a) Serotonin + acetyl coenzyme A $\xrightarrow[\text{transferase}]{N\text{-acetyl}}$ *N*-acetylserotonin

(b) *N*-Acetylserotonin + ^{3}H-methyl-*S*-adenosylmethionine

↓ Hydroxyindole-*O*-methyltransferase

^{3}H-Melatonin + *S*-adenosylhomocysteine

In the first reaction serotonin is *N*-acetylated, and in the second reaction *N*-acetylserotonin is *O*-methylated. *S*-Adenosylmethionine, with a ^{3}H-labeled methyl group, is utilized in the second reaction. The ^{3}H-labeled melatonin that is formed as a result of the two reactions is extracted into an organic solvent and the radioactivity is measured in a liquid scintillation counter.

The *N*-acetyltransferase enzyme used in the first reaction is isolated from rat liver. The hydroxyindole-*O*-methyltransferase used in the second reaction is isolated from bovine pineal glands. It is not necessary to separate serotonin (either by solvent extraction or by column chromatography) prior to performing the assay. The limit of sensitivity of the procedure is 50 pg of serotonin. A modification of the procedure has also been described (505) in which serotonin is chemically *N*-acetylated and only the second reaction is enzymatic. The sensitivity of this modified procedure is reported to be 5 pg of serotonin.

Although this procedure is very sensitive, it has not been widely used. It shares the disadvantages of many enzymatic procedures that the enzymes are intrinsically unstable and are not readily available. In addition, the high cost of the radiolabeled *S*-adenosylmethionine makes the assay expensive.

High Performance Liquid Chromatography

High performance liquid chromatography (HPLC) can be used for the measurement of 5-hydroxyindoles since these are easily oxidizable and are electroactive. In practice the 5-hydroxyindoles are separated on an HPLC column. When eluted from the column the 5-hydroxyindoles are oxidized at an electrode surface under controlled potential conditions. The current resulting from the transfer of electrons across the interface of the electrode and the solution is directly proportional to the concentration of 5-hydroxyindole eluted. The retention time of the analyte on the column is characteristic of the compound and gives some specificity to the method. In addition, the electrochemical detector is selective in that only compounds that are electroactive at a given potential are oxidized. For a discussion of the principles of electrochemical detection the reader is referred to the review by Kissinger, Bruntlett, and Shoup (570).

Several methods have been described for the measurement of 5-hydroxyindoles by HPLC with electrochemical detection (28,720,782,1124). Tissue samples require a minimum of preparation before extracts are injected on the HPLC. Typically, tissues are simply homogenized and the proteins precipitated. Aliquots (usually 20–50 μl) of the protein-free extract are injected directly onto the column (28,611,887). The three major 5-hydroxyindoles can be measured in the same extract after separation on a suitable HPLC column. However, a number of investigators prefer to use a "clean-up" procedure prior to the HPLC separation. In the clean-up procedures either ion-exchange or solvent partition methods are used to separate serotonin from interfering compounds. The methods used for the clean-up (448,580) are the same as those used prior to the measurement of 5-hydroxyindoles by fluorometric procedures. An advantage of using a clean-up procedure is that it lends specificity to the method. In addition, it is often possible to concentrate the analyte during the preliminary separation. This concentration is particularly useful when serotonin is to be measured in perfusates (448), in serum, or in cerebrospinal fluid (580). Prior clean-up of samples usually has the advantage of prolonging the life of the HPLC columns. A disadvantage of using a preliminary clean-up is that this usually precludes measuring all of the three major 5-hydroxyindoles in the same sample using a single injection on the HPLC.

Two important variables in the separation of 5-hydroxyindoles using HPLC are the stationary phase and the mobile phase. Two stationary phases are in common use for serotonin separations: strong cation-exchange resins, and a microparticular reversed phase matrix (octadecyl silane, or C_{18}), with the reversed phase separations being the most frequently used. Recently separations have been described using smaller particle size for reversed phase separations, and as a general rule the smaller the particle size the better the resolution achieved (602). Mobile phases that are used vary with regard to ionic strength, pH, polarity, and temperature. For a discussion of the effects of changes in these variables, the reader is referred to reviews by Mefford (719) and Meek (718).

The sensitivity of HPLC with electrochemical detection for the quantitation of 5-hydroxyindoles is high. Amounts of serotonin as low as 5 pg can be detected by this technique (720). In addition, the method is rapid and all three major 5-hydroxyindoles can be quantitated in less than 15 min (611).

HPLC with electrochemical detection has been supplanting other techniques for the quantitation of serotonin and other 5-hydroxyindoles. This technique has the advantages of speed, high sensitivity and specificity, of requiring a minimum of preparation of samples, and of relatively low cost for the instrumentation and reagents.

Other Methods

HPLC combined with fluorometric methods for quantitation has been used for measurement of serotonin concentrations (223,307). This method is not as sensitive as HPLC with electrochemical detection and, in addition, the instrumentation is more expensive.

A radioimmunoassay method for serotonin is available which is specific and sensitive to 1 ng of serotonin (847). It has not been widely used for quantitation but rather for immunocytochemical localization of serotonin in tissues (323).

BIOSYNTHESIS OF SEROTONIN

Overview

Serotonin is synthesized from the essential amino acid L-tryptophan by two enzymatic reactions (Fig. 1–1). The biosynthesis of serotonin represents a minor metabolic route for tryptophan, and normally not more than 2% of ingested tryptophan is utilized for the daily synthesis of about 10 mg serotonin (630). The rate-limiting step in the synthesis of serotonin from tryptophan is the first reaction catalyzed by tryptophan hydroxylase, an enzyme which occurs only in cells synthesizing serotonin. The second enzyme, L-aromatic amino acid decarboxylase, is widely distributed in tissues. In addition, L-aromatic amino acid decarboxylase is not specific and will catalyze the decarboxylation of all aromatic amino acids, including L-3,4-dihydroxyphenylalanine, the precursor of catecholamines.

Tryptophan Hydroxylase

Tryptophan hydroxylase activity has been detected in almost all tissues and cells that contain high concentrations of serotonin, including brain (359,1047), enterochromaffin cells of the gastrointestinal mucosa (625), serotonergic neurons in the gut (392), carcinoid tumor cells (661), bovine pineal gland (255), and neoplastic

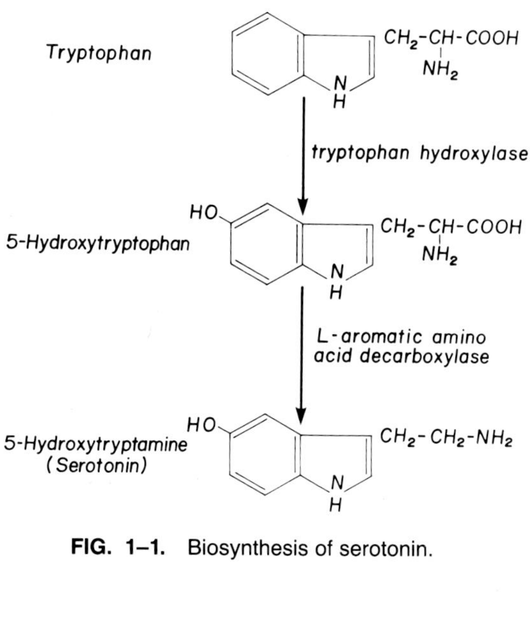

FIG. 1–1. Biosynthesis of serotonin.

mouse mast cells (662), but not platelets (753). In the brain this enzyme is localized uniquely in serotonergic neurons and its presence has been used as a marker for these neurons (600). Within cells the enzyme is localized in the cytosol (661), although some activity is also associated with subcellular organelles (425,534).

Tryptophan hydroxylase has been purified extensively and antibodies to it have been produced (534). This has made possible the immunohistochemical localization of the enzyme in the brain (534) and also in the periphery (392).

The reaction catalyzed by tryptophan hydroxylase is represented by the equation below:

$$\text{L-Tryptophan} + \text{tetrahydrobiopterin} + O_2 \longrightarrow$$
$$\text{5-hydroxytryptophan} + H_2O + \text{quinonoid dihydrobiopterin}$$

The quinonoid dihydrobiopterin is reduced back to tetrahydrobiopterin in a reaction catalyzed by dihydrobiopterin reductase, using reduced nicotinamide adenine dinucleotide or reduced nicotinamide adenine dinucleotide phosphate as cofactors (359).

The Michaelis constant (K_m) of tryptophan hydroxylase for tryptophan is relatively high (3×10^{-5} M) (1047). It is probable that under normal circumstances the concentration of tryptophan in cells that synthesize serotonin is below the K_m of tryptophan hydroxylase so that the enzyme is not saturated *in vivo*. Tryptophan hydroxylase is inhibited by concentrations of tryptophan higher than 2×10^{-4} M (359).

The K_m for oxygen of tryptophan hydroxylase is about 18 mm Hg when assayed under standard conditions. This suggests that it is not fully saturated with oxygen under the range of concentrations which occurs *in vivo* (425,561). In addition, the K_m for tryptophan and the K_m for oxygen show some interdependence (561).

The K_m of tryptophan hydroxylase from rabbit hindbrain for tetrahydrobiopterin (3×10^{-5} M) is probably higher than the tissue concentration of this compound (359). Various synthetic pteridines have frequently been used in enzyme assays instead of tetrahydrobiopterin. However, the kinetic properties of the enzyme, notably the K_m for tryptophan, are much higher when these synthetic pteridines are used than they are when tetrahydrobiopterin is the cofactor (359).

Tryptophan hydroxylase is stimulated in the presence of Fe^{2+} (359). The enzyme is easily inactivated on exposure to molecular O_2, and sulfhydryl compounds such as dithiothreitol are usually added to enzyme assay mixtures to protect the hydroxylase from this inactivation (604).

Tryptophan hydroxylase is inhibited both *in vivo* and *in vitro* by the amino acid *p*-chlorophenylalanine (533). This compound is a competitive inhibitor *in vitro*, but *in vivo* it produces a long-lasting inhibition of the enzyme and concomitant reduction in the levels of serotonin in the tissues. Tryptophan hydroxylase is also inhibited after *in vivo* administration of *p*-chloroamphetamine (935), although this drug probably has multiple other actions affecting the turnover of serotonin (367).

L-Aromatic Amino Acid Decarboxylase

L-Aromatic amino acid decarboxylase catalyzes the decarboxylation of several amino acids, including tryptophan, tyrosine, phenylalanine, dihydroxyphenylalanine, and 5-hydroxytryptophan. This enzyme is widely distributed in many tissues, with the kidney having especially high activity. Within the cell the enzyme is localized mainly in the cytosol, although some activity may be associated with particulate fractions (908).

Although it has been suggested that different enzymes catalyze the decarboxylation of 3,4-dihydroxyphenylalanine and 5-hydroxytryptophan, it is likely that the same enzyme catalyzes the decarboxylation of both amino acids (204). Pyridoxal phosphate is a coenzyme for L-aromatic amino acid decarboxylase, but excess of this coenzyme (greater than 6×10^{-7} M) inhibits decarboxylation (119).

The kinetic parameters of L-aromatic amino acid decarboxylase have been determined for nonpurified enzyme from cat brain, liver, and adrenals. The apparent K_m of the enzyme for 5-hydroxytryptophan was in the range of 8×10^{-6} M (119). The decarboxylase activity is lower in the brain than in the liver or the adrenals. Following long-term treatment with 5-hydroxytryptophan or with dihydroxyphenylalanine the L-aromatic amino acid decarboxylase activity increases in brain but decreases in liver (119) and in human erythrocytes (1035).

L-Aromatic amino acid decarboxylase is inhibited by a number of amino acid analogs, notably α-methyldihydroxyphenylalanine (α-methyldopa) and N'-DL-seryl-N^2-(2,3,4-trihydroxybenzyl)hydrazine (R0 4–4602). Many other miscellaneous compounds also inhibit the enzyme (see 385).

Regulation

Since the concentration of tryptophan in tissues is probably lower than the K_m of tryptophan hydroxylase for the precursor, the availability of the latter may control the rate of synthesis of serotonin. Diets poor in tryptophan affect tissue concentrations of serotonin (203). On the other hand, administration of tryptophan will increase the concentration of serotonin in tissues (742). Little information is available on the uptake of tryptophan from plasma by peripheral cells that synthesize serotonin. There are two saturable systems for uptake of tryptophan by brain cells, one with high affinity and the other with lower affinity for tryptophan (682). The significance of these two uptake systems is not yet understood.

The synthesis of serotonin in brain cells is regulated by mechanisms that activate or inhibit tryptophan hydroxylase (449). Thus the rate of synthesis appears to increase when the amine is released by nerve stimulation (697). This increase in serotonin synthesis is caused probably by a sequence of events, the first of which is the influx of Ca^{2+} into the neuron. This increase in Ca^{2+} concentration triggers a specific protein kinase capable of phosphorylating tryptophan hydroxylase. The phosphorylation of tryptophan hydroxylase results in increase in the maximal velocity (V_{max}) of the enzyme. Tryptophan hydroxylase is inhibited by an intraneuronal pool of serotonin (449), although serotonin does not inhibit the enzyme *in vitro* (422).

METABOLISM OF SEROTONIN

Overview

The major pathway of metabolism of serotonin is by oxidative deamination with the production of the unstable intermediate 5-hydroxyindoleacetaldehyde (Fig. 1–2). This aldehyde can undergo oxidation catalyzed by the enzyme aldehyde dehydrogenase with the formation of 5-hydroxyindoleacetic acid, or it can undergo reduction catalyzed by aldehyde reductase with the formation of 5-hydroxytryptophol. Under normal circumstances, the predominant pathway is oxidation to 5-hydroxyindoleacetic acid.

Minor metabolic pathways for serotonin include conjugation with sulfate or glucuronide, and *N*-acetylation. *O*-Methylation of serotonin and of *N*-acetylserotonin occurs, and *N*-methylation of serotonin has also been described.

Monoamine Oxidase, Aldehyde Dehydrogenase, and Aldehyde Reductase

Monoamine oxidase is a ubiquitous enzyme, with particularly high activity in the liver. The enzyme is present both intra- and extraneuronally (145). It is a flavoprotein and is localized on the surface of mitochondria (831).

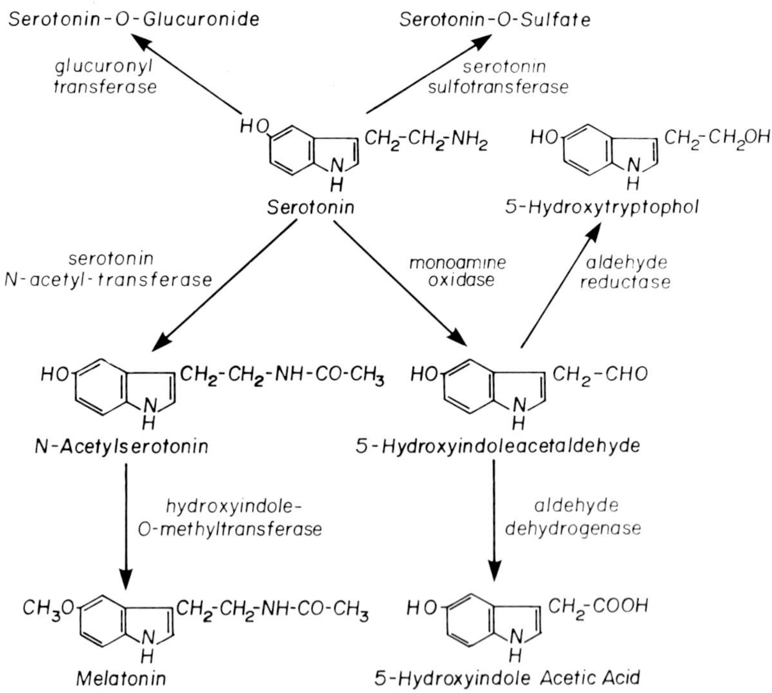

FIG. 1–2. Metabolism of serotonin.

The reaction catalyzed by monoamine oxidase is represented by the equation below:

$$\text{Serotonin} + O_2 + H_2O \rightarrow \text{5-hydroxyindoleacetaldehyde} + H_2O_2 + NH_3$$

The enzyme will catalyze the oxidative deamination of a wide range of amines including dopamine, norepinephrine, tyramine, phenylethylamine, and tryptamine. Monoamine oxidase activity can be differentiated into two major forms (A and B) on the basis of selective inhibition by different inhibitors (538) as well as separation by immunoaffinity chromatography (267).

Monoamine oxidase A is responsible for the oxidative deamination of serotonin and norepinephrine. The B form of the enzyme preferentially deaminates phenylethylamine (1172). The type A enzyme can be localized both intra- and extraneuronally, whereas the type B enzyme is predominantly, or perhaps exclusively, extraneuronal (145). The A form of the enzyme is inhibited by low concentrations of clorgyline (538), whereas the B form is preferentially inhibited by deprenyl (576).

The apparent K_m for serotonin is 7×10^{-5} M for monoamine oxidase A solubilized from mitochondria of the human brain (831). The apparent K_m of monoamine oxidase from rat brain for oxygen is 10^{-3} M; enzyme activity can be increased by approximately 50% when assays are done in 100% oxygen rather than in air (775).

5-Hydroxyindoleacetaldehyde, formed from serotonin by oxidative deamination, is rapidly oxidized to 5-hydroxyindoleacetic acid by a reaction catalyzed by the enzyme aldehyde dehydrogenase. This enzyme has been detected in many tissues, including the brain (321,622). In subcellular fractions of the bovine brain, the highest activity is detected in the mitochondrial preparations but lesser activity is also present in the cell supernate (321). The enzyme catalyzes the oxidation of aldehydes formed from norepinephrine and dopamine as well as 5-hydroxyindoleacetaldehyde. Nicotinamide adenine dinucleotide is necessary as a coenzyme and no activity is detected when nicotinamide adenine dinucleotide phosphate is substituted. The K_m for 5-hydroxyindoleacetaldehyde is low (2.5×10^{-6} M) (321).

5-Hydroxyindoleacetaldehyde can be reduced to the alcohol 5-hydroxytryptophol by a reaction catalyzed by the enzyme aldehyde reductase. In normal circumstances this pathway is of lesser importance than the oxidation to 5-hydroxyindoleacetic acid, but after the ingestion of ethyl alcohol there is a shunting from the oxidative to the reductive pathway (224,1055). However, 5-hydroxytryptophol can also be detected in peripheral tissues (1057) and in brain (270) under normal conditions.

Aldehyde reductase is present in the cell supernate from the pig brain. Nicotinamine adenine dinucleotide phosphate is the required coenzyme. The enzyme had a high affinity for 5-hydroxyindoleacetaldehyde with an apparent K_m of 4×10^{-6} M (1053). A similar aldehyde reductase has also been detected that catalyzes the reduction of the aldehydes formed by oxidative deamination of norepinephrine and dopamine (1016).

Minor Metabolic Pathways for Serotonin

Sulfation

An enzyme serotonin-sulfotransferase that catalyzes the formation of serotonin-*O*-sulfate has been detected in liver, lung, kidney, and brain (477). This enzyme is localized in the supernatant fraction of cells (478). It utilizes 3′-phosphoadenosine,5′-phosphosulfate as sulfate donor. When serotonin is *O*-sulfated it loses biological activity, and the formation of serotonin-*O*-sulfate may represent a mechanism for temporary inactivation of serotonin. Indeed, in many tissues arylsulfatase enzymes can cleave serotonin-*O*-sulfate with the regeneration of free serotonin (568).

Serotonin-*O*-sulfate has been detected in superfusates of rat spinal cord and in ventriculocisternal perfusates of the cat (446). The amounts of serotonin-*O*-sulfate produced increase only modestly when monoamine oxidase is inhibited (446,447). Serotonin-*O*-sulfate has also been detected in the urine of normal individuals (225).

A similar direct *O*-sulfation of catecholamines is catalyzed by the enzyme phenolsulfotransferase (532). It appears, however, that different enzymes are responsible for the sulfation of serotonin and the catecholamines, since phenolsulfotransferase preparations active towards catecholamines are relatively inactive towards serotonin (892).

Glucuronidation

Serotonin can be conjugated with glucuronic acid by a reaction catalyzed by the enzyme uridine diphosphate glucuronyltransferase. Glucuronidation is a reaction undergone by many compounds, including those that are foreign to the body, and often glucuronidation represents a detoxification mechanism for potentially toxic compounds. The enzyme glucuronyltransferase is found in many tissues, but the liver has the highest activity. Serotonin-*O*-glucuronide is a major metabolite of serotonin in the isolated perfused rat liver (1057). When monoamine oxidase is inhibited in the liver, there are large increases in the amounts of serotonin glucuronide, and glucuronidation easily compensates for the blocked pathway (1057). Serotonin-*O*-glucuronide has also been detected in the urine of patients with the carcinoid syndrome (712).

N-*Acetylation*

In the pineal body serotonin is a precursor of the hormone melatonin (Fig. 1–2); the first step in this biosynthesis is the *N*-acetylation of serotonin. *N*-Acetylation of serotonin has also been described in brain (828) and liver (1057). The enzyme responsible is serotonin *N*-acetyltransferase, and acetyl coenzyme A is utilized in the reaction. Within the cell serotonin *N*-acetyltransferase activity is localized in the cytosol (828). The activity of serotonin *N*-acetyltransferase in the pineal is 15 times greater during the night than during the day, and it is the activity of this enzyme that regulates the synthesis of melatonin (573). The concentration of melatonin in pineal body also shows a striking diurnal rhythm which is in phase with the rhythm of serotonin *N*-acetyltransferase.

O-*Methylation and* N-*Methylation*

In the pineal body *N*-acetylserotonin is *O*-methylated in a reaction catalyzed by the enzyme hydroxyindole-*O*-methyltransferase. *S*-Adenosylmethionine is the methyl donor in this reaction, and the product of the reaction is melatonin. Most hydroxyindole-*O*-methyltransferase activity is localized in the pineal. It has, however, been shown that serotonin can itself be *O*-methylated; this reaction has been detected in the hypothalamus with the formation of 5-methoxytryptamine (424).

The *N*-methylation of serotonin has been described in the lung (48) and the brain (750) of the rabbit. The enzyme involved also uses *S*-adenosylmethionine as methyl donor. *N*-Methylated amines are psychotoxic (48).

ACKNOWLEDGMENT

This work was supported in part by grants NS 9143 and NS 17858.

— 2 —

Peripheral Sources of Serotonin and Serotonin-Binding Proteins

*Michael D. Gershon and **Hadassah Tamir

*Departments of *Anatomy and Cell Biology, and **Psychiatry, Columbia University, College of Physicians and Surgeons, **New York State Psychiatric Institute, New York, New York 10032*

Serotonin is stored in both peripheral endocrine and neuronal cells that can release it into the circulation. As serotonin is a powerful amine, the circulation of free serotonin may be dangerous to an organism. Mechanisms that have evolved to deal with this potential danger involve the binding of serotonin to proteins. A variety of these proteins exist, and two have been well characterized.

PERIPHERAL SOURCES

Enterochromaffin Cells

Relatively little is known about the endocrine storage of serotonin. The enteroendocrine (enterochromaffin) cells of the gastrointestinal wall contain the bulk of the serotonin of the body (317) and probably account for the majority of the serotonin reaching the blood. The enterochromaffin cells release serotonin when stimulated by acetylcholine (347,567), noradrenergic nerve stimulation (12), rises in intraluminal pressure in the intestine (136), or a decline in intestinal pH (565,893). The release of serotonin by these cells might activate intrinsic enteric, primary afferent nerve fibers to trigger the peristaltic reflex (136,137).

Serotonergic Neurons

Serotonin fulfills all of the criteria needed for the identification of an enteric neurotransmitter (386,390).

Serotonin is synthesized by enteric neurons from tryptophan (292), a substance for which these neurons have a high-affinity uptake mechanism (391). Enteric neurons also contain tryptophan hydroxylase, the rate-limiting enzyme in the synthesis of serotonin (392). The monoamine is stored within these neurons in high concentration and can be demonstrated there by formaldehyde-induced histofluorescence (903) or, more easily, by immunocytochemistry (194,374,924) (Fig. 2–1). The immunocytochemical localization reveals that most enteric serotonergic neurons reside in the myenteric plexus. From the ganglia of this plexus they project within it in both oral and anal directions, but also to follower neurons in the submucosal plexus (374). The concentration of serotonin within the longitudinal muscle–myenteric plexus layer of the gut approximates about 100 ng/g; however, as this layer is composed mainly of muscle the content of serotonin of the myenteric plexus itself may be 10-fold higher (1023).

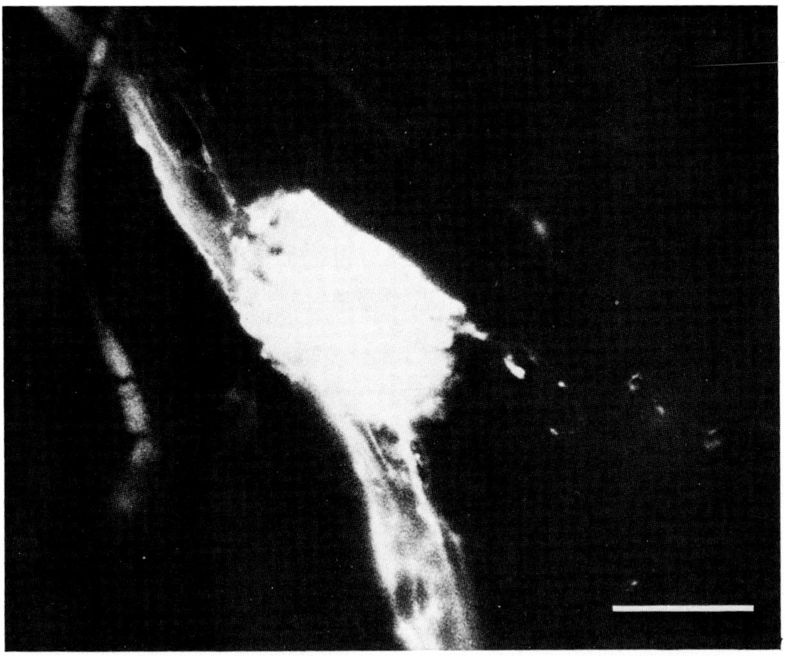

FIG. 2–1. Immunocytochemical demonstration of serotonin in the myenteric plexus of a mouse. The tissue was fixed with formaldehyde and the longitudinal muscle with adherent myenteric plexus were dissected and processed as whole mounts. A monoclonal antibody to serotonin was used and serotonin immunoreactivity was located using a tetramethylrhodamine isothiocyanate-labeled secondary antibody. A thick tract of fibers and a single neuronal perikaryon exhibit intense serotonin immunofluorescence. The tract is running through a nonfluorescent ganglion and is overlaid by the fluorescent serotonergic neuron. Marker = 30 μm.

When the enteric nerves are stimulated electrically, serotonin is released, together with neurectodermal serotonin-binding protein, by a Ca^{2+}-dependent mechanism (541). Newly taken up ^{3}H-serotonin is released preferentially, but endogenous serotonin is released as well (402,490). The neural release of serotonin can be detected by monitoring its efflux from the serosal surface of the gut (402). The release of serotonin by mucosal or enterochromaffin cells leads to the appearance of the monoamine in the lumen of the bowel or in the portal blood (566). The gut contains a barrier to the free transmural passage of serotonin from the mucosa to the serosa (402). This barrier is probably located at the junction between the mucosa and submucosa or within the mucosa itself. When applied to the mucosal surface of the intestine, serotonin stimulates primary afferent nerve fibers that in turn trigger the peristaltic reflex; however, when it is applied to the serosal surface of the gut the amine blocks peristalsis by depolarizing enteric ganglia (135). The normal function of neurons depends on brief and discrete applications of the transmitter. Thus, the intraenteric barrier preventing the free diffusion of serotonin may serve to insulate the enteric ganglia from being overwhelmed by the release of serotonin from the mucosal enterochromaffin cells. Actually, the stimulation of mucosal afferents by serotonin causes the release of serotonin from neurons within the myenteric plexus (390). Thus, the serosal release of serotonin is detected when the gut displays a high degree of propulsive activity (432).

Enteric neurons not only synthesize, store, and release serotonin, they also possess a specific uptake mechanism for the amine (388,396,903). The uptake of serotonin is Na^{+}- and Ca^{2+}-dependent and requires energy. Aerobic metabolism of glucose is primarily responsible for providing this energy, but anaerobic metabolism can take over. The uptake of serotonin is saturable. As it is inhibited by ouabain, it must involve a carrier molecule linked to Na^{+}/K^{+}-adenosine triphosphatase (ATPase). The uptake mechanism for serotonin is not inhibited by up to a thousandfold excess of norepinephrine or by chemical sympathectomy with 6-hydroxydopamine; therefore, it is distinct from the neuronal carrier for catecholamines. It is not present in noradrenergic axons. The sensitivity of the enteric uptake of serotonin to inhibitory drugs is quantitatively similar to that of central serotonergic neurons, with an order of potency: chlorimipramine>imipramine> desmethylimipramine. The newer inhibitors, such as fluoxetine (540), zimelidine, and trazodone, also block enteric uptake of serotonin. Zimelidine is the most effective of the group in distinguishing uptake of serotonin from that of norepinephrine *(unpublished observations)*. The uptake of serotonin is effective enough to remove the neurotransmitter from the junctional cleft. The uptake of labeled serotonin allows the radioautographic detection of serotonergic elements by means of light (Fig. 2–2) and electron microscopy (Fig. 2–3) (386,903), and yields the same results as obtained with antibodies to serotonin. The use of the radioautographic technique has demonstrated that enteric serotonergic neurons are at least primarily, and possibly entirely, interneurons (Fig. 2–3).

Serotonin probably mediates the descending wave of inhibition of vagal excitation that accompanies the peristaltic reflex (545). The monoamine causes the slow

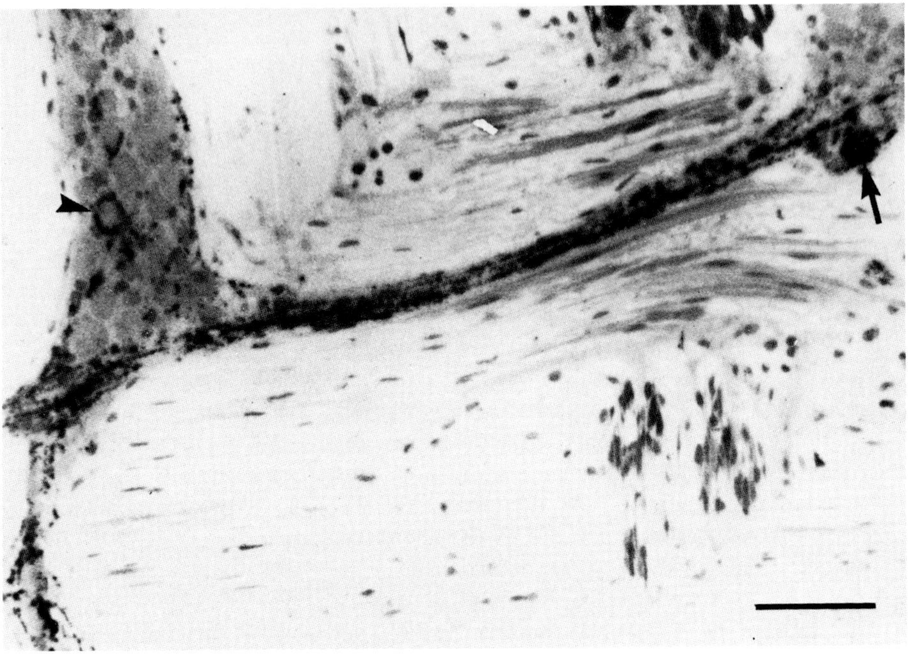

FIG. 2–2. Radioautographic demonstration of the uptake of ^{3}H-serotonin by serotonergic elements of the myenteric plexus. The longitudinal muscle with adherent myenteric plexus was dissected from the small intestine of a guinea pig and incubated with ^{3}H-serotonin (0.5 μM). The tissue was fixed with aldehydes and osmium tetroxide, embedded in plastic, and sectioned at 1 μm for radioautography. A densely labeled serotonergic tract runs through a connective string between the ganglion at the *upper right* and the ganglion at the *left* of the field. There is one labeled cell body in the ganglion at the *right (arrow).* The dense accumulations of silver grains in apposition to neurons in the ganglion at the *left* represent serotonergic synapses. One cell in the ganglion is ringed by such labeling *(arrowhead).* Compare the patterns of labeling with the immunofluorescence of serotonin shown in Fig. 2–1. Marker = 100 μm.

excitatory postsynaptic potential (epsp) in one of the categories of myenteric ganglion cell (type II/AH neuron) (1167); substance P may mediate this response in other cells (751). Every type II/AH neuron that shows a slow epsp in response to fiber tract stimulation is covered by serotonergic synapses, whereas such synapses are rare in cells that do not show the response (312). The enteric neuronal serotonergic receptor has recently been defined and specific antagonists, dipeptides of 5-hydroxytryptophan, have been developed (124,401). These antagonists block both the effects of serotonin on type II/AH neurons and the slow epsp.

→

FIG. 2–3. Demonstration of the location of serotonergic and noradrenergic terminals in the myenteric plexus. The dissected longitudinal muscle and the adherent myenteric plexus were incubated with ^{3}H-serotonin and then fixed with aldehyde and $NaMnO_4$. This procedure preserves ^{3}H-serotonin in tissue while at the same time specifically marking the small (50 nm) dense-cored vesicles of noradrenergic axons *(filled arrows).* Note the mixed population of vesicles in the serotonergic varicosities *(open arrows).* Serotonergic varicosities form synapses on processes and cell bodies **(inset)** of other neurons. Noradrenergic terminals are not labeled by ^{3}H-serotonin.

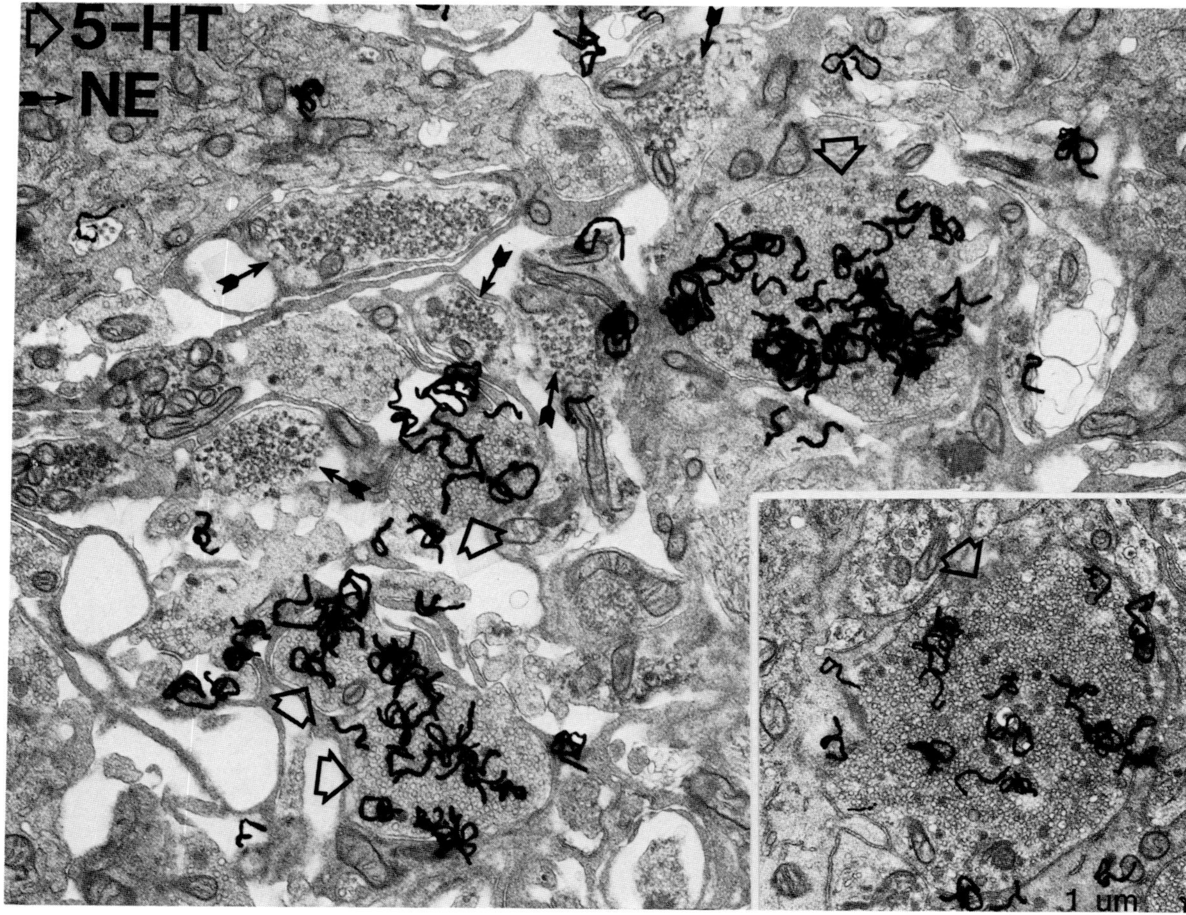
5-HT
NE
1 um

Serotonin can evoke a rapid as well as a slow depolarization of myenteric ganglion cells. It also inhibits the release of acetylcholine prejunctionally (788). In the submucosal plexus, serotonin activates all ganglion cells (479). The multiplicity of these actions of serotonin makes it difficult to interpret pharmacological experiments in which it is applied to segments of whole gut. Nevertheless, it seems probable that enteric serotonergic neurons promote propulsive activity of the bowel. This supposition is supported by the observation that specific destruction of enteric serotonergic neurons with the indolic neurotoxin 5,7-dihydroxytryptamine (62,399,400) slows intestinal motility in mice *(unpublished observations)*.

Since the identification of serotonin as an enteric neurotransmitter, the amine has been found in other peripheral neural sites. These locations include cerebral arteries (299,426), the superior cervical ganglion (1112), the nodose ganglion (381), and the adrenal medulla (491,1111).

SEROTONIN-BINDING PROTEINS

Potential Danger of Circulating Free Serotonin

Serotonin increases vascular permeability, especially in postcapillary venules (674,811). The potentially devastating action of serotonin is illustrated by anaphylactic shock in mice where it is the lethal mediator (397). Serotonin characteristically causes the endothelial cells of postcapillary venules to contract. This contraction opens the intercellular junctions between endothelial cells, permitting the leakage of plasma out of the venules (43). Thus, serotonin reduces the barrier between the blood and connective tissue, although the coarse filter represented by the basal lamina of the affected venules remains intact. This filter holds back cellular elements of the blood, but not fluid, salts, or even macromolecules. The defenses that have evolved to prevent the circulation of free serotonin also attest to the danger inherent in the molecule. Platelets take up serotonin by a system of active transport (113); the liver catabolizes the amine (825); pulmonary endothelial cells take up serotonin (53,153). All of these processes have the common effect of rapidly removing free serotonin from the circulation.

Serotonin is active in many organs and tissues. Among other actions, it (a) stimulates platelet aggregation (63,733); (b) acts in delayed hypersensitivity reactions (389); (c) activates vascular (317), bronchial (317), and intestinal smooth muscle (317,379); and (d) stimulates sensory nerve endings (341,948,1154), sympathetic nerve terminals (288,351), and intrinsic enteric motor neurons (132,379,788,1167). This partial listing makes it apparent why the action of serotonin must, for the survival of the organism, remain well localized after its release. Given all of the considerations that would seem to make the circulation of free serotonin an end to be avoided, the number of peripheral (outside the brain and spinal cord) sites of storage for the monoamine (Table 2–1), which all have the potential to deliver serotonin into the systemic circulation, may seem surprising.

One physiological mechanism for preventing the actions of serotonin is to specifically bind it to macromolecules as soon as it appears in free form. A number

TABLE 2–1. *Peripheral sites where serotonin has been localized*

Tissue	Uptake mechanism	Synthesis	Reference
Enterochromaffin cells	+	+	317,328,625,1039
Platelets	+	−	218,856
Mast cells	+	+	70,603
Parafollicular cells	−	+	327,395,793
Adrenal medulla	+	?	398,491,1111
Pineal	+	+	84,792,986
Pituitary	+	+	795,830,930
Pancreatic islet	+	+	326,582,794
Enteric nervous	+	+	194,293,373,386,388,396

of such serotonin-binding proteins exist, which vary in their properties depending on their source. The two best-known are the serotonin-binding protein of serotonergic neurons (1022,1027–1030) and serotonectin, a protein present in blood (853,1024,1031,1032). Other less well-characterized serotonin-binding proteins are present in glial cells (474), enterochromaffin cells of the gastrointestinal mucosa (412), and mast cells of the rat (1033).

Neurectodermal Binding Proteins

Neurectodermal serotonin-binding protein is concentrated in synaptosomes of the central nervous system and is released as a soluble protein when synaptosomes rupture (1028). Lesions made in the serotonergic nuclei of the median raphe (601) cause the loss of neurectodermal serotonin-binding protein from the forebrain, indicating that it is present in serotonergic neurons (1030). Neurectodermal serotonin-binding protein is present in the myenteric plexus of the gut, where its appearance during ontogeny correlates with the appearance of enteric serotonergic neurons detected by other means (542). The protein can be found in one additional site outside the central nervous system, namely the thyroid gland. Here the distribution of neurectodermal serotonin-binding protein correlates well with the distribution of serotonin, and both correlate with the distribution of parafollicular cells (81,82). This then suggests that the parafollicular cells contain the neurectodermal serotonin-binding protein, as confirmed by affinity chromatography studies that demonstrate the presence of serotonin and neurectodermal serotonin-binding protein in such cells isolated from sheep thyroids (82). Parafollicular cells have a neurectodermal origin, as do central and peripheral serotonergic neurons. They develop from precursor cells that emigrate to the thyroid from the neural crest (620,834).

Although the neurectodermal serotonin-binding protein was first recovered from the $100{,}000 \times g$ supernatant fraction of brain homogenates, it probably is a component of synaptic vesicles. In common with other particulate elements of neurons (421), neurectodermal serotonin-binding protein is transported proximo-distally in serotonergic axons by fast axonal transport (1026). Thus, when the thoracic spinal cords of rabbits are transected, neurectodermal serotonin-binding protein accumulates above the lesion and is lost distal to the cut. The rate of accumulation of the proteins in the proximal cord is compatible with a rate of transport of approximately 100 mm/day, which probably is an underestimate due to the relatively long segments of spinal cord required for the assay of neurectodermal serotonin-binding protein. This indirect evidence suggesting that neurectodermal serotonin-binding protein is stored in synaptic vesicles is supplemented by two lines of direct evidence that lead to the same conclusion. Stimulation of the enteric nervous system *in vitro* produces not only a release of serotonin (402,490), but also a Ca^{2+}-dependent release of neurectodermal serotonin-binding protein (541). There is also a resting release of neurectodermal serotonin-binding protein but, unlike the stimulated increment in the release of the protein, its resting release is Ca^{2+} independent. In contrast to neurectodermal serotonin-binding protein, the cytosolic marker protein, lactic acid dehydrogenase, is not released by nerve stimulation. Exocytosis is the only known mechanism that accounts for the Ca^{2+}-dependent, stimulus-coupled release of neurectodermal serotonin-binding protein. This, in turn, implies that this protein must be stored in synaptic vesicles that can fuse with the plasma membranes of nerve terminals at the active zones of synapses. Careful rupture of brain synaptosomes, followed by isolation of synaptic vesicles by differential centrifugation in K^+-containing media, allows the detection of neurectodermal serotonin-binding protein in the vesicular fraction. Neurectodermal serotonin-binding protein is four times more concentrated in synaptic vesicles than in the synaptosomal supernatant (1026).

If neurectodermal serotonin-binding protein is a component of synaptic vesicles, the questions arise as to what its function is and whether or not it forms a complex with serotonin. Exogenous labeled serotonin was used to investigate this issue. The cerebral ventricles of rats were perfused with ^{3}H-serotonin for 3 hr. ^{3}H-Serotonin is taken up and concentrated by serotonergic terminal axons under these conditions (162). The radioautographic analysis of the distribution of ^{3}H-serotonin reveals that the radioactive compound is distributed nonuniformly and preferentially labels axon terminals and some cell bodies in the raphe nuclei close to the ventricular system. The brains labeled with ^{3}H-serotonin were homogenized and centrifuged, and the supernatant was subjected to molecular sieve chromatography. A significant amount of ^{3}H-serotonin was eluted with the protein fraction. When subjected to polyacrylamide gel electrophoresis this labeled protein migrates, as does extracted neurectodermal serotonin-binding protein (394). There are two forms of neurectodermal serotonin-binding protein, one of 56 kilodaltons (kd), the other of 45 kd. When neurectodermal serotonin-binding protein is extracted from brain, the 56-kd form of the protein predominates; however, when the complex of ^{3}H-serotonin and protein is formed within intact neurons and is subsequently recovered as described, the 45-kd form predominates. Since the process of ventricular perfusion with ^{3}H-serotonin preferentially labels terminals, the 45-kd form of neurectodermal serotonin-binding protein may be characteristic of terminal axons, while the 56-kd form may be most abundant in cell bodies and preterminal axons. If so, the 56-kd protein may be the precursor of the 45-kd form of neurectodermal serotonin-binding protein. This would explain why the 45-kd form accumulates in preparations of 56-kd neurectodermal serotonin-binding protein during storage, and why antibodies prepared against the 56-kd form cross-react with the 45-kd material. ^{3}H-Serotonin can also be used to label stores of neurectodermal serotonin-binding protein within intact, enteric serotonergic neurons, which take up ^{3}H-serotonin (396). Homogenization and analysis of ^{3}H-serotonin-labeled protein derived from enteric neurons demonstrate that the protein, like that derived from central neurons, is predominantly of the 45-kd type. In the case of the gut, however, the 45-kd form of neurectodermal serotonin-binding protein is normally more abundant than the 56-kd form (394).

One can only speculate on the exact role of neurectodermal serotonin-binding protein; however, the characteristics of the protein (Table 2–2) suggest a possibility. Neurectodermal serotonin-binding protein binds serotonin very tightly under conditions similar to those encountered inside the cells. The dissociation constants of the complex between serotonin and the binding protein are in the range of 10^{-10} M and 10^{-8} M, in buffers containing K^+ and no Na^+ or Ca^{2+}. Both Na^+ and Ca^{2+} antagonize the binding of serotonin by neurectodermal serotonin-binding protein, and at concentrations of Na^+ and Ca^{2+} found in extracellular fluid the protein loses its affinity for serotonin. The binding of ^{3}H-serotonin is also enhanced by hydrophobic conditions that might prevail in synaptic vesicles. For example, exposure to gangliosides or mild detergents stimulates the binding of ^{3}H-serotonin by neurectodermal serotonin-binding protein (1025). Thus, the protein could act as an intravesicular store for the amine. Binding of several molecules of serotonin to one macromolecule of neurectodermal serotonin-binding protein would reduce the in-

TABLE 2–2. *Properties of proteins that bind serotonin isolated from neurectoderm-derived tissue and from human blood*

	Brain neurectodermal serotonin-binding protein	Serotonectin (human or rat)
Apparent molecular weight (SDS-PAGE)	45 + 56 kd	200 kd
Carbohydrate content	None	12%
K_D (nM)	4×10^{-10} 0.3×10^{-7}	0.04×10^{-10} 0.1×10^{-7}
Enhancement of serotonin binding by Fe (fold)	15	2–3
Enhancement of binding by gangliosides (fold)	3	None
Inhibition of binding by reserpine (5 μm)	50%	0
Inhibition of binding by Na^+-Ca^{2+} (Krebs solution)	95%	5%

travesicular osmotic pressure below that which would exist if serotonin were in free solution. The extremely tight, high-affinity binding of serotonin by neurectodermal serotonin-binding protein would assure that there would be very little free serotonin within synaptic vesicles until the neurectodermal serotonin-binding protein was saturated. Nevertheless, at the time of exocytosis the vesicular contents are exposed to increased concentrations of Na^+ and Ca^{2+}. This would cause neurectodermal serotonin-binding protein to lose its affinity for serotonin and the amine would then be able to diffuse away from its binding protein to reach serotonergic receptors across the synaptic gap. Such a mechanism would help to prevent swelling of synaptic vesicles despite their very high content of serotonin (651). Neurectodermal serotonin-binding protein may play another role. The final step of the synthesis of serotonin occurs in the cytosol of serotonergic axon terminals (see Chapter 1). This cytosolic store of newly synthesized serotonin will be supplemented by serotonin transported into the serotonergic terminals by the Na^+-dependent, energy-requiring uptake mechanism of the neuronal cell membrane (388,903). The cytosolic serotonin, taken up or newly synthesized, must be loaded into synaptic vesicles for subsequent release from the neurons by exocytosis. A variety of mechanisms may contribute to this transport of serotonin into synaptic vesicles. However, the binding of the monoamine to a macromolecule localized within the interior of synaptic vesicles would greatly facilitate the process. Indeed, by removing free serotonin entering the vesicles, it would tend to favor a continued inwardly directed cytosol-to-vesicle concentration gradient of the serotonergic transmitter.

An unresolved problem for this formulation, which assigns a critical role to neurectodermal serotonin-binding protein in the physiology of serotonergic synaptic vesicles, is how to explain the recycling of the latter. Although it is not known whether or not serotonin-containing synaptic vesicles, or their membranes, can be reused, other types of neuronal vesicles are recycled (475,736,1050). If this is the case for serotonergic vesicles, it is not clear whether they can retain or refill their stores of neurectodermal serotonin-binding protein. Possibly, the protein could be retained despite the release of soluble vesicle contents during exocytosis if it were bound as a peripheral protein to the inner surface of the membrane of synaptic vesicles. If so, then only a portion of the total vesicular content of neurectodermal serotonin-binding protein might be lost as a result of transmitter release, and the remaining bound portion would be retrieved by endocytosis along with the vesicular membrane during the recovery phase following stimulation.

It is uncertain whether or not a nonvesicular store of neurectodermal serotonin-binding protein is present in the neuronal cytosol. Whenever brain or enteric nervous tissues are homogenized, there is always a considerable amount of neurectodermal serotonin-binding protein in the final high-speed supernatant (1026,1028). Unfortunately, the supernatant also contains an equally large amount of serotonin and it is impossible to distinguish material derived from broken or leaking synaptic vesicles from true components of the cytosol. Nevertheless, serotonin is protected from monoamine oxidase when bound to neurectodermal serotonin-binding protein

(1023); thus, one of the forms of the protein may be truly cytosolic and function to protect cytosolic serotonin.

After dissociation from neurectodermal serotonin-binding protein, the free amine can bind to its receptor, it can bind to the uptake site on the presynaptic neuron, or, if it is released from a peripheral neuron and diffuses away, it can reach the circulation.

Serotonectin

Once serotonin enters the circulation it comes into contact with plasma proteins that instantaneously bind it. For example, although the binding of serotonin by albumin is not of high affinity (853), the large amount of albumin in the blood accounts for a considerable fraction of the protein-bound serotonin in circulation. The blood also contains another protein, serotonectin, that binds the monoamine with a relatively high affinity (1024,1031,1032). Serotonectin is a glycoprotein and differs in this respect from neurectodermal serotonin-binding protein. Serotonectin, moreover, is not found in synaptosomes, whereas neurectodermal serotonin-binding protein has not been detected in the circulation. These points illustrate major differences between the two molecules.

Serotonectin was first found to be associated with rat platelets; however, it could be washed away from the plasma membrane of platelets with physiological salt solutions without disrupting the lipid bilayer of that membrane (1031). Serotonectin thus behaves as a peripheral membrane protein, bound to the outer face of the plasma membrane of the platelets. The platelet-bound and free circulating forms of serotonectin probably are in equilibrium. Rat serotonectin has been purified and antibodies have been obtained against it. The use of these antibodies for immunoprecipitation permitted the demonstration that serotonectin is synthesized in blood-forming organs such as the spleen and the bone marrow. Immunocytochemical localization of serotonectin by means of these antibodies confirms, at the light and electron microscopic levels, that serotonectin is associated with the plasma membrane of platelets, but not with that of other blood constituents or of endothelial cells. The immunocytochemical studies reveal no internal serotonectin-like immunoreactivity in the cytoplasm of the platelets. Within the plasma membrane itself, serotonectin seems to be associated only with the most peripheral part of the platelet membrane in direct contact with the blood, since the surface-connecting system of the platelets (an invagination of the plasma membrane) does not bind antibodies against serotonectin. Exposure of platelets to antibodies to serotonectin antagonizes the uptake of serotonin; however, serotonectin itself is not the molecule that transports serotonin through the platelet plasma membrane. Indeed, ^{3}H-imipramine, a marker for the transport system, does not bind to serotonectin and it continues to bind to platelets even when serotonectin has been removed from platelet surfaces.

Serotonectin is present in man (1032). Human serotonectin, like rat serotonectin, is a glycoprotein with a molecular weight of about 200 kd (Table 2–2). It shares

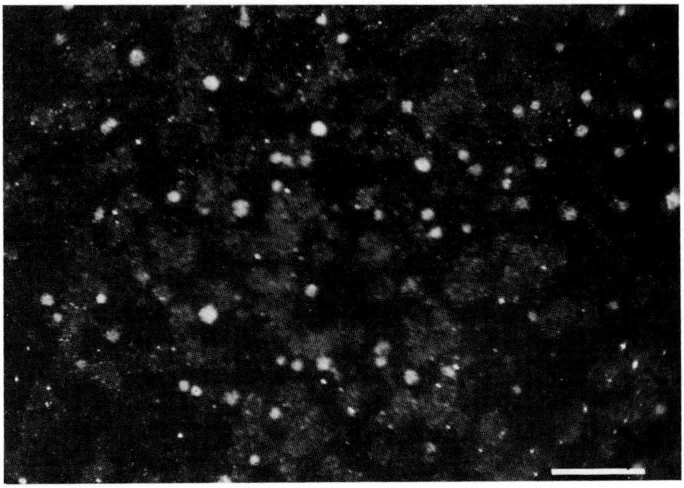

FIG. 2–4. Immunofluorescence of serotonectin on human platelets. A smear of human peripheral blood was fixed and processed for the immunocytochemical demonstration of human serotonectin. The cells were exposed to a monospecific polyclonal antiserum and the serotonectin immunoreactivity was located using a tetramethylrhodamine isothiocyanate-labeled secondary antiserum. Platelets demonstrate bright immunofluorescence while red cells in the field do not. Marker = 30 μm.

many characteristics with rat serotonectin, such as its affinity for serotonin and platelets (Fig. 2–4) and its inhibitory effect on the uptake of serotonin. Human serotonectin has also been purified and used to generate antibodies. One significant difference between human and rat serotonectin is that the human protein binds to more cells than that of the rat. Human serotonectin also binds to granulocytes and mononuclear cells, although not as tightly as it binds to platelets. In the bone marrow, human serotonectin is associated with eosinophilic metamyelocytes and megakaryocytes, indicating a possible site of synthesis of the protein in the marrow.

Serotonectin, as well as albumin, will tend to reduce the concentration of free serotonin in the blood. In addition, the binding of serotonectin to platelet surfaces might facilitate the uptake of serotonin. This interpretation is supported not only by the ability of antibodies to serotonectin to interfere with the uptake of serotonin, but also by the higher efficacy of platelet uptake in the presence of plasma containing serotonectin than in that of physiological salt solution, in which serotonectin is washed away from the surface of the platelets. Platelet aggregation induced by serotonin is not affected by antibodies against serotonectin. It is unknown what abnormality might result from a deficiency in serotonectin or from abnormal binding of serotonectin to its acceptor sites.

ACKNOWLEDGMENT

This work was supported by NIH grants NS12969, NS15547, and BNS NIMH 37575.

— 3 —

Peripheral Metabolism of Serotonin

C. Norman Gillis

Departments of Pharmacology and Anesthesiology, Yale University School of Medicine, New Haven, Connecticut 06510

SEROTONIN REMOVAL FROM BLOOD

Most of the serotonin in the body is found in enterochromaffin cells of the intestine, the central nervous system, and blood. About 90% of the amine in blood is present in a functionally inactive form in small, electron-dense granules within the platelets (1048). The remainder, which under physiological conditions can be as much as 20 ng/ml, is free in plasma (197). Normally, this concentration of free amine is too low to produce measurable cardiovascular effects. However, under certain conditions, notably the carcinoid syndrome (131), experimental pulmonary embolization (914), or endotoxin shock (998), free serotonin concentrations in blood may increase to the point where significant cardiovascular or other actions might be anticipated. Two processes, however, will tend to limit any increase in the concentration of free serotonin in the blood. These are (a) rapid removal of the amine from blood by endothelial cells of blood vessels (especially those of the lung), and (b) enzymatic degradation by the A isozyme of monoamine oxidase, particularly in the liver. The purpose of this chapter is to describe these processes which, together, can restore the blood serotonin level to normal. In addition, consideration is given to possible consequences to the organism as a whole that might follow disruption of these mechanisms, as is postulated to occur in certain diseases.

If serotonin is released from enterochromaffin cells or carcinoid tumors of the intestine into portal blood, it first passes through the liver, which can bring about a substantial inactivation of the amine. The serotonin gradient between portal and hepatic veins in the anesthetized dog normally averages 30%; however, when serotonin is continuously infused into the portal vein to mimic intestinal release of amine in the carcinoid syndrome, the liver extraction increases to as much as 80% (289). Thomas and Vane (1038) confirmed this observation, measuring hepatic venous blood concentrations of serotonin by means of a blood-bathed bioassay technique. Removal of serotonin by the liver is largely a reflection of enzymatic deamination by monoamine oxidase with formation, predominantly, of 5-hydroxyindoleacetic acid.

Free serotonin that escapes deamination in the liver enters the hepatic vein and is then transported to the lung. Within seconds, most of the amine is removed by an avid uptake process and is then rapidly oxidized by monoamine oxidase. In fact, this type of serotonin removal is one of the nonventilatory functions of the lung (409,546). Serotonin can also be taken up by other means. For example, it has been shown in experiments with mice (117), rats, and guinea pigs (49,931) that intravenously administered serotonin can be accumulated and stored by the lung in structures that protect it from monoamine oxidase. This form of storage, to some extent, could reflect margination of platelets that take up serotonin and are then trapped within the lung circulation. Support for this suggestion is found in experiments (117) in which fluoxetine, which inhibits uptake of serotonin by platelets, decreased the retention by the lung of unchanged serotonin, 15 min or 2 hr after its intravenous administration. In addition, some accumulation may occur in adrenergic nerve endings (1105), mast cells, or neuroepithelial bodies (546) within the lung. These processes, while they may contribute to storage of administered amine, occur at too slow a rate to account for the rapid disappearance of serotonin following intravenous injection or infusion, when over 90% is removed during the 3 to 6 sec taken for a single passage through the pulmonary circulation (1038). Rather, such rapid removal of free serotonin is believed to be mediated by a transport process at the luminal membrane of endothelial cells in the pulmonary microvasculature.

The ability of lung to remove serotonin was first observed about 60 years ago when Starling and Verney (996) reported that they could not adequately perfuse an isolated canine kidney with defibrinated blood because of slowly developing vasoconstriction in the preparation. In contrast, however, they found that the kidneys could be perfused effectively if the blood first passed through a heart-lung preparation. Accordingly, Starling and Verney concluded that a vasoconstrictor substance was removed from blood during its passage through the lungs. It remained for Gaddum, Hebb, and Silver (378), 30 years later, to identify the vasoconstrictor substance as serotonin. In addition to serotonin, it now seems likely that other blood-borne vasoactive substances that are removed by the lung [for example, certain prostaglandins (54)] also may have contributed to the responses observed by Starling and Verney. In 1967 Thomas and Vane reported that lungs of anesthetized dogs removed up to 90% of intravenous infusions of serotonin. Several reviews of the removal of free serotonin by the lung (Table 3–1) are available

TABLE 3–1. *Properties of serotonin removal by the lung*[a,b]

K_m (μM)	2–5	Monoamine oxidase inhibition	0
Low sodium	↓	6-Hydroxydopamine	0
Hypothermia	↓	Reserpine	0
High potassium	↓	After removal	Deamination followed by release of 5-hydroxyindole acetic acid
Ouabain	↓		
Cocaine	↓		
Imipramine, desmethyl imipramine	↓		

[a]Symbols: ↓, Inhibition; 0, no effect.
[b]Material assembled from the following review articles: Bakhle and Vane, Junod, Steinberg and Fisher, Gillis and Pitt, Block and Stalcup (54,102,409,546,999).

(54,102,409,546,999); the interested reader is referred to these overview papers for more detailed citations than can be presented here.

The removal of serotonin by the lung is a very rapid process, which inactivates the amine within seconds of its intravenous injection or release from storage sites. The characteristics of the removal process suggest a carrier-mediated mechanism. Thus, it is temperature-sensitive, energy-dependent, and saturable in isolated perfused lungs of dog, rat, rabbit, and guinea pig. The apparent K_m, determined by several different techniques, is about 2 μM. Furthermore, removal is inhibited by hypothermia, low sodium or high potassium, cardiac glycosides, cocaine, and imipramine. This spectrum of properties seems similar to those displayed by the well-documented uptake and storage of norepinephrine [or serotonin (117,1105)] by adrenergic neurons. However, several important differences also exist. Thus, after its removal by the lung, serotonin is extensively deaminated by monoamine oxidase. Since this oxidation is inhibited by harmaline or clorgyline, both of which specifically inhibit the A isozyme of monoamine oxidase, apparently this form of the enzyme is involved (922) (Fig. 3–1). The primary metabolite formed is 5-hydroxyindoleacetic acid, although small amounts of 5-hydroxytryptophol could also be detected in the effluent medium from rabbit lungs perfused with serotonin (201; see Chapter 1). Even within the duration of a single transpulmonary transit *in vivo*, there is substantial formation of 5-hydroxyindoleacetic acid, which can be detected in carotid arterial blood (152). In fact, enzymatic degradation is so rapid that when isolated lungs are perfused for only 5 min with ^{14}C-serotonin, virtually all of the radioactivity remaining in the lungs is in the form of the deaminated product, 5-hydroxyindoleacetic acid (546). Unlike the metabolism of serotonin in the brain, no methylation of the amine can be detected in the lung, even when *S*-adenosyl methionine is provided as a methyl donor (201). Pretreatment of animals with 6-hydroxydopamine (516), which causes chemical denervation of adrenergically innervated tissues, does not alter the removal of serotonin, implying that the process is independent of adrenergic neurons in pulmonary blood vessels. Interestingly, however, in lungs from rabbits treated with 6-hydroxydopamine, the vasoconstrictor

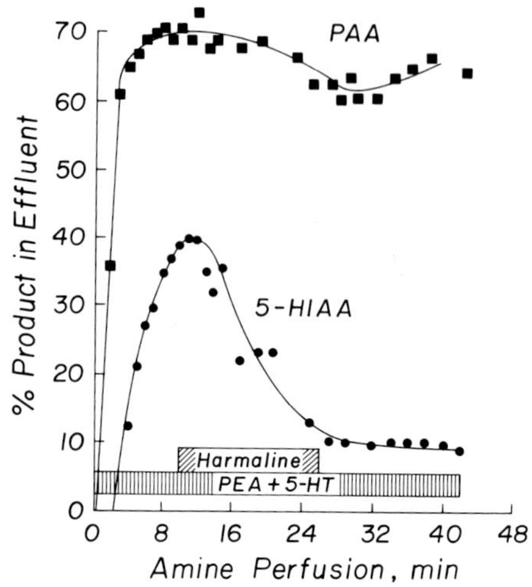

FIG. 3–1. Formation and subsequent appearance of deaminated metabolites from the rabbit lung perfused simultaneously with 0.34 μM ^{14}C-phenylethylamine and ^{3}H-5-hydroxytryptamine. Harmaline (8 μM), an inhibitor of the type A isozyme of monoamine oxidase, was present in the perfusion medium during the time interval shown by the *upper horizontal bar.* Note the inhibition of 5-hydroxyindoleacetic acid (5-HIAA) formation from 5-hydroxytryptamine and the relative *absence* of harmaline-induced inhibition of phenylacetic acid (PAA) from phenylethylamine. (Redrawn from ref. 922.)

response to infused serotonin was significantly greater than in lungs from control animals (516). The reason for this potentiation is unknown. Lungs from reserpine-treated animals also take up and metabolize serotonin normally (546). As long as 1 week after its intravenous injection, serotonin can still be found in lungs of mice and rats (49,117). This form of uptake is unrelated to endothelial cell properties and probably reflects a small amount of serotonin which is taken up and stored in adrenergic nerves (1105).

By means of fluorescence histochemistry and autoradiography, several groups have demonstrated that the main site for rapid removal of serotonin is the endothelial cell of capillaries and small blood vessels. However, these demonstrations require that monoamine oxidase be inhibited in order to avoid the normal deamination of serotonin after its transport into endothelial cells. Apparently, therefore, the A isozyme of monoamine oxidase resides in these cells (922). Concomitant with its removal and deamination, the biological activity of serotonin is reduced by over 90% in a single passage through the lungs of the rat and the dog (15,1038). These observations emphasize that the pulmonary removal and the subsequent deamination of serotonin protect the left heart and the systemic circulation from monoamine released from platelets or other tissue stores into venous blood. Although inhibitors of monoamine oxidase prevent intrapulmonary deamination of serotonin after its uptake, they do not affect the rate of amine removal (409,1038). This suggests that transport of serotonin into endothelial cells, rather than intraendothelial deamination, is the rate-limiting step in overall removal. The removal of serotonin by the lung shows some degree of structural specificity since it is able to distinguish between serotonin and structurally related compounds. For example, *N*-acetyl-5-methoxytryptamine (melatonin) is not taken up by lung after intravenous injection (585). Also, alpha-methyl serotonin has about two-thirds of the affinity of serotonin for removal by the rat lung (15). Furthermore, no enzymatic degradation of alpha-methyl serotonin occurs, because the alpha-methyl group makes this substance an unsuitable substrate for monoamine oxidase.

Lungs of all mammalian species studied so far have the ability to remove serotonin. Recently, indicator dilution techniques (Fig. 3–2) have been used extensively to study kinetic features of the removal of serotonin by lungs of dogs, anesthetized rabbits, and patients undergoing cardiopulmonary bypass for myocardial revascularization (406,408,897). Like those of other species, lungs of anesthetized humans remove between 80 and 90% of a bolus injection of serotonin given into the right atrium via the proximal port of a Swan-Ganz catheter. The multiple indicator dilution technique has proved to be very useful in studying the possible physiological and pathophysiological roles of nonrespiratory lung functions (406). Specifically, this technique has permitted repeated measurement of serotonin removal in the same animal before and after certain drugs or other treatments. In this way, it has been demonstrated that the mechanism of pulmonary removal of serotonin in the anesthetized animal is similar to that seen in experiments carried out with perfused lungs *in vitro*. Thus, removal of serotonin *in vivo* is inhibited by cocaine or imipramine, is saturable, and is followed by intrapulmonary oxidative deamination (152,153).

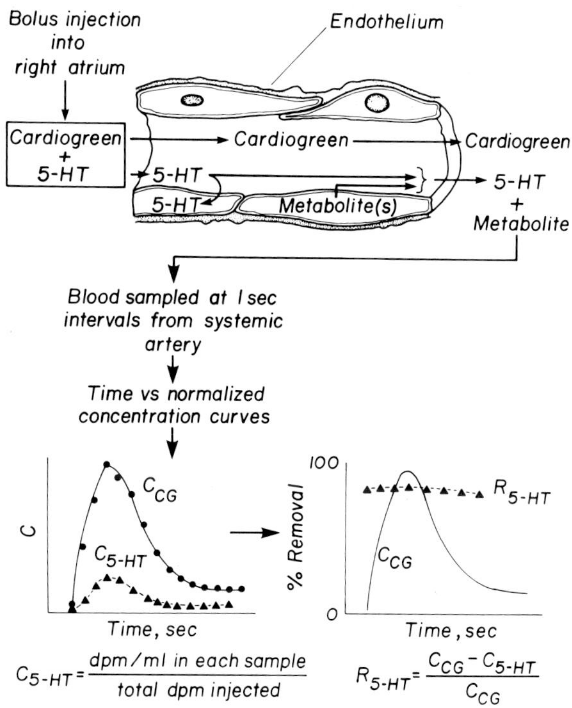

FIG. 3–2. Principles underlying the application of multiple indicator dilution techniques to measure the removal of serotonin by the lung. For further details see refs. 406 and 897.

In view of the ability of platelets to sequester serotonin, it is reasonable to question whether this process is responsible for some of the amine removal. The mean transit time of blood through the pulmonary vasculature of rabbits, dogs, and humans normally lies within the range of 4 to 10 sec. Within this time, the uptake of serotonin by platelets is very small. Thus, in the dog there is no significant uptake of serotonin by platelets during the time taken for a single passage of blood through the lung (897). Similar results are obtained in the rabbit *(unpublished observations)*. Thus, although platelets take up serotonin from plasma, the process is too slow relative to the uptake by endothelial cells to explain single-pass removal of free serotonin. In fact, Thomas and Vane (1038) reported that the half-time of serotonin in dog blood (reflecting platelet uptake) was 60 to 120 sec. Therefore, over longer time intervals and many circulation times, uptake of serotonin by platelets will contribute significantly to overall removal, as will uptake into other structures, including mast cells and adrenergic nerves.

From the foregoing description it is clear that any serotonin released by, for example, platelet aggregation will be rapidly inactivated by endothelial transport and subsequent oxidative deamination in the lung. The endothelial cells in other vascular beds may also contribute to the removal of serotonin. Thus Thomas and Vane (1038) reported that about 45% of infused serotonin disappeared in a single passage through the peripheral vasculature of the dog. In their experiments, serotonin was infused into the ascending aorta and detected by superfusion of mixed venous (right ventricular) blood over rat stomach strips, which served as a bioassay organ. In agreement with these studies, 50% of an intraarterial bolus injection of serotonin disappeared in a single passage through the hindlimb vascular bed of anesthetized dogs (855). This process was imipramine-sensitive. It is unlikely that this reflects inhibition of serotonin uptake by adrenergic neurons, in view of the rapidity of removal. The inhibitory effect of imipramine establishes that the amine is not merely diffusing from the vascular space. Therefore, although there is no direct evidence (for example, by histochemical localization of serotonin), it seems probable that the disappearance of the amine reflects uptake and metabolism by the endothelial cells of the skeletal muscle vascular bed. Work carried out on isolated endothelial cells supports this interpretation. Thus, endothelial cells isolated from bovine aorta (980), porcine aorta and pulmonary artery (547), human umbilical vein (980), rat epididymal fat pads (905), and rat brain (451) take up serotonin by a saturable process which is inhibited by hypothermia, ouabain, cocaine, or imipramine. In view of this sensitivity to inhibitors of active uptake, and the saturability of the process in the extrapulmonary vascular beds and in the various endothelial cells studied, it is probable that serotonin removal occurs in all such cells. However, it has been argued that the very high surface area/volume ratio of the pulmonary microvasculature endows the endothelium of this vascular bed with a quantitatively more significant role in serotonin removal than any other organ of the body (405,546). Indeed, the strategic anatomic location of the lung between right and left heart, coupled with the fact that the entire cardiac output passes through the pulmonary circulation several times a minute, places the lung in a unique position to serve as a metabolic filter (409).

Several other lung structures are capable of taking up and storing serotonin. These include mast cells, Kultschitzy cells, neuroepithelial bodies (999), and adrenergic nerve endings of blood vessels (117,1105). It seems quite likely that any serotonin that escapes deamination in endothelial cells might well be transferred to these cells and thus account for the prolonged retention of unchanged amine (49,117). However, as mentioned earlier, these processes are too slow to account for the rapid inactivation of free serotonin in blood.

At first sight, the proposal that the lung plays perhaps as significant a role in metabolizing circulating serotonin as does the liver may seem unusual. However, it should be recalled that drug clearance (i.e., the volume of blood cleared of drug per unit time) is a function of both the rate of transport (or metabolism) *and* the blood flow (or rate of drug delivery) to the organ. Since blood flow to the liver averages 30% of the cardiac output, which is, of course, essentially the blood flow through the lungs, it follows that the lung could have much less intrinsic ability to transport or metabolize a drug and yet be as effective as the liver in this regard. For drugs that are metabolized under first-order conditions by liver enzymes, the intrinsic clearance (i.e., the maximum ability of the liver to remove the drug from blood or perfusion medium in the absence of flow limitation) can be calculated from the ratio V_{max}/K_m. The kinetic constants can be derived by standard techniques *in vitro* (872). Wiersma and Roth (1150) provided evidence that the liver of the rat, perfused at a normal flow rate, clears serotonin at roughly the rate calculated from the ratio of V_{max} and K_m for liver monoamine oxidase. In marked contrast, however, the clearance of serotonin by the lung of the same animal, also perfused at physiological flow for that organ, was over three times that predicted from the kinetic constants for lung monoamine oxidase. Clearly, therefore, the removal of serotonin by the lung depends on processes other than deamination by monoamine oxidase, and in fact reflects the very high rate of removal into pulmonary endothelial cells. The absence of a similar process in the liver may be due to the fact that the endothelial lining of the hepatic vascular spaces is discontinuous and thus probably allows more direct access to the monoamine oxidase of the liver.

SEROTONIN REMOVAL: RELATION TO CARDIOVASCULAR FUNCTION

The predominant mechanism whereby serotonin removal by the pulmonary endothelium might influence cardiovascular function presumably is by changing the concentration of serotonin in the blood and thus in the environment of vascular or cardiac receptors. Since serotonin causes or facilitates platelet aggregation, removal of the amine by the lung normally may prevent systemic arterial thrombogenesis (409). Certainly, quite small changes in the magnitude of removal by the lung could produce dramatic changes in the serotonin concentration of the systemic arterial blood. For example, about 95% of the serotonin is normally inactivated by the lung and only 5% reaches the left atrium. However, a modest decline in the degree of inactivation, perhaps to 85%, could result in a threefold increase in concentration of the amine in the left atrial and hence the systemic arterial blood (409). Thus, given only diminished endothelial removal of serotonin, the systemic vascular response to this substance could be dramatically increased or prolonged (409). In view of the central role of the pulmonary endothelium in this process, it is appropriate now to consider some situations in which drugs or other experimental procedures known to affect these cells have been studied.

Drug-Induced Changes in Endothelial Function

The alkaloid monocrotaline is known to cause both portal and pulmonary hypertension in several species (713). Following treatment with monocrotaline, endothelial cells of rat lungs are swollen and the vasculature shows evidence of thrombosis, as well as lesions in the arterial media. It might therefore be expected that metabolic functions of the endothelial cells would be affected. In fact, the pulmonary removal of serotonin is reduced by about 50% (406). Elevated levels of circulating serotonin might increase pulmonary pressure *in vivo* (812). Therefore, an impaired ability to remove and inactivate serotonin may be related to the pulmonary and perhaps also to the portal hypertension produced by monocrotaline. Administration of this alkaloid also causes interstitial edema in the lungs (713). In this context, it is interesting to note that serotonin widens interendothelial cell gaps (674) and perhaps therefore increases the permeability of pulmonary capillaries. In support of this proposal is the observation that serotonin increased lung lymph flow in sheep (128).

The use of aminorex, an amphetamine derivative formerly used as an anorectic agent, was associated with an unusually high incidence of primary pulmonary hypertension. Subsequently it was shown that aminorex interfered with the removal of serotonin by the lung (960). Not surprisingly, therefore, the pulmonary hypertension was treated successfully with a serotonergic antagonist (999).

Endothelial Function and Hypertension

Lungs of spontaneously hypertensive rats have a reduced capacity to inactivate serotonin (923). Furthermore, platelets of hypertensive patients also have a much reduced ability to take up serotonin; and the platelet turnover is enhanced in hypertensive patients. Both of these effects could lead to increased levels of circulating serotonin (1072,1075; Chapter 12).

Serotonin Removal and Pulmonary Embolization

The free concentration of serotonin in the blood is elevated significantly during pulmonary embolization (914). Thus, experimental pulmonary embolization produced by intravenous infusion of autologous blood clots is associated with a release of serotonin from platelets that aggregate and adhere to the clot (914). It is clear that enough serotonin is released from aggregating platelets to contract canine pulmonary blood vessels (708). Thus it is likely that pulmonary hypertension following embolization could be related to production of serotonin from aggregating platelets. Such embolization decreases lung compliance and increases air flow resistance, effects which are consistent with narrowing of airways (998,1037). Since these effects can be prevented by treatment of the dogs with methysergide, a serotonergic inhibitor, the likely mediator affecting the lung mechanics is serotonin released from platelet aggregates. Plasma concentrations of serotonin are increased threefold after intravenous administration of autologous clots. Coincident with this effect is a 29% decrease in platelet count. The concentration of serotonin in platelets, however, was unchanged. The increased plasma level of serotonin is correlated, to a significant degree, with the increase in pulmonary vascular resistance, but not with effects of the amine on the airways (1064). This suggests that prostaglandins, which are also released during embolization, act to inhibit the endothelial removal of serotonin in the lung. In confirmation of this proposal, the increase in vascular resistance is inhibited by the serotonergic antagonist cyproheptadine, which does not affect the abnormalities in gas exchange associated with embolization (1064). That embolization specifically reduces removal of serotonin is confirmed by the findings that glass bead embolization reduces the uptake of serotonin but not that of norepinephrine by the lung (340). Since there is significant mechanical obstruction and some resultant loss of surface area from glass bead embolization, the effect of progressive pulmonary lobar ligation on removal also has been examined. The removal of serotonin decreases only when all but one lobe of the lung are ligated, although the decrease is less than that produced by glass beads (which probably cause less mechanical obstruction and reduction of vascular surface area). Concomitantly, under these conditions of ligation, the removal of norepinephrine by the lungs also is reduced. These observations indicate that a simple reduction of surface area has a relatively nonspecific effect on the removal of biogenic amines; furthermore, heterogeneity of perfusion may also play a role in this effect. It is also likely that embolization may raise the local concentrations of serotonin sufficiently to cause saturation of the uptake mechanism at endothelial sites.

Serotonin Removal and Cardiovascular Shock

Endotoxin and intestinal ischemic shock have been linked to increased concentrations of serotonin in the blood (524,998). In this case also, the source of the amine probably is platelet aggregation. Radioimmunoassay procedures do not detect a difference in levels of serotonin between venous blood of control rats and of those subject to 30 min of occlusion of the superior mesenteric artery, if measured 15 min after the arterial clamp is released (443). However, by that time any elevation of circulating serotonin must have been dissipated by endothelial uptake as well as by deamination by the liver.

There are other conditions in which there is evidence of impaired removal of serotonin, probably resulting from endothelial damage. These include treatment with paraquat or bleomycin and exposure to high oxygen tension (102,406). In these cases, increased blood levels of serotonin (and other vasoactive hormones normally inactivated by the lung) may be present, and the cardiovascular consequences might attend such an effect on circulating free serotonin.

ACKNOWLEDGMENT

The author wishes to express his appreciation to Dr. Bruce R. Pitt of Yale University for helpful discussion and comments, and to Ms. Suzanne Abrams for her careful secretarial assistance during the preparation of this manuscript. Studies from the author's laboratory referred to in this chapter were supported by grants HL-13315, HL-23245, and HL-27207 from the National Institutes of Health.

— 4 —

Uptake and Storage of Serotonin by Platelets

J. F. Stoltz

Centre Régional de Transfusion et d'Hématologie, Vandoeuvre-lès-Nancy, France

Janeway et al. (523) discovered that a vasoconstrictor appears during coagulation and platelet lysis. Rapport et al. (874) identified the substance as serotonin. The platelets of all mammals contain serotonin at concentrations that vary according to the species. Although originally it was stated that platelets did not possess the enzymes required for synthesizing the amine (1183), more recent investigations have revealed low-level enzyme activity. However, the bulk of the serotonin contained in platelets originates in the enterochromaffin cells of the gastrointestinal tract (115,317,856).

Serotonin is taken up by platelets during their lifetime in the circulating blood. The only blood containing measurable amounts of serotonin in plasma is that in the portal vein, probably because of release of the amine from the enterochromaffin cells of the intestinal mucosa (see Chapter 20). Although platelets contain monoamine oxidase, the amount of serotonin metabolized to 5-hydroxyindoleacetic acid and 5-hydroxytryptophol under normal conditions is very low (856). The serotonin in platelets is not exposed to the action of monoamine oxidase because the amine is localized in dense granules (213). The dense granules also contain nucleotides, cations, proteins, etc. These organelles (1,000 to 2,000 Å in diameter) are surrounded by a single membrane and differ from other subcellular platelet structures such as α-granules, mitochondria, and glycogen particles. The storage organelles for serotonin are formed in the megakaryocytes, where they do not contain high amounts of the monoamine. The concentration of serotonin in platelets varies considerably from one species to another (Table 4–1). The storage of serotonin by

TABLE 4–1. *Serotonin levels in platelets of different animal species*

Species	$\mu g/10^9$ Cells	μmol/mg Protein
Rabbit	11.2[a]	86,000[d]
	7.5[b]	
Cow	5.0	
Pig	3.6[a]	
	2.2[b]	
Dog	1.7[b]	
Rat	1.1[a]	9,000[d]
	0.2–0.4[b]	
Cat	0.9	
Man	0.5[a]	
	0.5–1.5[b]	2,200[d]
	0.15–0.6[c]	
Guinea pig	0.35[a]	820[d]
	0.20–0.4[b]	

Data from refs. [a]295, [b]502, [c]198, [d]214.

platelets is not a simple process, but involves at least three steps: (a) transport through the platelet membrane; (b) transport through the storage-granule membrane; and (c) formation of high-molecular-weight complexes with adenine nucleotides and divalent cations. Equilibrium studies with prolonged incubation periods only indicate the overall contribution of the three steps and do not provide the means of studying the individual components.

PLATELET MEMBRANE

The investigation of the uptake process at the platelet membrane must be conducted under experimental conditions which allow measurement of the initial uptake rate. Incubation periods lasting less than 10 sec should be applied (295). Even then, no certainty exists that only the first step in the uptake of serotonin is assessed. The suspending medium used for the experiments has to be defined accurately. Sneddon (985) emphasized the importance of carrying out the measurements in an artificial medium rather than in plasma to allow control of pH and avoid possible interference by plasma proteins. However, the use of platelet preparations washed in artificial medium may introduce artifacts such as morphological or biological changes. Thus, it is difficult to obtain accurate quantitative data on the mechanisms of serotonin uptake by the platelets. This uptake involves both passive and active transport.

Passive Transport

Passive transport is (a) not connected to energy production; (b) only effective when the extracellular concentration of serotonin is higher than 10 μM; (c) highly temperature-dependent (decreasing the temperature from 37 to 32°C decreases passive transport by 50%); and (d) proportional to the serotonin concentration in the external medium. Passive transport is not affected by agents such as chlorpromazine and dismethylimipramine. Low concentrations of ADP that do not cause platelet aggregation appear to facilitate passive diffusion of serotonin (680).

Active Transport

Active transport is the dominant process when the extracellular concentration of serotonin is below 5 μM. In view of the low levels of serotonin in plasma, this mechanism plays a physiologically important role. Its major characteristics are as follows: (a) It occurs against a high concentration gradient between platelets and plasma (e.g., 1000:1). (b) The relationship between the initial rate of uptake and the substrate concentration obeys Michaelis kinetics (452,502). (c) It is a saturable process. (d) It shows structural specificity, albeit not absolute since platelets also accumulate 5-hydroxy-α-methyltryptamine and 5,6-dihydroxytryptamine. (e) It depends on the extracellular sodium concentration. In the absence of Na^+ there is no transport of serotonin, but as Na^+ is added back to the suspending medium serotonin uptake is stimulated. A linear relationship exists between the Na^+ concentration and the initial rate of uptake, which suggests that Na^+ acts as a cosubstrate with serotonin in the transport process (985). (f) It is decreased or inhibited by some metabolic inhibitors. (g) It is inhibited by a variety of drugs, of which the tricyclic antidepressants are the prototypes (Table 4–2).

TABLE 4–2. *Concentrations of various drugs causing 50% inhibition (IC_{50} in μM) of the effects of serotonin on platelets*

Drugs	5-HT receptor binding	Uptake	Platelet aggregation	Platelet shape change
Chlorimipramine	0.1[a], 0.05[b], 0.5[c]	0.3[a], 0.1[b], 0.2[c], 0.03[d]	50[b]	50[b], 0.4[c], 0.6[d], 0.2[e]
Methysergide	200[a], 0.002[c]	100[a], 310[d]	0.02[b]	0.1[b], 0.005[c], 0.02[d], 0.01[e]
Imipramine	0.01[b], 0.1[c], 0.02[f]	5[b], 0.7[c]	500[b]	500[b], 0.25[c], 0.8[e]
Fluoxetine	0.01[a], 0.1[b]	0.01[a], 5[b]	50[b]	50[b]
Nortriptyline	0.2[a,f]	1[a]		
Chlorpromazine	5[a], 0.02[c]	5[a]		0.03[c,e]
Mianserin	50[a]	80[a]		
Pizotifen	100[a], 0.001[c]	500[a]		0.001[c]
Tryptamine	0.5[b]	50[b]	50[b]	0.5[b], 1.4[e]
Cinanserin	0.01[b], 0.001[c]	500[b]	5[b]	5[b], 0.003[c], 0.05[e]
Cyproheptadine	0.01[b], 0.001[c]	500[b]	0.05[b]	0.5[b], 0.003[c]
Atropine	0.75[c]			1.6[c]
Dopamine		145[d]		192[d]
d-LSD	0.001[c]			0.006[c], 0.004[e]
Desipramine	4.5[c]	7.5[c]		0.4[c]
Tyramine		12[d]		39[d]
Spiroperidol				0.002[e]

Data from refs. [a]848, [b]945, [c]295, [d]614, [e]617, [f]253.

In addition to the physicochemical and biological parameters that control the uptake of serotonin by platelets, the rheological parameters of blood flow also affect it. Thus, in platelets submitted to a shear stress of more than 100 dyne/cm^2 for 5 min the uptake of ^{14}C-serotonin is decreased by more than 40%, and in platelets exposed to a shear stress of approximately 200 dyne/cm^2 by more than 60% (27,1004,1005).

UPTAKE IN DENSE GRANULES

The dense granules are surrounded by a single membrane. They primarily contain serotonin and 5′-di- and triphosphonucleotides (411). The uptake of serotonin into the dense granules is temperature dependent but is unaffected by metabolic antagonists such as iodoacetate or ouabain or by glucose depletion, indicating it may not be an active process. It can be inhibited pharmacologically with reserpine (215).

STORAGE

The content of serotonin in the dense granules, measured after ultracentrifugation separation, varies considerably depending on the species studied (858). For example, granules from platelets of the rabbit contain 21.0 ± 1.7 nmol serotonin/μg protein but those of the guinea pig 0.6 ± 0.2 nmol/μg protein. These differences can be observed by comparing electron microscopic photographs of platelets from different species in specimens fixed with glutaraldehyde and osmium tetroxide. For example, the majority of the dense granules of platelets of the rabbit are highly osmiophilic, whereas a large number of human granules appear empty or only partially filled. These observations indicate that the storage granules in man are not saturated. Similar conclusions have been reached in the guinea pig, where the serotonin level could be increased 10-fold by daily injections of serotonin. The mechanism for storing serotonin in the granules is not completely clear. It probably involves complex formation with ATP and divalent cations (856; see Chapter 7).

β-Adrenoceptor blocking drugs such as alprenolol, atenolol, or propranolol can induce liberation of ^{14}C-serotonin from rat platelets by a mechanism other than the classic release reaction (1051). It is possible that these drugs, owing to their physicochemical properties, interact with phospholipids of the platelet membrane resulting in changes in its microviscosity. Propranolol (1 μM) inhibits the initial rate of transport of serotonin by approximately 50% but does not affect the steady-state level of accumulation (927). Other molecules, such as suloctidil, may cause the leakage of serotonin from the dense granules.

BINDING PROTEINS

Binding proteins exist that contribute to the storage and/or uptake processes of serotonin in various cells. The prototype of these proteins is the specific protein known as serotonectin, which is found on the outer surface of platelets (387). It is a membrane-bound glycoprotein that can be extracted with salt solutions (1031). It is synthetized in blood-forming organs and is probably in equilibrium between plasma and thrombocytes. It does not bind endothelial cells. Since antiserotonectin antibodies interfere with the uptake of serotonin, it can be assumed that serotonectin contributes to the uptake mechanism (see Chapter 2). The protein nature of the receptors that take part in the active transport of serotonin is indicated by the finding that small concentrations of trypsin, which do not induce release, inhibit serotonin uptake in rat platelets (985).

UPTAKE SITES, BINDING SITES, AND SEROTONERGIC RECEPTORS

When the uptake of serotonin by the platelet membrane is determined while active transport is inhibited, the number of binding sites revealed is approximately 10^4 per cell (112). These results were obtained by incubating platelet suspensions at 0°C with radioactive serotonin added in increasing concentrations. Because of the low specific activity of the radiolabeled serotonin available, the receptors involved could not be characterized. In an attempt to reveal the existence of different receptors, Born et al. (114) investigated the specificity of the reactions of serotonin with platelets by comparing the inhibitory effects of 17 analogs and two known antagonists (imipramine and methysergide) on aggregation and uptake. No correlation emerged between the inhibitory potencies of the analogs on uptake and aggregation. With increasing aliphatic substitution on the amino group, the derivatives became increasingly potent in inhibiting aggregation. This study also revealed the different effects of imipramine (uptake inhibitor) and methysergide (aggregation inhibitor). These results, in particular the differences in inhibitory potencies of imipramine and methysergide on the two processes, indicated that two types of binding sites exist on the platelet membrane: one site mediating uptake and the other aggregation (114).

Gaddum and Picarelli (379) originally classified serotonergic receptors as D and M receptors (see Chapter 5). The receptor mediating aggregation induced by serotonin corresponds to the D type (732). Sialic acid may be one of its components since serotonin causes aggregation of platelets enriched with *N*-acetylneuraminic acid (728). However, this interpretation is hard to reconcile with the observation that reduction in platelet surface charge, as caused by sialic acid, increases aggregation induced by both serotonin and ADP. The sialic acid groups are not the only negative ionized groups on the surface of the platelets that contribute to the apparent charge of the cell membrane (1003) and possibly play a role in the response to serotonin. Furthermore, the sialic acid component of the serotonergic receptors mediating aggregation is undoubtedly not a specific feature of these receptors, since uptake is also accelerated with the incorporation of *N*-acetylneuraminic acid, and since incubation of rabbit and human platelets with neuroaminidase reduces the uptake.

On the basis of a Scatchard analysis of ^{3}H-serotonin binding as a function of the concentration of unlabeled serotonin, the cell membrane of rat platelets must contain three different serotonergic interaction sites (site A: 2,500/platelet, site B: 12,500/

platelet, site C: 46,000/platelet). Low concentrations of serotonergic antagonists [e.g., *d*-lysergic acid diethylamide (LSD)] inhibit binding at the high-affinity site without affecting the affinity of the other two (295). The affinity constant for the site with intermediate affinity is comparable with that of the uptake process. However, an analysis based on a Scatchard plot does not allow definite conclusions when used alone. Thus, Shick and McKean (945), using a similar experimental procedure, described only one site for human platelets, but Peters and Grahame-Smith (848) identified two whereby the low-activity site was apparently connected with uptake. This confirms the results of Laubscher and Pletscher (617) who, by comparing a series of neuropsychotropic drugs, also suggested the presence of different receptors mediating the uptake of and the shape changes induced by serotonin. The receptor mediating aggregation can be subtyped further as an S_2-serotonergic receptor (637,841) since the effect of serotonin on platelet aggregation is inhibited by S_2-serotonergic antagonists such as ketanserin.

— 5 —

Serotonergic Binding Sites

J. E. Leysen

Department of Biochemical Pharmacology, Janssen Pharmaceutica, Beerse, Belgium

As early as 1878, Langley defined the receptor concept to explain mutual interaction of drugs on isolated tissues (615). For almost a century, the characterization and classification of receptors was based on observations of the pharmacological effects elicited by drugs. A first classification of serotonergic receptors was proposed by Gaddum and Picarelli (379) who studied serotonin-induced contractions in isolated segments of the ileum and defined M and D receptors. Contractions antagonized by morphine were classified as a phenomenon mediated by an M receptor. Since morphine prevented contractions of the gut by inhibiting the release of acetylcholine following stimulation by serotonin, M receptors were referred to as neuronal serotonergic receptors. D receptors were characterized as the receptors mediating the contraction of enteric smooth muscle induced by serotonin and antagonized by dibenzylamine. In the intact organism serotonergic antagonists were found to cause profound behavioral effects but did not allow a particular classification of serotonergic receptors (518).

The advent of highly labeled radioactive drugs opened new perspectives for the study of receptors at the molecular level, and radioligand receptor binding became a powerful research tool. Before a radioligand binding site is classified as a receptor a role of the binding site in pharmacological or physiological effects must be demonstrated. Unfortunately, this prerequisite, inherent to the receptor concept, is often neglected or not fulfilled.

The first attempts to label serotonergic receptors in homogenates of brain tissue used ^{3}H-lysergic acid diethylamide (LSD) as ligand and an equilibrium dialysis method (332). The breakthrough came with the use of the rapid filtration technique for separating membrane-bound and free radioactivity. Bennett and Snyder (74,75) studied serotonergic receptor binding in mammalian brain using both ^{3}H-LSD and ^{3}H-serotonin and detected high-affinity saturable binding sites with both ligands. The notable differences between the ^{3}H-LSD and ^{3}H-serotonin binding led to the suggestion of a "two-state" receptor-model, with ^{3}H-serotonin labeling the agonist state and ^{3}H-LSD labeling the antagonist state. One year later it was found that ^{3}H-spiperone, a neuroleptic used as dopaminergic ligand, labeled a population of sites in the rat frontal cortex showing the features of serotonergic binding sites. The sites were essentially similar to those labeled with ^{3}H-LSD. Among a variety of neurotransmitters, serotonin showed the highest binding affinity for the sites and the binding affinity of drugs correlated with their potency to antagonize the excitatory behavior mediated by serotonin in rodents (641,642). This was the first demonstration of a relationship between affinities for a binding site and a response to serotonin. Peroutka and Snyder (841) challenged the earlier proposed two-state receptor model and supplied evidence that cortical binding sites labeled by ^{3}H-serotonin and by ^{3}H-spiperone are distinct molecular entities. The notation 5-HT$_1$ (S$_1$-serotonergic) and 5-HT$_2$ (S$_2$-serotonergic) sites was introduced, predominantly based on their binding properties. ^{3}H-LSD appeared to label both the S$_1$- and the S$_2$-serotonergic sites. These binding sites, in particular the S$_1$-serotonergic types, have been further subclassified. This chapter reviews the properties and the distribution of the S$_1$- and S$_2$-serotonergic binding sites. Emphasis is placed on the relationship between the binding and actions of serotonin as identified by *in vivo* and *in vitro* pharmacological studies.

METHODS FOR BINDING STUDIES *IN VITRO*

Dissected tissues are homogenized in ice-cold buffered medium and membranes are isolated and washed by centrifugation procedures. The final membrane pellets are suspended in buffer in a high dilution (2–10 mg original wet weight of tissue/ml). Small aliquots (1–5 ml) of the membrane suspension are incubated with a low concentration (nanomolar order) of a radioactive ligand. The incubation is stopped by rapid filtration of the incubation mixture over glass fiber filters, to separate membrane-bound and free radioactivity. Membrane-bound radioactivity, retained on the glass fiber filters, is counted in a liquid scintillation spectrometer.

The specific binding of the radioactive ligand is defined as the difference in bound radioactivity in assays in the absence and the presence of excess of an unlabeled competitor. The total number of binding sites (B_{max} value) and the binding affinity of the radioactive ligand, given by the equilibrium dissociation constant (K_D value), are derived from radioligand binding-concentration curves. These are often analyzed in Scatchard plots (940), which are a mathematical transformation of the law of mass action: linear Scatchard plots indicate a simple interaction of the labeled ligand with a single binding site, while curvilinear Scatchard plots point to complex interactions probably involving multiple binding sites.

Binding affinities of unlabeled drugs are derived from binding-inhibition curves, obtained through incubating the radioactive ligand in the presence of increasing concentrations of the unlabeled drug. The IC_{50} value is the drug concentration causing 50% inhibition of the specific binding of the radioactive ligand. The equilibrium inhibition constant (K_I value) is calculated according to Cheng and Prusoff (163) as $K_I = IC_{50}/[1 + C/K_D]$ where C is the concentration and K_D the equilibrium dissociation constant of the radioactive ligand. Analysis of the shape of the inhibition curves, either graphically (e.g., in Hill plots; 954) or with a computer (259,260), reveals whether the labeled and unlabeled drugs compete for a single binding site or whether multiple binding sites may be involved.

DETECTION OF SEROTONERGIC BINDING SITES

S_1-Serotonergic Sites

S_1-Serotonergic sites are labeled by ^{3}H-serotonin in nanomolar concentrations in membrane preparations of mammalian brain. For optimal assay conditions freshly purified ^{3}H-serotonin must be used and 4 mM $CaCl_2$ must be present in the incubation mixture, as well as an optimal concentration of ascorbic acid. Variations in assay conditions have often been the cause of inconsistent findings (see refs. 634,757,758). Methodological problems have hampered the quantification of S_1-serotonergic binding sites and the assessment of the binding affinity of ^{3}H-serotonin; variations in K_D values between 3 and 12 nM have been reported. There is, however, fair agreement about the general binding properties of the S_1-serotonergic sites labeled with ^{3}H-serotonin.

^{3}H-Serotonin binding is potently inhibited by serotonin-like indole derivatives. Most ergot derivatives, which have an indole moiety in their structure, bind with high affinity to the S_1-serotonergic sites. Known serotonergic antagonists, lacking the indole structure, show very low or no binding affinity for the sites, except for methiothepin (Fig. 5–1). Drugs that are referred to as serotonergic antagonists are not selective for serotonergic binding sites (Table 5–1). The lack of specificity of most of these drugs can be appreciated from the K_I values measured in different binding tests (Table 5–2). For example, methiothepin is an extremely nonselective ligand.

Besides the fact that most antagonists only poorly inhibit the binding of ^{3}H-serotonin, certain compounds, in particular spiperone, produce shallow inhibition curves. This was interpreted by Pedigo et al. (836) as indicating multiple binding sites. They introduced a subclassification of S_1-serotonergic sites into 5-HT_{1A} sites and 5-HT_{1B} sites. The 5-HT_{1A} sites, detected in the upper part of the inhibition curves, have been characterized by high-affinity binding for serotonin and for

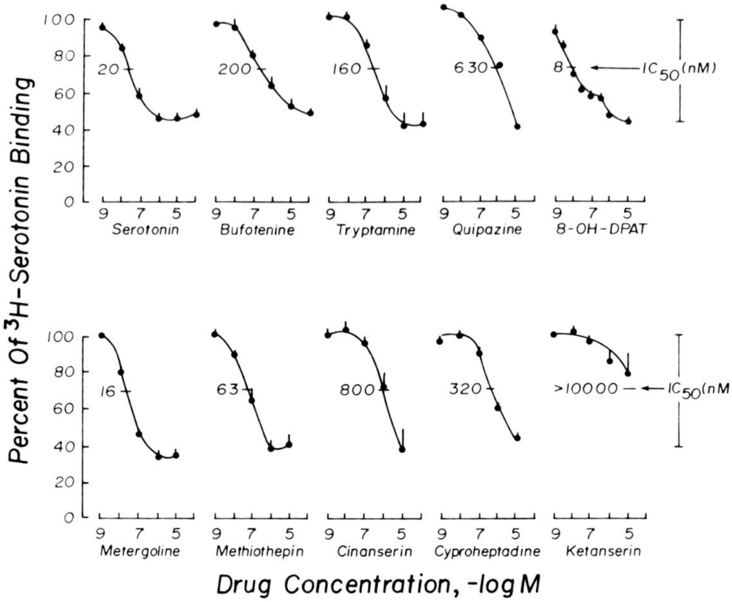

FIG. 5–1. Inhibition by serotonergic agonists and antagonists of ^{3}H-serotonin binding in rat hippocampal membranes. Assays were run at 37°C using 3 nM ^{3}H-serotonin in Tris-HCl buffer (pH 7.6), containing 4 mM $CaCl_2$. Specific binding was defined as the portion of binding inhibited by 1 μM LSD. Rapid filtration was used to harvest labeled membranes (634). IC_{50} values indicate concentrations inhibiting 50% of the specific binding of the radioligand.

serotonergic antagonists. The 5-HT_{1B} binding sites, represented in the lower part of the inhibition curves, show a high affinity for agonists but a low one for antagonists. The heterogeneity of S_1-serotonergic binding sites was further substantiated by investigations with the serotonergic agonist 8-hydroxy-2(di-*n*-propylamino)tetralin (8-OH-DPAT; chemical abstract name: 7-(dipropylamino)-5,6,7,8-tetrahydro-1-naphthalenol). 8-OH-DPAT was the first agonist without indole moiety which potently inhibited ^{3}H-serotonin binding; however, the inhibition curve revealed an intermediary plateau (Fig. 5–1). Middlemiss and Fozard (735), using purportedly specific binding assay conditions for measuring 5-HT_{1A} and 5-HT_{1B} sites, found high-affinity binding of 8-OH-DPAT to 5-HT_{1A} sites only, which suggested that the compound could be a useful tool to explore the significance of the 5-HT_{1A} binding sites. Unlike 5,6-dihydroxy- or 6,7-dihydroxy 2-aminotetralins and 5-, or 6-, or 7-monohydroxy 2-aminotetralins, which possess dopaminergic agonistic properties and bind to dopaminergic sites (144), 8-OH-DPAT may be a selective S_1-serotonergic agonist (480), as also appears from its receptor-binding profile (Tables 5–1 and 5–2). The compound can elicit a serotonergic behavioral syndrome; however, in contrast to the behavioral changes induced by 5-hydroxytryptophan (518), the effect of 8-OH-DPAT cannot be antagonized by serotonergic antagonists such as cyproheptadine and metergoline (480). ^{3}H-8-OH-DPAT used as radioligand labels sites in hippocampal and striatal tissue which show somewhat

TABLE 5–1. *Potencies of compounds, belonging to different pharmacological classes, to interact with biochemical models for serotonergic receptors*

	S_2-Serotonergic[a] ^{3}H-Ketanserin Frontal cortex	S_1-Serotonergic[b] ^{3}H-Serotonin Hippocampus	Serotonin-stimulated adenylate cyclase[c]
Serotonergic antagonists			
Metergoline	0.28	20	18,100
Ketanserin	0.39	>>1,000	—
Methiothepin	0.39	62	11,900
Cyproheptadine	0.44	700	11,500
Pipamperone	0.78	5,000	—
Methysergide	0.94	99	—
Mianserin	1.4	1,100	27,600
Cinanserin	2.0	3,100	21,500
LSD	2.5	20	—
Dopaminergic antagonists			
Spiperone	0.53	160	3,800
Chlorpromazine	3.3	3,100	—
Haloperidol	22	7,900	—
Histaminergic antagonists			
Astemizole	5.3	7,900	—
Pyrilamine	>>10,000	>>10,000	—
Adrenergic antagonists			
Phentolamine	187	7,900	—
Propranolol	590	400	—
Yohimbine	660	210	—
Prazosin	>>10,000	>>10,000	—
Amine uptake blockers			
Amitriptyline	4.2	1,200	—
Imipramine	37	>>10,000	—
Citalopram	1,660	—	—
Serotonergic agonists			
Bufotenine	118	62	—
Quipazine	235	1,600	—
Serotonin	296	7.8	—
Tryptamine	1,500	197	—
Mescaline	5,900	70,000	—
8-OH-DPAT	>>10,000	4	—
Dopaminergic agonists			
Dopamine	93,500	20,000	—
5,6-di-OH-DPAT	23,500	>10,000	—
Various transmitters			
Epinephrine	47,000	—	—
Norepinephrine	>>100,000	200,000	—
Histamine	>>100,000	—	—
Acetylcholine	120,000	>>100,000	—

>>*x:* No inhibition of binding at *x*, the highest concentration tested.
[a]K_i values (nanomolar) for inhibition of specific ^{3}H-ligand binding (643).
[b]K_i values (nanomolar) for inhibition of specific binding (637,643).
[c]IC_{50} values for inhibition of serotonin-stimulated cyclic AMP formation (774).

TABLE 5–2. K_I *values (nanomolar) for the interaction with various neurotransmitter receptor-binding models*

Receptor: Ligand: Tissue:	H_1-Histaminergic ^{3}H-Pyrilamine Cerebellum	α_1-Adrenergic ^{3}H-WB 4101 Forebrain	α_2-Adrenergic ^{3}H-Clonidine Cortex	D_2-Dopaminergic ^{3}H-Haloperidol Striatum	Muscarinic ^{3}H-Dexetimide Striatum
Serotonergic antagonists					
Metergoline	1,100	38	380	23	>>10,000
Ketanserin	10	10	>>10,000	220	>>10,000
Methiothepin	4.9	0.47	48	4.0	>>10,000
Cyproheptadine	2.7	100	760	31	19
Pipamperone	>>10,000	54	610	120	>>10,000
Methysergide	>>10,000	2,300	2,600	200	>>10,000
Mianserin	2.9	82	60	620	>>10,000
Cinanserin	1,200	1,200	>10,000	1,600	>>10,000
LSD	>>10,000	160	58	20	>>10,000
Dopaminergic antagonists					
Spiperone	>>10,000	10	>>10,000	0.16	>>10,000
Chlorpromazine	6.9	1.7	1,000	18	162
Haloperidol	>>10,000	8.0	>>10,000	1.2	>>10,000
Histaminergic antagonists					
Astemizole	0.80	6.6	>>10,000	>>10,000	>>10,000
Pyrilamine	0.87	3,500	>>10,000	>>10,000	>>10,000

Adrenergic antagonists					
Phentolamine	880	2.8	5.2	2,800	>>10,000
Propranolol	>>1,000	12,000	>>10,000	>>1,000	>>10,000
Yohimbine	>>1,000	410	150	590	>>10,000
Prazosin	>>1,000	0.58	4,700	>>1,000	>>10,000
Amine uptake blockers					
Amitriptyline	0.74	30	740	210	20
Imipramine	11	72	>1,000	390	120
Citalopram	300	820	>>10,000	5,000	8,000
Serotonergic agonists					
Bufotenine	800	3,600	3,000	>>10,000	>>10,000
Quipazine	630	1,300	7,400	40,000	25,000
Serotonin	>>10,000	19,000	33,000	12,000	>>10,000
Tryptamine	9,000	13,000	17,000	40,000	>>10,000
Mescaline	>>1,000	>>1,000	>>1,000	>>10,000	>>10,000
8-OH-DPAT	>10,000	1,030	480	2,200	7,800
Dopaminergic agonists					
Dopamine	>>100,000	17,000	140	140	>>10,000
5,6-di-OH-DPAT	16,000	2,200	32	13	39,000
Various transmitters					
Epinephrine	>>100,000	440	9	620	>>10,000
Norepinephrine	>>100,000	820	7.4	2,200	>>10,000
Histamine	14,000	>>100,000	19,000	>>100,000	>>10,000
Acetylcholine	>>100,000	>>100,000	>>100,000	>>100,000	39,000

Values for serotonergic and dopaminergic antagonists are from ref. 637. Values for other compounds are unpublished data, determined as described in ref. 637.

different binding properties (420,685). In hippocampal tissue a higher number of sites are present, with a K_D value of about 4 nM, whereas in the striatum the K_D value is 11 nM. Most serotonergic agonists and antagonists are less potent inhibitors of the binding of ^{3}H-8-OH-DPAT in the striatum than in the hippocampus. The binding properties of the sites labeled with ^{3}H-8-OH-DPAT in the hippocampus show close similarities with those of S_1-serotonergic sites labeled by ^{3}H-serotonin; the correlation with the properties of the striatal binding sites for ^{3}H-8-OH-DPAT is not as good (Table 5–3). Because of its particular pharmacological properties, ^{3}H-8-OH-DPAT is perhaps the most appropriate substance to trace the role of S_1-serotonergic binding sites.

S_2-Serotonergic Sites

S_2-Serotonergic sites were first identified using ^{3}H-spiperone in frontal cortex tissue (641,642). The sites were characterized by a high binding affinity, in the nanomolar range, for known serotonergic antagonists belonging to different chemical classes. Serotonin itself has a relatively low affinity (micromolar range) for S_2-serotonergic binding sites, although it still binds at least 100-fold more potently to these sites than other neurotransmitters. The use of ^{3}H-spiperone in the S_2-serotonergic binding studies suffered from the disadvantage that the ligand also labels dopaminergic receptors (Tables 5–1 and 5–2). This hampered the study of S_2-serotonergic sites in areas rich in dopaminergic receptors (633). Several other radioligands have been tried to study S_2-serotonergic sites. ^{3}H-LSD labels not only both S_1- and S_2-serotonergic binding sites, but also dopaminergic receptors [reviewed by Leysen (634)]. In the frontal cortex, which is rich in S_2-serotonergic receptors, binding sites labeled with ^{3}H-LSD and those labeled with ^{3}H-spiperone have the same properties (634). Another isotope of LSD, ^{125}I-LSD, is more selective for S_2-serotonergic sites, and because of its very high specific radioactivity it may be used for autoradiographic studies (309,457). ^{3}H-Mianserin, which is a tetracyclic, mixed serotonergic and histaminergic antagonist, labels both S_2-serotonergic and H_1-histaminergic binding sites (844; see Tables 5–1 and 5–2). To render this ligand suitable for S_2-serotonergic binding studies, the addition of a selective H_1-histaminergic receptor blocker (e.g., triprolidine) is required.

^{3}H-Ketanserin, a serotonergic antagonist that primarily binds to S_2-serotonergic receptors, lacks interaction with S_1-serotonergic binding sites (637,643). Actually, nanomolar concentrations ($\leq$5 nM) of this ligand label solely S_2-serotonergic sites. H_1-Histaminergic and α_1-adrenergic sites, for which ketanserin shows a moderate affinity (Table 5–2), are apparently not labeled at the low nanomolar concentrations of ^{3}H-ketanserin which are used in the S_2-serotonergic binding studies.

The K_D values of ^{3}H-ketanserin for S_2-serotonergic sites, assayed at 37°C, range from 0.4 to 1 nM. The specific ^{3}H-ketanserin binding, as defined by the displacement by a 1,000-fold excess of methysergide, follows regular equilibrium binding kinetics, which, according to the law of mass action, conforms with the interaction at a single site. Bound ^{3}H-ketanserin can be displaced by serotonergic agonists and

TABLE 5–3. *Correlation between drug potencies to inhibit the binding of various radioligands to S_1- and S_2-serotonergic sites; K_i values of at least 18 drugs in a wide activity scale are considered*

	Spearman rank correlation coefficients								
	I	II	III	IV	V	VI	VII	VIII	IX
S_1-Serotonergic sites									
(I) ^{3}H-Serotonin, rat hippocampus		$r_s = .912$ $p < .001$***	$r_s = .869$ $p < .001$***	$r_s = -.095$ $p > .05$	$r_s = -.017$ $p > .05$	$r_s = -.017$ $p > .05$	$r_s = -.154$ $p > .05$	$r_s = -.017$ $p > .05$	$r_s = -.018$ $p > .05$
(II) ^{3}H-8-OH-DPAT, rat hippocampus			$r_s = .860$ $p < .001$***	$r_s = .214$ $p > .05$	$r_s = .400$ $p > .05$	$r_s = .400$ $p > .05$	$r_s = .564$ $p > .05$	$r_s = .400$ $p > .05$	$r_s = .378$ $p > .05$
(III) ^{3}H-8-OH-DPAT, rat striatum				$r_s = .143$ $p > .05$	$r_s = .417$ $p > .05$	$r_s = .417$ $p > .05$	$r_s = .564$ $p > .05$	$r_s = .417$ $p > .05$	$r_s = .270$ $p > .05$
S_2-Serotonergic sites									
(IV) ^{3}H-Spiperone, rat frontal cortex					$r_s = .960$ $p < .001$***	$r_s = .967$ $p < .001$***	$r_s = .924$ $p < .001$***	$r_s = .984$ $p < .001$***	$r_s = .907$ $p < .001$***
(V) ^{3}H-Ketanserin, rat frontal cortex						$r_s = .949$ $p < .001$***	$r_s = .887$ $p < .001$***	$r_s = .979$ $p < .001$***	$r_s = .879$ $p < .001$***
(VI) ^{3}H-Ketanserin, rat striatum							$r_s = .890$ $p < .001$***	$r_s = .973$ $p < .001$***	$r_s = .893$ $p < .001$***
(VII) ^{3}H-Ketanserin, human frontal cortex								$r_s = .890$ $p < .001$***	$r_s = .953$ $p < .001$***
(VIII) ^{3}H-Ketanserin, cat platelets									$r_s = .958$ $p < .001$***
(IX) ^{3}H-LSD, human platelets									

antagonists from various chemical classes. Serotonin itself has an IC_{50} of 1,000 nM ($K_I = 300$ nM); bufotenine is two and a half times more active, and tryptamine five times less active than serotonin. 8-OH-DPAT shows no appreciable binding affinity for S_2-serotonergic sites. Other neurotransmitters, such as histamine and norepinephrine, do not bind to S_2-serotonergic sites up to millimolar concentrations; dopamine and epinephrine are at least 100 times weaker than serotonin. In contrast, all the serotonergic antagonists have an IC_{50} value in the nanomolar range, although they do not necessarily interact selectively with serotonergic receptors (Tables 5–1 and 5–2). Furthermore, compounds known for a different pharmacological property can bind potently to S_2-serotonergic sites. For example, amitriptyline, which blocks the uptake of serotonin (see Chapter 4), has a binding affinity for S_2-serotonergic sites in the nanomolar range.

The S_1- and S_2-serotonergic sites are unrelated (Figs. 5–1 and 5–2; Tables 5–1 and 5–3). However, S_2-serotonergic sites detected with different ^{3}H-ligands in different tissues from different species have similar binding properties. This is illustrated by the highly significant correlations between the binding affinities of drugs for S_2-serotonergic sites labeled with ^{3}H-ketanserin in membrane preparations of different tissues, such as rat and human prefrontal cortex, rat striatum, and cat platelets, with ^{3}H-spiperone in membranes from the rat frontal cortex, or with ^{3}H-LSD in membranes from human platelets (Table 5–3).

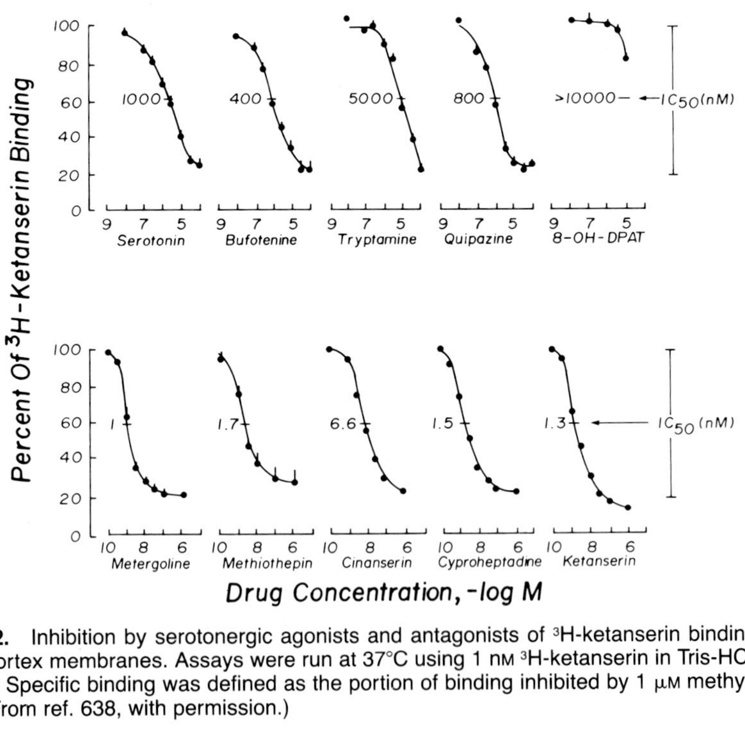

FIG. 5–2. Inhibition by serotonergic agonists and antagonists of ^{3}H-ketanserin binding in rat frontal cortex membranes. Assays were run at 37°C using 1 nM ^{3}H-ketanserin in Tris-HCl buffer (pH 7.7). Specific binding was defined as the portion of binding inhibited by 1 μM methysergide (643). (From ref. 638, with permission.)

Serotonin-Stimulated Adenylate Cyclase

In the colliculi of newborn rats micromolar concentrations of serotonin stimulate adenylate cyclase activity. The physiological role of this cyclase is thus far unknown. Serotonergic antagonists inhibit this effect poorly or not at all, thus suggesting that binding sites that show decreased affinity for agonists in the presence of guanyl nucleotides are coupled to an adenylate cyclase (906). Hence S_1-serotonergic binding sites were supposed to be linked to serotonin-stimulated adenylate cyclase (838,839). However, no correlation exists between the potency of drugs to inhibit ^{3}H-serotonin binding to S_1-serotonergic sites and that to reduce serotonin-stimulated cyclic AMP production (Table 5–1). The S_1-serotonergic binding sites and the serotonin-stimulated adenylate cyclase exhibit a distinct regional and subcellular distribution. This refutes the possibility of a coupling between the two systems (634,644).

LOCALIZATION OF SEROTONERGIC BINDING SITES

In the brain, S_1-serotonergic binding sites are most numerous in the hippocampus and the striatum, but S_2-serotonergic sites are concentrated in the prefrontal cortex, the tuberculum olfactorium, and the nucleus accumbens (Table 5–4). The distribution of S_1- or of S_2-serotonergic sites does not correspond to the regional distribution of specific serotonin uptake or of serotonin content in the brain areas (644,645). This absence of correspondence between density of binding sites and

TABLE 5–4. *Distribution of S_1- and S_2-serotonergic binding sites and of serotonin uptake in subcortical areas of the rat brain*

	S_2	S_1	Uptake
Prefrontal cortex	100	100	100
Tuberculum olfactorium	56	67	156
Nucleus accumbens	40	—	92
Striatum	36	150	93
Substantia nigra	15	—	131
Thalamus	10	82	132
Hypothalamus	9	—	109
Medulla oblongata	10	—	87
Hippocampus	12	198	67
Spinal cord	1	—	48
Cerebellum	0	0	17

One hundred percent values in prefrontal cortex tissue represented: for S_2-serotonergic binding sites (using 2 nM ^{3}H-ketanserin): 21 ± 6 fmol/mg wet weight tissue (N = 12); for S_1-serotonergic binding sites (using 3 nM ^{3}H-serotonin): 13 ± 1 fmol/mg wet weight tissue (N = 9); and for uptake (using 10 nM H-serotonin): 150 ± 30 fmol/mg wet weight tissue, 5 min.

degree of serotonergic innervation could be used as an argument against a role of the binding sites in neurotransmission (58). However, the ratio between functional and spare receptors and the relationship between receptor density, receptor function, and intensity of elicited effect still are unknown factors. The distribution of S_2-serotonergic binding sites is similar in the brains of various mammalian species, including humans (645,952). Moreover, the ontogenetic development of the S_2-serotonergic binding sites in rat brain reveals that the sites are detectable from the sixteenth day after gestation. A marked development occurs during the second postnatal week. This parallels the development of the neuronal system and that of receptor sites for various other neurotransmitters (133). These observations favor a probable role of S_2-serotonergic sites in neurotransmission. In contrast, the number of S_1-serotonergic sites is high during early embryonic development but decreases substantially during the first 2 weeks of life. Since serotonin may play a role in cell differentiation, the decrease in the number of S_1-serotonergic sites shortly after birth could be linked to a reduced involvement of serotonergic sites in this process (1065).

Information on the cellular localization of the serotonergic binding sites in the brain is obtained from studies of neuronal lesions in rats. Regarding the S_1-serotonergic sites, different findings were made with different ^{3}H-ligands. Destruction of catecholaminergic neurons following local injection of 6-hydroxydopamine does not affect the binding of ^{3}H-serotonin in the forebrain. This suggests that the number of S_1-serotonergic sites on dopaminergic or noradrenergic neurons is too low to be detected. Lesions of neuronal cell bodies in the striatum, by the local application of kainic acid, decrease the high-affinity binding of ^{3}H-serotonin, but apparently not that of ^{3}H-8-OH-DPAT in the striatum. However, the binding of ^{3}H-8-OH-DPAT in the hippocampus decreases after injection of kainic acid in this area. Destruction of serotonergic neurons by 5,6- or 6,7-dihydroxytryptamine (see Chapter 6) does not alter the binding of ^{3}H-serotonin in the forebrain nor that of ^{3}H-8-OH-DPAT in the hippocampus; however, binding of ^{3}H-8-OH-DPAT in the striatum seems decreased. The S_1-serotonergic binding sites affected by kainic acid could be postsynaptic, whereas the striatal sites to which ^{3}H-8-OH-DPAT binds probably are localized on serotonergic nerve terminals (420,634).

S_2-Serotonergic binding sites labeled by ^{3}H-ketanserin in the frontal cortex and the striatum are not affected by the destruction of either catecholaminergic or serotonergic neurons, which indicates that these sites are not located on the terminals of these neurons (639,645). "Up-regulation" of S_2-serotonergic binding sites does not occur after denervation of the serotonergic neurons. This would imply that the serotonergic system is regulated differently than other neuronal systems, such as the dopaminergic system where denervation supersensitivity is observed commonly. High doses of kainic acid in the frontal cortex cause a decrease in S_2-serotonergic binding sites accompanied by a decrease in the activity of glutamic acid decarboxylase, the enzyme contained in γ-aminobutyric acid (GABA)-ergic neurons. Hence, S_2-serotonergic binding sites may be associated with GABA-ergic neurons (645).

In peripheral tissues the concentration of serotonergic binding sites is very low. Thus, all attempts to detect S_2-serotonergic binding sites in vascular tissues have remained virtually negative, although there is ample evidence for a functional role of S_2-serotonergic receptors in this tissue. On platelet membranes, however, S_2-serotonergic binding sites have been detected (383,640).

It can be inferred from the localization studies that data on the distribution of binding sites or on the effects of neuronal lesions, *per se*, do not provide sufficient proof for either attributing or denying a functional role to serotonergic binding sites. On the one hand, this is the consequence of our lack of information regarding the ratio between functionally active and so-called spare receptor sites. On the other hand, it appears that binding studies still are too much prone to artefacts and such findings need extremely critical evaluation, a point that is often neglected.

ROLE OF SEROTONERGIC BINDING SITES

Correlations and the Investigation of the Functional Role of Binding Sites

A crucial issue in radioligand binding research is the demonstration of a functional role of the discovered binding sites. The most readily available means to attempt this is the determination of a correlation between binding affinities of drugs and their potencies in functional tests. The value of correlations has been much criticized, and they are often misused. Nevertheless, such correlations can allow acceptable conclusions provided large series of drugs are considered spanning a wide activity scale and comprising compounds of different chemical classes. When *in vitro* binding is compared with *in vivo* pharmacological potency, factors such as resorption, distribution, and metabolism of the drugs must be kept in mind and the deviation of a single compound from a well-established correlation line should not lead to hypothesizing receptor subclasses. However, when *in vitro* binding is compared with *in vitro* pharmacological observations both the rank order of potencies and the absolute activities of drugs must correspond in both tests.

The investigation of the role of S_2-serotonergic binding sites has been much more successful than that of S_1-serotonergic sites. This is because all known serotonergic antagonists, independent of their chemical class, bind to the S_2-serotonergic sites, and the binding affinity of the individual drugs varies from the subnanomolar to the micromolar range. In contrast, very few compounds bind potently to S_1-serotonergic sites, and most are indole derivatives with agonistic properties.

In Vivo Effects of Serotonergic Agonists

In rodents, the administration of precursors of serotonin or of serotonergic agonists elicits multiple behavioral effects (Table 5–5). The effects reflect excitation (head twitches, forepaw treading, myoclonus, wet dog shakes), impairment of blood circulation (cyanosis, gastric lesions), and subjective perceptions of the drugs by the animals (discriminative stimulus). Studies involving large series of serotonergic antagonists show that the affinities of the drugs for S_2-serotonergic binding sites

TABLE 5–5. *Effects of serotonergic agonists* in vivo

Serotonergic agonist	Observation	Receptor site	References
5-Hydroxytryptophan + monoamine oxidase inhibitor	Serotonin syndrome, head twitches, forepaw treading, cyanosis, hindlimb abduction, . . . , gastric lesions, dead (rats, mice)	S_2-Serotonergic	*a,b*
5-Hydroxytryptophan	Head twitches (rats, guinea pigs)	S_2-Serotonergic	*c*
Mescaline	Head twitches (rats, mice)	S_2-Serotonergic	*d,e*
Tryptamine	Forepaw treading (rats)	S_2-Serotonergic	*f*
5-Hydroxytryptophan	Myoclonus (guinea pig)	Multiple serotonergic receptors?	*g*
5-Hydroxytryptophan, LSD, quipazine	Discriminative stimulus (rats)	S_2-Serotonergic Component	*h* *i,j*
Compound 48/80	Gastric lesions (rats)	S_2-Serotonergic	*k*
Serotonin	Paw edema (rats)	S_2-Serotonergic	*b*

Data from refs. [a]518, [b]812, [c]839, [d]643, [e]779, [f]642, [g]669, [h]186, [i]410, [j]360, [k]47.

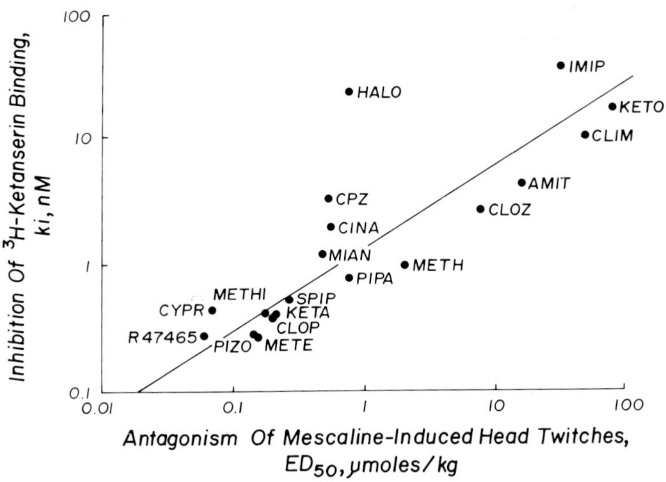

FIG. 5–3. Correlation between the affinity of drugs for S_1-serotonergic binding sites (K_I values measured in ^{3}H-ketanserin binding tests using rat frontal cortex membranes) and their potency to antagonize mescaline-induced head twitches in rats. (Data from ref. 779.) Abbreviations are the first letters of the generic names of the drugs as described in ref. 643. (From ref. 643, with permission.)

correlate with their potencies to antagonize the behavioral excitation induced by serotonin (Fig. 5–3 and Table 5–5). The binding affinities are correlated similarly with the potencies to antagonize serotonin-induced gastric lesions (47) or serotonin-induced edema of the paw (810). This suggests that S_2-serotonergic binding sites play a role in the behavioral excitation, the circulatory impairment, and the inflammation caused by the monoamine. Studies with a limited number of drugs suggest the involvement of multiple, not further identified, serotonergic receptors in 5-hydroxytryptophan-induced myoclonus in the guinea pig (669). Discriminative stimuli induced in the rat by serotonergic agonists also have an S_2-serotonergic component (Fig. 5–4; 186,360,410). The subjective feelings (anxiety, neurotic depression, tension) perceived by animals following treatment with drugs such as LSD and quipazine are probably mediated in part by S_2-serotonergic binding sites. Likewise, serotonin may be involved in sexual behavior (7), sleep (458), and pain perception (512); studies related to these effects have used only a few serotonergic agonists and antagonists, which does not allow proper identification and characterization of the subtype of binding sites mediating the response.

Reliable proof for a role of S_1-serotonergic binding sites in the behavioral effects of serotonin is not available.

In Vitro Effects of Serotonergic Agonists on Smooth Muscle

Serotonin has multiple effects on isolated smooth muscle of various organs (Table 5–6). M and D receptors have not yet found a corresponding radioligand-binding model (635,643). Thus, no significant correlation exists between the affinity of various drugs for S_1- and S_2-serotonergic binding sites and their potency to antagonize serotonin-induced contractions of the fundus of the rat, which probably is the best model for D receptors (Figs. 5–5 right and 5–6 right). Since very few drugs act on M receptors, they probably are different from both S_1- and S_2-serotonergic binding sites. By contrast, a close correspondence is found between the affinity of drugs for S_2-serotonergic binding sites and their potency to antagonize serotonin-induced contractions of the caudal artery of the rat (Fig. 5–6; 634,643).

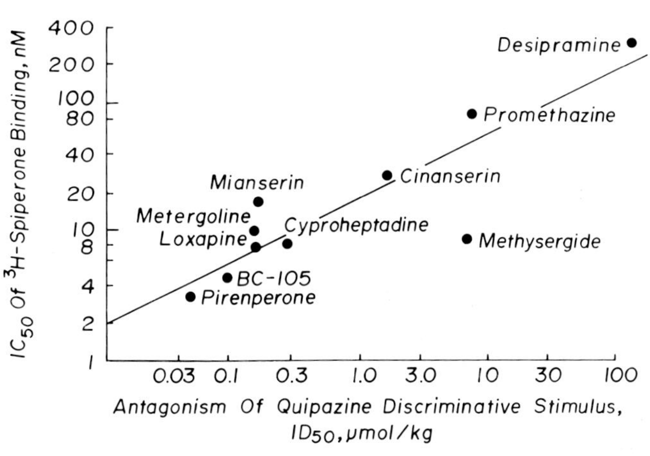

FIG. 5–4. Correlation ($r = 0.87$) between the affinity of drugs for S_2-serotonergic binding sites (IC_{50} values measured in ^3H-spiperone binding tests using rat frontal cortex membranes) and their potency to antagonize quipazine-induced discriminative stimuli in rats. (From ref. 360, with permission.)

TABLE 5–6. *Serotonin-induced effects in isolated tissues*

Tissue and serotonin-induced effect	Typical antagonists	Receptor site	References
Gastrointestinal smooth muscle			
Contraction of ileum	Morphine	M receptor	*a*
Contraction of ileum, fundus	Phenoxybenzamine	D receptor	*a,b*
Vascular smooth muscle			
Contraction of			
Rat caudal artery	Ketanserin	S_2-Serotonergic	*c,d*
Rabbit femoral artery	Cyproheptadine		*e,f,g*
Dog basilar artery	Methysergide		
Contraction of			
Dog saphenous vein	Phentolamine	α_1-Adrenergic	*c*
Rabbit ear artery	Prazosin		
Relaxation of contracted vessels			
Dog coronary artery	Methiothepin Methysergide	?	*f,g*
Trachea, guinea pig	Ketanserin Cyproheptadine Methysergide	S_2-Serotonergic	*h*

Data from refs. [a]379, [b]1069, [c]1093, [d]643, [e]1092, [f]760, [g]1095, [h]1097.

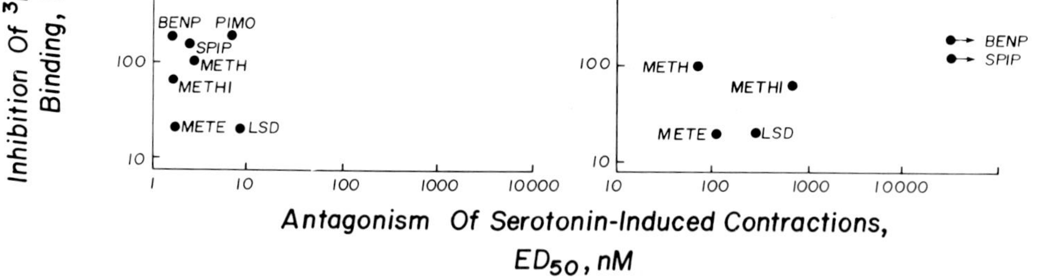

FIG. 5–5. Absence of correlation between the affinity of drugs for S_1-serotonergic binding sites (K_I values measured in ^{3}H-serotonin binding tests using rat hippocampal membranes) and their potency to antagonize **(left)** serotonin-induced contractions of isolated caudal arteries of the rat (data from ref. 1093) or **(right)** serotonin-induced contractions of the isolated rat fundus (data from ref. 1094). (From ref. 635, with permission.)

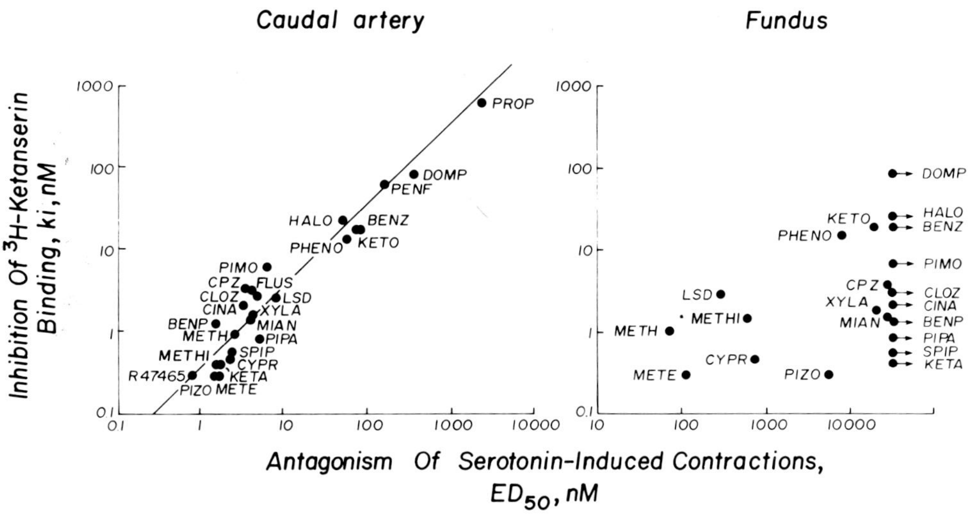

FIG. 5–6. Correlation between the affinity of drugs for S_2-serotonergic binding sites (K_i values measured in ^{3}H-ketanserin binding tests using rat frontal cortex membranes) and their potency to antagonize **(left)** serotonin-induced contractions in isolated rat caudal artery (1093) and **(right)** serotonin-induced contractions of the isolated fundus of the rat (1094). (From ref. 635, with permission.)

Serotonin-induced contractions of the femoral artery of the rabbit apparently are also mediated by S_2-serotonergic receptors (1092). To elicit contraction of isolated blood vessels, concentrations of 10^{-7} to 10^{-6} M serotonin are required (1088). These concentrations correspond to the affinity of serotonin for S_2-serotonergic binding sites (Fig. 5–2). Thus, a role of S_2-serotonergic binding sites as true receptors is most likely in vascular smooth muscle. The failure to label S_2-serotonergic binding sites in homogenates of blood vessels probably reflects a low receptor density despite a high functional activity. S_1-Serotonergic binding sites are not involved in the contractions of the caudal artery of the rat induced by serotonin (Fig. 5–5 left). In the canine basilar artery Peroutka et al. (840) suggested that the contractile response to serotonin is mediated by S_1-serotonergic receptors. However, the action of antagonists on contractions caused by serotonin in this blood vessel is complex and does not obey the criteria of competitivity (1095). Thus, a more thorough analysis suggested that S_2-serotonergic receptors contribute to the serotonin-induced contraction of the basilar artery (760,1095).

S_2-Serotonergic binding sites also mediate the amplifying effect of serotonin on the contractions of isolated blood vessels induced by various stimuli (1094,1096; see Chapter 9).

Serotonin at higher concentrations can interact with α_1-adrenergic receptors in certain vascular tissues (Table 5–6). These effects of the monoamine are antagonized by α_1-adrenergic blockers.

Effects of Serotonin on Neurotransmitter Release

S_1-Serotonergic sites may play a role in the modulation by serotonin of the release of several neurotransmitters, but none of the proposed hypotheses have been substantiated irrefutably. Thus, the suggested involvement of S_1-serotonergic sites in an apparent modulation by serotonin of the release of ^{3}H-dopamine evoked by potassium from striatal slices (311) has been refuted (644). Activation of S_1-serotonergic binding sites may contribute to the attenuation by serotonin of its own release in cerebral tissue ("autoreceptor" inhibition; 308,692). Martin and Sanders-Bush (692) found that the potency of a limited number of serotonergic agonists to inhibit potassium-stimulated release of ^{3}H-serotonin from hypothalamic synaptosomes correlated with their binding affinity for S_1-serotonergic sites. However, the relative affinity of antagonists to S_1-serotonergic binding sites (metergoline>methiothepin>methysergide) (644,774,841) did not correspond to their potency to reverse inhibition of ^{3}H-serotonin release by agonists. Engel et al. (308) favored the interpretation that the autoreceptors are S_1-serotonergic binding sites based on the finding of a rank correlation between the affinity of indoleamines for S_1-serotonergic binding sites and the inhibition of electrically evoked release of ^{3}H-serotonin from cortical tissue. However, no correlation was present when serotonergic antagonists were considered. Engel et al. (308) further suggested that the inhibition by serotonin of norepinephrine release from sympathetic nerve endings in the canine saphenous vein is mediated by S_1-serotonergic sites. This interpretation awaits confirmation by studies using appropriate antagonists.

Effects of Serotonin on Neuronal Firing

Serotonin exerts both excitatory and inhibitory effects on neuronal firing. From electrophysiological studies a subclassification of serotonergic receptors has been proposed (909) which is quite distinct from that derived from binding studies. Peroutka et al. (839) suggested that the excitatory serotonergic receptors defined electrophysiologically correspond to S_2-serotonergic binding sites and that neuronal inhibition by serotonin could be mediated by S_1-serotonergic sites. Neither assumption is substantiated by direct comparison of drug potencies in binding assays and electrophysiological tests. From the limited data available, few similarities are obvious between the electrophysiologically defined receptors and receptor-binding sites.

S_2-Serotonergic Binding Sites and Depression

Peroutka and Snyder (842,843) observed that chronic treatment of rats with certain antidepressant agents, in particular tricyclic drugs, causes a down-regulation of S_2-serotonergic and β-adrenergic binding sites. The slow development of the down-regulation was thought to correspond to the slow onset of the therapeutic action of these drugs. Hence, it was suggested that S_2-serotonergic binding sites may play a role in the etiology of depression. However, the mechanism by which

S_2-serotonergic sites are down-regulated is not fully understood (29). Down-regulation is observed only with a limited number of antidepressant drugs (842). Potent, and selective, blockers of the uptake of serotonin, such as citalopram, are ineffective (514). Repeated electroshock apparently causes up-regulation of S_2-serotonergic binding sites (564). No information seems available on the effect of specific S_2-serotonergic antagonists in the treatment of depression. However, preliminary clinical studies indicate that ritanserin [a selective, potent, and long-acting S_2-serotonergic antagonist (39,156,780,894)] possesses anxiolytic properties.

THE PLATELET AS AN INTEGRAL MODEL FOR THE STUDY OF S_2-SEROTONERGIC RECEPTORS

Serotonin affects platelet function (235,242; see Chapters 4, 7, and 15). S_2-Serotonergic binding sites were demonstrated in platelet membranes of cat and human with radioligand-binding techniques, using ^{3}H-ketanserin and ^{3}H-LSD (383,640). The specific binding sites on the platelet membranes display the same properties as the S_2-serotonergic sites in the brain (Fig. 5–7 and Table 5–3). In platelets of the cat the potency of serotonin to induce aggregation ($K_m = 600$ nM) matched that to inhibit ^{3}H-ketanserin binding ($K_I = 300$ nM). The potency of antagonists to inhibit the serotonin-induced aggregation closely corresponds to their affinity for S_2-serotonergic binding sites (Fig. 5–8) (242). The inactivity, in both tests, of a specific serotonin uptake blocker such as citalopram demonstrates that the effect of serotonin on aggregation is unrelated to its uptake into the platelets (242,640; see Chapter 4). The affinity of drugs for S_2-serotonergic binding sites, detected with ^{3}H-LSD in membranes of human platelets, correlates with their potency to antagonize serotonin-induced shape changes (383). The changes in the

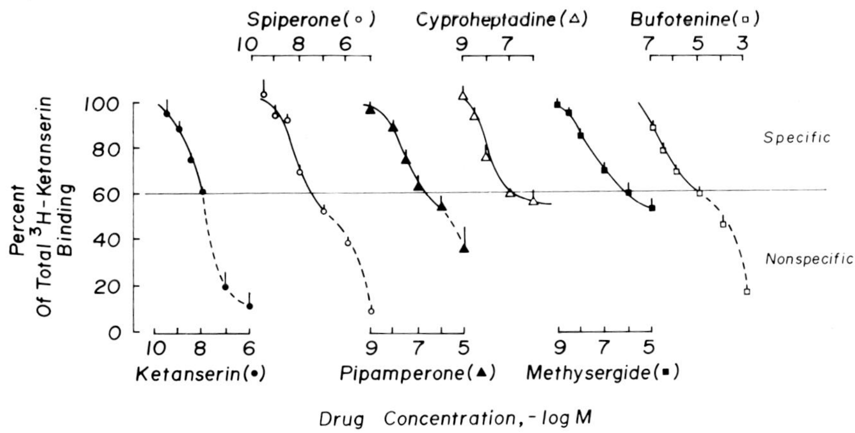

FIG. 5–7. Inhibition by various drugs of ^{3}H-ketanserin binding (2 nM) to membrane preparations of cat platelets. Forty percent of the ^{3}H-ketanserin binding is inhibited by the various serotonergic antagonists and agonists, which represents specific binding to S_2-serotonergic sites. Another 40% is inhibited only by compounds that show structural similarity to ketanserin; it represents non-specific binding (see 636). (From ref. 640, with permission.)

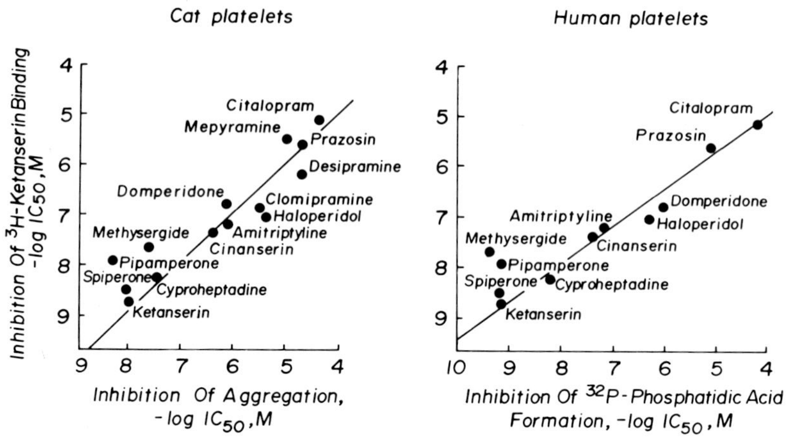

FIG. 5–8. Correlation between the affinity of drugs for S_2-serotonergic binding sites (IC_{50} values measured in ^{3}H-ketanserin binding tests using cat platelet membranes) and their potency to inhibit serotonin-induced **(left)** aggregation of cat platelets (data from ref. 243); or **(right)** increases in ^{32}P-phosphatidic acid formation in human platelets (data from ref. 229). (From ref. 638, with permission.)

turnover of phospholipids induced by serotonin and the subsequent phosphorylation of 20K and 40K proteins also are triggered by receptors with characteristics similar to those of S_2-serotonergic binding sites (Fig. 5–8; 229,230,782). This provides evidence that the phospholipid turnover in which Ca^{2+}, diacylglycerol, and presumably myo-inositol D-1,4,5-trisphosphate are second messengers (339) forms part of the signal-transducing system coupled to S_2-serotonergic receptor sites (638).

ACKNOWLEDGMENT

I am much indebted to my colleagues for providing the data from their studies: Dr. F. Awouters and C. Niemegeers, *in vivo* pharmacological studies; Dr. F. Colpaert, behavioral studies on drug discrimination; Dr. D. de Chaffoy de Courcelles, biochemical studies on phospholipid turnover; Dr. F. De Clerck, studies on platelet function; Dr. J. Van Nueten, *in vitro* pharmacological studies using isolated tissues. Only because of close cooperation with them has identification of the role of receptor-binding sites been possible. Part of this work was supported by a grant from I.W.O.N.L.

6

Serotonin and Central Regulation of Arterial Blood Pressure

William A. Wolf, Donald M. Kuhn, and Walter Lovenberg

Section on Biochemical Pharmacology, National Institutes of Health, National Heart, Lung, and Blood Institute, Bethesda, Maryland 20205

The distribution of serotonin-containing cell bodies and their projections in the central nervous system parallel those of central pathways subserving cardiovascular regulation, which suggests a role for serotonergic neurons in the control of cardiovascular function (158,606). The present chapter focuses on the neuroanatomical and functional integration of the central serotonergic systems which affect cardiovascular activity. For the discussion of central mechanisms mediating cardiovascular activity in general, the reader is referred to specialized reviews (e.g., 210,589).

SEROTONIN BIOSYNTHESIS AND NEUROANATOMY

Biosynthesis

Serotonin does not readily cross the blood-brain barrier; therefore, *de novo* synthesis must occur within neurons from the precursor amino acid L-tryptophan (see Chapter 1). The presence of tryptophan hydroxylase determines the serotonergic identity of the neuron, and this enzymatic conversion is usually the rate-limiting step in the synthesis of serotonin. By contrast, aromatic L-amino acid decarboxylase is not restricted to serotonergic neurons, since catecholaminergic neurons, endothelial cells, and peripheral organs all contain high levels of it. This is an important point, considering that systemic administration of 5-hydroxytryptophan is one strategy by which cerebral serotonin levels can be enhanced. However, serotonin will be formed in any location where decarboxylase is found.

The brain can be selectively depleted of serotonin by inhibiting tryptophan hydroxylase using the irreversible enzyme inhibitor *p*-chlorophenylalanine. In addition, two structural analogs of serotonin, 5,6-dihydroxytryptamine and 5,7-dihydroxytryptamine, are cytotoxic agents taken up into neurons by the amine uptake mechanisms. When administered after a selective noradrenergic uptake inhibitor such as desmethylimipramine, uptake of these neurotoxins is limited to serotonergic neurons, thereby enhancing the specificity of the destruction.

Overview of Neuroanatomy

Dahlstrom and Fuxe (207) originally described the existence of serotonergic neurons in the central nervous system of the rat using fluorescence histochemistry. Nine distinct nuclei of serotonin-containing cell bodies were labeled B1–B9 according to their caudal-rostral location. Most of these cell groups (all but B9 and several aspects of B1, B3) coincide with the medially located raphe nuclei. Since then, extensive analyses of the projections of these raphe nuclei have been performed. One of the first techniques involved the autoradiographic localization of ^{3}H-amino acids which are transported in an anterograde direction after microinjection of these substances into cell body sites such as a raphe nucleus (50,189,851). Subsequently, the use of markers such as horseradish peroxidase which are transported in a retrograde direction provided information as to the site of origin of projections to a specific terminal field (79,1020). While these studies provided valuable information concerning the neuroanatomical distribution of neuronal projections from a given raphe nucleus, in most cases no knowledge was available as to whether these projections were serotonergic. This is an important point in light of the fact that the raphe nuclei are not comprised solely of serotonin-containing cells (334,1151), and may help explain some of the continuing controversy surrounding the role of the central serotonergic system in the regulation of arterial blood pressure. More recent immunohistofluorescent techniques provide a much more sensitive and specific means of identifying serotonin-containing neurons in the central nervous system (1000). With the advent of retrogradely transported fluorescent markers, double labeling studies, which map the projections of fibers of an immunohistochemically identified neuron to a specific terminal field, provide a powerful tool in the analysis of neurochemically defined pathways in the brain (938). In general, while projections from the raphe nuclei are extensive, certain major pathways predominate in the serotonergic control of arterial blood pressure. These can be classified into ascending systems and descending systems.

Ascending Serotonergic Neurons

The dorsal and median raphe nuclei [B7 and B8, respectively, of Dahlstrom and Fuxe (207)], which are located in the mesencephalic tegmentum, send predominantly ascending fibers via the medial forebrain bundle to forebrain structures.

While many forebrain regions are innervated by serotonergic fibers from these nuclei (e.g., cortex, hippocampus, striatum, habenula, etc.), the region of greatest interest with respect to cardiovascular function is the hypothalamus.

In the rat, cat, and monkey, autoradiographic, biochemical, and immunohistofluorescent studies demonstrate the existence of a dense serotonergic innervation to many regions of the hypothalamus (50,851,930,949,1000). In the lateral hypothalamus this innervation originates from both the dorsal and median raphe nuclei. Physiological data also confirm that the hypothalamus is an important site of central serotonergic influence on arterial blood pressure.

Descending Serotonergic Neurons

Midbrain-pontine nuclei

The brainstem raphe nuclei send descending projections to the spinal cord. Double labeling studies using horseradish peroxidase and serotonin immunohistofluorescence (67) as well as autoradiographic studies (851) describe projections from the dorsal raphe, median raphe, and B9 to the raphe magnus in the brainstem. In addition, the dorsal raphe nucleus projects to the cervical and lumbar regions of the spinal cord in the rat, cat, and monkey (684,851). While these projections may not be serotonergic, double labeling studies demonstrate minor serotonergic projections from B9 and the dorsal raphe nucleus to the cervical spinal cord in the rat (122). Furthermore, autoradiographic studies show projections from the dorsal raphe nucleus to the locus ceruleus, nucleus tractus solitarius, and various other brainstem regions (189,851).

Medullary raphe nuclei

Most of the serotonergic innervation of the spinal cord is derived from the medullary raphe nuclei: the raphe pallidus (B1), raphe obscurus (B2), raphe magnus (B3), and a cluster of cells situated in the ventrolateral medulla considered to be lateral extensions of B1 and B3 (207,1000). Briefly, the raphe obscurus is distributed in the medulla as a paramedial, bilateral column of cells extending rostrally from the pyramidal decussation to the level of the inferior olive. The raphe pallidus is a small, densely packed cluster of cells located ventromedially between the pyramidal tracts and extending rostrally from the inferior olive to the level of the facial nucleus. In addition, a dense cluster of cells is distributed just lateral to the pyramidal tracts at the ventral surface of the medulla, extending rostrally from the inferior olive to the facial nucleus. The more caudal aspects of this cluster are considered part of B1 while the more rostral aspects are considered part of B3, but they will be collectively termed "ventrolateral B1, B3." Finally, the raphe magnus is a large cluster of cell bodies located in the medial medulla extending rostrally from the inferior olive to the level of the superior olive.

The location of these nuclei coincides with the distribution of regions involved in the regulation of sympathetic activity and blood pressure (472,589,656), suggesting an important role for medullary serotonergic mechanisms. Further evidence in support of this hypothesis exists in the numerous autoradiography (105,653,655), fluorescence, and immunohistochemistry studies (122,123) showing that these medullary nuclei project to various levels of the spinal cord.

The ventrolateral B1, B3 region, the raphe pallidus, and the raphe obscurus all send prominent serotonergic projections to the intermediolateral cell column of the spinal cord. The intermediolateral cell column is located in the thoracic and upper lumbar regions of the spinal cord and contains the preganglionic sympathetic neurons which play a major role in cardiovascular control (589). In comparison, the raphe magnus sends projections to the lumbosacral cord as well as the cervical spinal cord, although perhaps as few as 10% of the total population of the cells of the raphe magnus project to the spinal cord, and these may not all be serotonergic in nature (122,1145). In fact, the estimates of the proportion of serotonergic cells in the various medullary raphe nuclei range from approximately 5 to 15% in the raphe magnus, to 35% in the raphe obscurus, and to 50% in the raphe pallidus (1145,1151). Furthermore, the medullary raphe nuclei innervate a number of brainstem regions including those implicated in cardiovascular regulation, such as the nucleus tractus solitarius and the dorsal motor nucleus of the vagus (105,910). Finally, these raphe nuclei receive innervation from the nucleus tractus solitarius and other regions (67,654). Not all of these reciprocal projections are serotonergic.

CARDIOVASCULAR EFFECTS OF ALTERING SEROTONIN LEVELS IN THE BRAIN

The administration of L-tryptophan either in intravenous doses of up to 100 mg/kg, by infusion, or intraventricularly produces no conspicuous change in blood pressure in conscious or anesthetized normotensive rats (298,596,785). Sved and colleagues (1010), however, have demonstrated that the intraperitoneal administration of 50 to 100 mg/kg of L-tryptophan lowers blood pressure in conscious spontaneously hypertensive rats. Wolf and Kuhn (1165,1166) further investigated this effect in both spontaneously hypertensive and normotensive rats and demonstrated that 100 mg/kg of L-tryptophan lowers blood pressure in hypertensives by approximately 30 mm Hg, but elevates it slightly in the normotensive rats. While doses of 25 to 50 mg/kg of L-tryptophan maximally increase serotonin levels in a comparable manner in both strains, only 100 mg/kg of L-tryptophan lowers blood pressure in the hypertensive rat. The administration of 100 mg/kg of the dextrorotary isomer of tryptophan to spontaneously hypertensive rats increases brain serotonin and its turnover to the same extent as an equimolar dose of L-tryptophan, because D-tryptophan is converted to L-tryptophan in the periphery (1173). However, D-tryptophan has no effect on blood pressure (1164). These studies suggest that the modest cardiovascular effects of L-tryptophan are not a result of enhanced serotonergic activity in the brain.

The administration of 5-hydroxytryptophan generally produces depressor effects in conscious or anesthetized rats, cats, or dogs (see 606). In addition, sympathetic activity is depressed and cardiovascular reflexes (e.g., increase in arterial blood pressure during bilateral carotid occlusion) are attenuated. The involvement of serotonin in the physiological responses to 5-hydroxytryptophan has been studied with peripherally acting or peripherally and centrally acting inhibitors of the enzyme decarboxylase. However, the interpretation of such experiments is difficult. Indeed, large amounts of 5-hydroxytryptophan circulate throughout the periphery and the brain allowing for conversion to serotonin throughout the body. While peripheral decarboxylase inhibitors may not alter the hypotensive action of 5-hydroxytryptophan, this does not necessarily determine whether this effect is due to enhanced central serotonergic activity or a direct peripheral effect of 5-hydroxytryptophan. In addition, the conversion of 5-hydroxytryptophan to serotonin throughout the brain, which can also lead to depletion of catecholamines, makes it impossible to ascertain if the enhancement of serotonergic mechanisms is involved in the cardiovascular actions of the precursor. Likewise, the results obtained with drugs causing depletion of serotonin are variable and difficult to interpret. Systemic administration of *p*-chlorophenylalanine can produce an increase, a decrease, or no change in blood pressure in normotensive or hypertensive rats, while intraventricular injections of the neurotoxins generally decrease blood pressure or produce no change (see 606).

CARDIOVASCULAR EFFECTS OF CENTRAL SEROTONERGIC NEURONAL ACTIVATION

Ascending Systems

Stimulation of the Midbrain Raphe

Kuhn and colleagues (605,1166) and others (904,984) have studied the cardiovascular effects of activation of the ascending serotonergic system (Fig. 6–1). Electrical stimulation of the dorsal or median raphe increases blood pressure while producing variable changes in heart rate. This pressor effect is localized to the dorsal and median raphe nuclei and is seen in both anesthetized normotensive and spontaneously hypertensive rats. Depletion of serotonin in the brain with *p*-chlorophenylalanine attenuates the pressor response; restoring serotonin levels in depleted animals by administering 5-hydroxytryptophan partially restores the pressor response to stimulation of the raphe. Intravenous administration of fluoxetine, an inhibitor of the uptake of serotonin, enhances and prolongs the pressor response to the stimulation. However, it is abolished by transections of the brain either rostral or caudal to the midbrain raphe, indicating that it is mediated by ascending projections to forebrain structures which then send descending projections caudal to the midbrain. Microinjection of the serotonergic antagonist 2-bromolysergic acid diethylamide (BOL) into the anterior hypothalamus (preoptic area), a region densely innervated by serotonergic fibers from the midbrain raphe, antagonizes the pressor response to stimulation of the raphe. An increase in sympathetic outflow mediates part of this pressor response, since pretreatment of rats with bretylium tosylate, which prevents the release of norepinephrine from sympathetic postganglionic

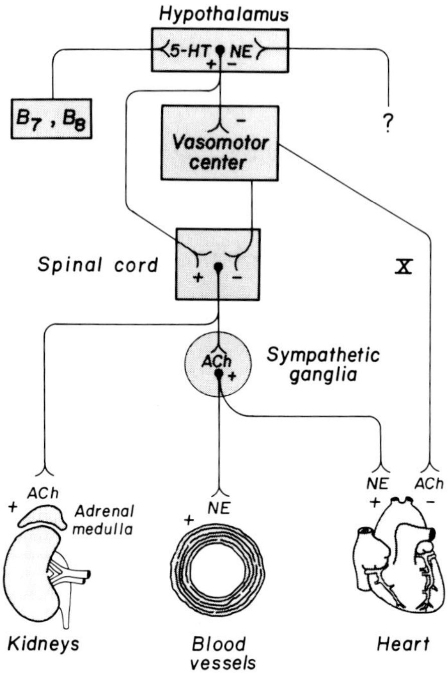

FIG. 6–1. Role of the ascending serotonergic system in the control of blood pressure. + Indicates excitatory influence, − indicates inhibitory influence. Abbreviations: 5-HT, serotonin; NE, norepinephrine; ACh, acetylcholine; B7, dorsal raphe; B8, median raphe; X, vagus nerve. Vasomotor center includes nucleus tractus solitarius and dorsal motor nucleus of the vagus.

nerves, attenuates it. These observations indicate that activation of an ascending serotonergic system by stimulation of the midbrain raphe produces an increase in blood pressure which is mediated in part by the hypothalamus and which requires transmission toward the spinal cord to enhance sympathetic activity.

Central Administration of Serotonergic Agonists and Antagonists

The central administration of serotonergic agents has variable cardiovascular effects (see 606). Some of the variability can probably be attributed to the diverse methodology employed. For example, even the intraventricular administration of serotonin provides a widespread distribution of serotonin throughout the brain and could simultaneously modify the activity of several serotonergic neuronal systems. Thus, in the anesthetized rat, the greatest increases in arterial blood pressure are observed after administration of serotonin into the lateral or the third ventricle; injection of the monoamine into the fourth ventricle produced small increases or no change in blood pressure, while administration into the cisterna magna actually produced depressor or biphasic (pressor-depressor) responses (595,613).

The pressor response to intraventricular serotonin or other serotonergic agonists (357) is not secondary to a decrease in respiratory activity (595); it is antagonized by intraventricular administration of the serotonergic antagonists BOL and methysergide. Transection of the spinal cord at the first or third, but not at the fifth or seventh, cervical levels prevents the serotonin-mediated increase in arterial blood pressure (595,613). Both adrenalectomy and treatment with bretylium tosylate attenuate, but do not abolish, the pressor effect of intraventricular injections of serotonin (595). Bretylium tosylate prevents the release of norepinephrine at sympathetic terminals, but not that of adrenal catecholamines. Repeated administration of nicotine, which induces a depolarization blockade of ganglionic transmission, prevents the pressor response to intraventricular serotonin (595). However, blockade of ganglionic nicotinic receptors with hexamethonium does not attenuate the serotonin-induced pressor response. This may suggest that the increased sympathetic activity is mediated by serotonergic ganglionic transmission (1123). The pressor response to intraventricular serotonin is antagonized by the systemic administration of α-adrenergic antagonists such as phenoxybenzamine and tolazoline (595). Thus, the intracerebral administration of serotonin produces an increase in arterial blood pressure, which is brought about by enhanced sympathetic activity resulting in neurogenic vasoconstriction and release of catecholamines from the adrenal medulla. The further localization of the central pressor effect of serotonin to a specific brain region was first described by Smits and Struyker-Boudier (983) and subsequently confirmed by others (904,1166). As little as 20 ng of serotonin injected into the anterior hypothalamus (preoptic region) produces significant increases in arterial blood pressure. The pressor responses to intrahypothalamic injection of serotonin can be attenuated by local or systemically administered metergoline, or intraventricular administration of methysergide. Wolf and colleagues (1166) extended these observations by demonstrating that a similar pressor response is evoked by the administration of serotonin in the anterior hypothalamus (preoptic region) of spontaneously hypertensive and normotensive Wistar-Kyoto rats; this effect is also antagonized by metergoline pretreatment (1166). The destruction of noradrenergic neurons in the anterior hypothalamus with 6-hydroxydopamine produces hypertension 5 to 20 days later (69). This hypertension is associated with a depletion of norepinephrine in the anterior hypothalamus while the levels of dopamine and serotonin remain unchanged. The hypertension was blocked by the neurotoxin 5,7-dihydroxytryptamine, given into the medial forebrain bundle 1 week prior to 6-hydroxydopamine. The neurotoxin decreased the levels of serotonin in both the anterior and posterior hypothalamus by approximately 80%, but did not alter arterial blood pressure. The local administration of methysergide also antagonized the hypertension induced by 6-hydroxydopamine. Thus, a decreased noradrenergic activity in the hypothalamus produces hypertension only in the presence of an intact serotonergic system, which suggests a functional interaction between these two hypothalamic neurochemical systems in the central cardiovascular effects of serotonin. The monoamine may not be solely responsible for the maintenance of arterial blood pressure, since the serotonergic neurotoxin has no effect on resting blood pressure. However, depletion of norepinephrine probably removes an inhibitory influence on the release of serotonin in the hypothalamus. The ensuing increase in release of the monoamine in turn raises arterial blood pressure.

Descending Systems

Serotonergic Neurons of the Brainstem and Arterial Baroreflex Activity

The increases in blood pressure produced by intravenous administration of pressor agents such as norepinephrine or epinephrine normally produce a reflex bradycardia mediated in part by neurons of the nucleus tractus solitarius. In both the cat (1019) and the rat (648), treatment with 5-hydroxytryptophan (into the fourth ventricle or intravenously) to augment the cerebral levels of serotonin attenuates this reflex bradycardia. This effect is dependent on the conversion of 5-hydroxytryptophan to serotonin in the brain. However, since the decarboxylase enzyme responsible for this conversion is ubiquitous, questions concerning the site and the specificity of the action of 5-hydroxytryptophan remain open. The effect of 5-hydroxytryptophan can be obtained in cats in which the spinal cord has been transected at the first cervical level (1019), which suggests that neurons in the brainstem, which alter the activity of the vagus nerve, can be inhibited by serotonin. In confirmation of this interpretation, 5,7-dihydroxytryptamine and *p*-chlorophenylalanine augment the reflex bradycardia caused by the injection of catecholamines in the rat (648). In the conscious rabbit, destruction of the serotonergic neurons of the brainstem by the intracisternal administration of 5,6-dihydroxytryptamine enhances the vagally mediated reflex bradycardia (461,591). These observations confirm that serotonergic neurons in the brainstem are involved in the transmission of baroreflex activity and serve to impair vagal responses.

Serotonergic inputs may also inhibit reflex responses mediated through sympathetic nerves. In this case, the brainstem and the hypothalamus may work in concert. The involvement of serotonergic neurons in the modulation of sympathetic reflexes is suggested by the alterations in the levels of serotonin and 5-hydroxyindoleacetic acid in the spinal cord, medulla, and hypothalamus after sinoaortic denervation (403,1156). It is uncertain whether these changes are the consequence or the cause of the hypertension induced by the denervation or whether they are unrelated to cardiovascular activity. In the rabbit, intracisternal administration of 5,6-dihydroxytryptamine prior to sinoaortic denervation prevents the development of hypertension, which suggests the need for intact bulbospinal serotonergic mechanisms in this response (1156). The intraventricular administration of serotonin to anesthetized cats or dogs not only reduces efferent sympathetic activity and arterial blood pressure, but also attenuates the increase in blood pressure and sympathetic activity caused by bilateral carotid occlusion and inhibits the depressor response to vagal afferent stimulation (61,89). Obviously, serotonin administered in this manner could act at a number of sites; the involvement of the hypothalamus in baroreflex transmission (210,589) makes this site a possibility.

In the anesthetized rat, the microinjection of serotonin into the nucleus tractus solitarius, which receives a dense serotonergic innervation, produces a dose-dependent increase in arterial blood pressure and in heart rate. These effects are enhanced by treatment with fluoxetine, but antagonized by administration of BOL or metergoline (1165). These findings suggest that the pressor response to medullary microinjections of serotonin is mediated by serotonergic receptors.

Microinjections of serotonin into the nucleus tractus solitarius of the anesthetized cat also increase arterial blood pressure and renal nerve activity (963). When serotonin is injected into the nucleus tractus solitarius of cats in which renal nerve activity is reflexly elevated by bilateral compression of the common carotid arteries, the nerve activity returns to baseline levels. Hence, serotonin may act to impair baroreflex transmission at the level of the nucleus tractus solitarius.

The hypotensive agent clonidine acts predominantly on bulbospinal mechanisms to reduce sympathetic outflow (590). It lowers blood pressure in part by increasing the sensitivity of the baroreflex mechanisms. Destruction of serotonergic neurons in the brainstem and the spinal cord abolishes the cardiovascular action of clonidine in the conscious rabbit (462). The release of serotonin is inhibited by α-adrenergic agonists (352,419). Thus, clonidine may act by suppressing the neuronal release of serotonin, which may normally inhibit transmission in the baroreflex circuitry.

Bulbospinal Serotonergic Mechanisms and Sympathetic Nervous Activity

Neurons within the intermediolateral cell column identified as preganglionic sympathetic neurons, as well as spinal neurons receiving visceral somatosensory afferents, are excited by iontophoretically applied serotonin and inhibited by serotonergic antagonists (191,254,548,704). Electrical stimulation of the medullary raphe nuclei can either decrease or increase arterial blood pressure and sympathetic nervous activity (3,190,472,1170) (Fig. 6–2). These different cardiovascular responses cannot be localized to one or two discrete areas (3,1170). However, stimulation of the more rostral or more caudal regions of the raphe magnus produces the greatest depressor effects. Pressor effects are more prominent during stimulation of the raphe obscurus and pallidus. When the ventrolateral B1, B3 region is stimulated, prominent pressor effects are seen (1170). This variability in response to electrical stimulation may be due to several factors. While electrical stimulation provides discrete localized activation of nerves, it does not necessarily distinguish between cell bodies and axons. More importantly, since the raphe nuclei contain a large proportion of nonserotonergic cells, it is uncertain in some cases whether the activation of serotonergic systems is responsible for the cardiovascular effects noted. In the stimulation studies mentioned above, no attempts were made to modify the physiological responses with serotonergic antagonists or with serotonergic neurotoxins. While serotonergic neurons appear to modulate a number of different physiological functions, the following studies suggest that activation of serotonergic pathways of the brainstem projecting to the intermediolateral cell column increases blood pressure.

Electrical stimulation of the ventrolateral B1, B3 region increases arterial blood pressure in the anesthetized rat. This effect is attenuated by specific destruction of the serotonergic processes to the intermediolateral cell column. In addition, pressor responses and increased renal nerve activity can be elicited by local application of

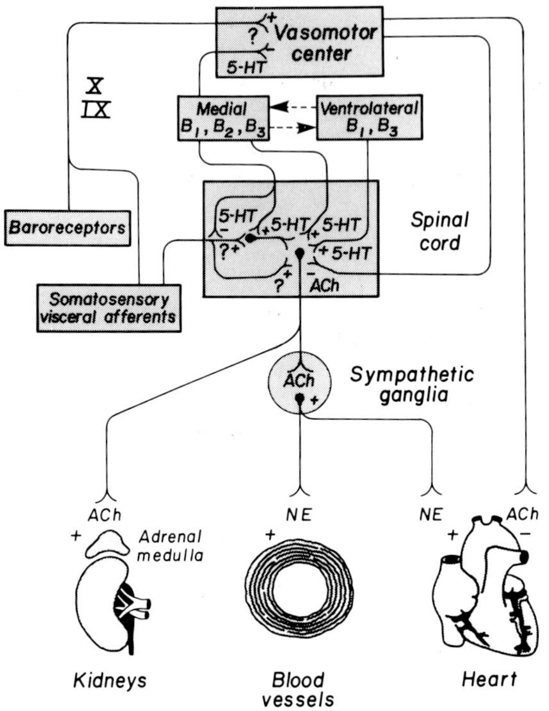

FIG. 6–2. Role of bulbospinal serotonergic systems in the control of blood pressure. + Indicates excitatory influence, – indicates inhibitory influence. Abbreviations: 5-HT, serotonin; NE, norepinephrine; ACh, acetylcholine; IX, glossopharyngeal nerve; X, vagus nerve; B1, raphe pallidus; B2, raphe obscurus; B3, raphe magnus. Vasomotor center includes nucleus tractus solitarius and dorsal motor nucleus of the vagus.

kainic acid or glutamate (499,701). These observations suggest that serotonergic neurons within the ventrolateral B1, B3 region mediate increases in blood pressure by increasing sympathetic nerve activity.

In addition, the electrical stimulation of selected regions of the midline medullary raphe nuclei produces either an increase or a decrease in sympathetic activity and arterial blood pressure. Subsequent systemic administration of the serotonergic antagonists methysergide and metergoline prevents only the increases in sympathetic activity (702). This suggests that the serotonergic pathways associated with the midline medullary raphe mediate increases in arterial blood pressure through enhanced sympathetic activity.

Studies with α-methyldopa and clonidine also suggest that some bulbospinal serotonergic systems regulate vasomotor tone via changes in sympathetic outflow. Thus, the hypotensive action of α-methyldopa, administered intraperitoneally to spontaneously hypertensive rats, can be attenuated by selective ablation of spinally descending serotonergic neurons (170). Furthermore, α-methyldopa locally applied to the ventrolateral B1, B3 region produces hypotensive effects that are also abolished by either intraventricular or intraspinal 5,7-dihydroxytryptamine (737). Intraventricular 6-hydroxydopamine has no effect on the hypotensive action of α-methyldopa. These studies implicate the ventrolateral B1, B3 serotonergic neurons as one possible site of action for the drug. Head and colleagues (462) have also shown that the hypotensive action of α-methyldopa and clonidine is dependent on the integrity of bulbospinal serotonergic mechanisms. These studies demonstrate that a bulbospinal serotonergic system plays a role in the maintenance of vasomotor tone and can increase or decrease it.

Depressor responses can be observed upon stimulation of the brainstem raphe (3,777,1170). The hypotensive effect of systemically administered 5-hydroxytryptophan may be due to an enhanced activity in a bulbospinal serotonergic pathway (871). However, without more direct pharmacological evidence (e.g., with the use of serotonergic antagonists), it is uncertain whether bulbospinal serotonergic pathways mediate a depressor effect. Such a role for serotonin seems possible, since evoked reflex activity in preganglionic sympathetic neurons can be attenuated by concurrent stimulation of the raphe pallidus or raphe obscurus (404). This raphe-mediated inhibition can be blocked by several serotonergic antagonists (e.g., methysergide, cinanserin). Furthermore, Franz and colleagues (353) have suggested that evoked spinal reflex activity is inhibited by serotonin, thus implying that sympathetic activity is under an inhibitory serotonergic control. These authors utilize 5-hydroxytryptophan, which by itself has direct effects on spinal reflex activity (444). However, blockade of the conversion of 5-hydroxytryptophan to serotonin did prevent the inhibitory action of the precursor on evoked reflex activity (777). Since these experiments were performed in spinalized cats as well, the possibility exists that activation of serotonergic receptors in the spinal cord may inhibit evoked sympathetic responses. Blockade of spinal transmission could be one mechanism by which the serotonergic system modulates nociception (481), and this may also play a role in associated cardiovascular responses (650). While there is as yet no direct evidence demonstrating that inhibitory spinal interneurons are stimulated by serotonin, this is one possible mechanism by which both excitatory and inhibitory efferent sympathetic responses could be seen (404). McCall and colleagues (703) describe such inhibitory interneurons modulating baroreflex activity. These interneurons are located within the intermediomedial nucleus of the spinal cord and receive innervation from both the nucleus tractus solitarius and sympathetic afferents of the inferior cardiac nerve. The effects of iontophoretically applied serotonin or stimulation of the raphe suggest that serotonin can either increase or decrease the activity of spinal cord interneurons (68).

7

Platelet Activation and Serotonin

Holm Holmsen

Department of Biochemistry, University of Bergen, N-5000 Bergen, Norway

Platelets are small anucleate cells that circulate in the blood of all animals and humans. They vary in size and life span among species. Platelets are shed from the cytoplasm of mature polyploid megakaryocytes in the bone marrow and sequestered by the reticuloendothelial system. In humans their average size is about 7 μm^3 (femtoliter) and their life span is 7 to 9 days. There are normally 1 to 3×10^8 platelets per milliliter of blood. Unstimulated (resting, circulating) platelets have a characteristic disc-like shape, which is believed to be maintained by a marginal bundle of microtubules; the most conspicuous ultrastructural feature of platelets, however, is the abundance of secretory storage granules, which consist of three types, besides scattered mitochondria and glycogen grains in the cytoplasm (Fig. 7–1). Another typical feature is the surface-connecting system (open canalicular system), a channel system that penetrates the entire cell, thus making open connections between the interior of the cell and its surface (Fig. 7–1). Several pieces of evidence indicate that this canalicular system transports the substances released from the storage granules to the platelets' environment.

The major function of platelets is in hemostasis. They are responsible for primary hemostasis, i.e., the formation of a platelet mass around a lesion in the vessel. Such activated platelets further stimulate formation of thrombin on their surface so that a meshwork of fibrin is formed among and on the platelets in the hemostatic plug and completes hemostasis. Under pathological conditions platelets participate in thrombus formation, particularly in arterioles and small arteries; many of the mechanisms that control platelet behavior in thrombosis are the same as those that

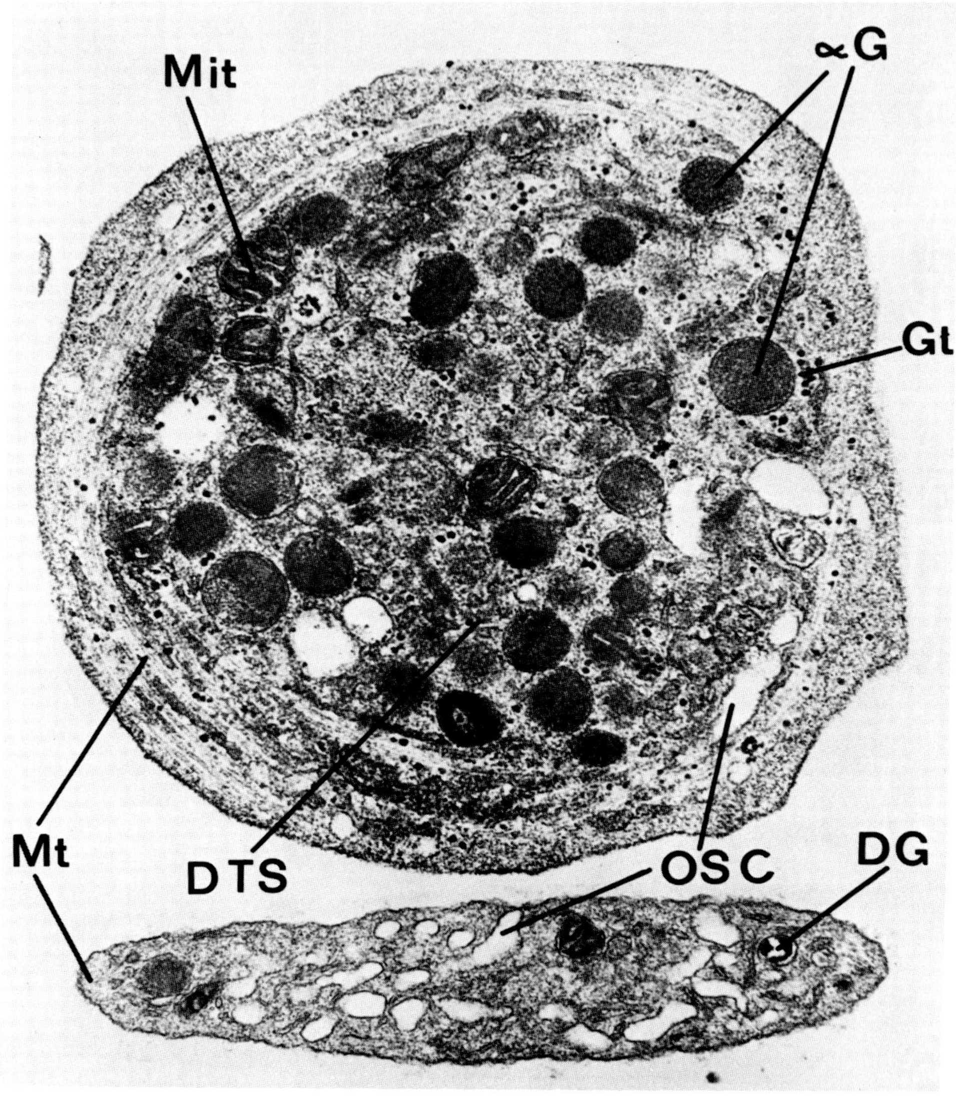

FIG. 7–1. Ultrastructure of a human platelet. Transection through the equatorial plane *(top)* and perpendicular to this plane *(bottom)*. Mit = Mitochondrion; αG = α-granule; Gl = glycogen; DG = dense granule (only remnants of contents can be seen); OSC = open canalicular system; DTS = dense tubular system; Mt = marginal bundle of microtubules.

operate in hemostasis. In addition to their primary role in hemostasis/thrombosis, platelets are unique in their capability to absorb serotonin from plasma and store it in their dense granules. Platelets may also help to control the growth of certain types of cells, including some cancer cells (276).

PLATELET RESPONSES AND THEIR ACTIVATION

Morphological and Biochemical Correlations

Light and electron microscopic studies of platelets during hemostasis *in vivo* reveal distinct stages in their transformation from free disc-like cells to packed (aggregated), degranulated cells of varying shapes in the hemostatic plug. Morphologically these stages are very similar to those occurring during platelet aggregation *in vitro*. The latter condition allows the determination of platelet responses to external stimuli, all of which depend on active metabolism in the cells (486). The individual responses observed *in vitro* are called shape change, aggregation, secretion (exocytosis of the contents of three different storage granules), and liberation of arachidonate.

Shape Change

A rapid (2–5 sec) transformation from the discoid to a spherical shape occurs during even minute stimulation of the platelets; philopods grow rapidly to give the platelet the aspect of a "spiny sphere." Shape changes are independent of extracellular Ca^{2+} and can be monitored by light transmission/scattering.

Aggregation

During aggregation the spherical platelets clump together. This process depends on extracellular Ca^{2+}, the presence of fibrinogen, and intact cellular energy metabolism. Fibrinogen does not bind to unstimulated (resting) platelets. However, stimulation of the cells causes two separate proteins (glycoproteins IIB and IIIA) to combine in the plasma membrane; this IIB-IIIA complex (fibrinogen receptor) has strong affinity for fibrinogen and binds it when extracellular Ca^{2+} is present (687,850). Platelets may also stick to foreign surfaces in a process called adhesion, the mechanism of which is poorly understood but which probably differs from that of aggregation.

Secretion

Platelets have three types of secretory storage organelles: the dense granules, the α-granules, and the acid hydrolase-containing vesicles. These storage organelles differ in ultrastructure (Fig. 7–1) and contents (Fig. 7–2). The substances stored in the granules are released to the surroundings of the platelets by exocytosis (platelet secretion) when the cells are stimulated, although different degrees of

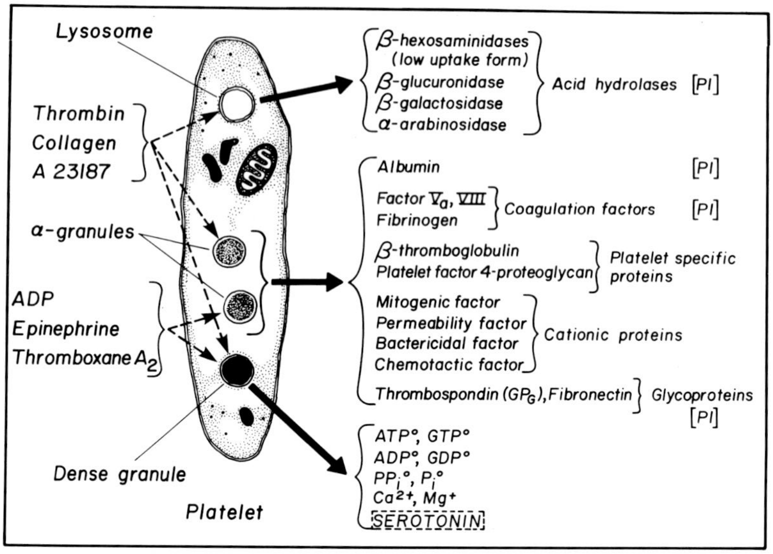

FIG. 7–2. Contents of storage granules in platelets. The contents are shown to be released by six agonists with different potencies (Table 7–1). A23187 is a divalent cationophore. ADP = Adenosine diphosphate; ATP = adenosine triphosphate; GDP = guanosine diphosphate; GTP = guanosine triphosphate; P_i = phosphatidylinositol; PP_i = phosphatidylinositol polyphosphates; [PI] = the substance in the granule is also present in plasma; ° = metabolic inertia. Serotonin is the only released substance that the platelet takes up rapidly from its surroundings.

stimulation are required for the discharge of the contents from each individual type of storage organelle (Fig. 7–2). With the exception of serotonin, secretable substances cannot be taken up by platelets.

Liberation of Arachidonate

There is no free arachidonic acid in resting platelets. Considerable amounts of the fatty acid are present esterified in the 2-position of membrane phosphoglycerolipids. The esters include phosphatidylcholine, phosphatidylethanolamine, phosphatidylserine, and phosphatidylinositol. Upon appropriate stimulation, phospholipase A_2 is activated which liberates free arachidonate from phosphatidylcholine, phosphatidylinositol, and phosphatidylethanolamine (130); phosphatidic acid and diacylglycerol are formed early after stimulation and are also potential sources of free arachidonate (94,776). The liberated free arachidonate is immediately oxygenated along the cyclo-oxygenase and lipoxygenase pathways. The prostaglandins and thromboxanes resulting from the action of cyclo-oxygenase can activate the platelets (see Table 7–1) and thus participate in a positive feedback regulation.

Platelet Agonists

Platelet agonists can be subdivided into groups depending on their potency (Table 7–1). The potency of an agonist is determined by how many different platelet

TABLE 7–1. *Human platelet agonists*

Weak	Intermediate	Strong
Vasopressin Serotonin Norepinephrine	ADP Epinephrine Thromboxane A_2 Prostaglandin endoperoxides Platelet-activating factor	Thrombin Collagen

Weak agonists produce shape change and reversible aggregation. Intermediate agonists produce shape change, irreversible aggregation, and secretion of the content of dense granules and α-granules as well as low but measurable amounts of arachidonate. Strong agonists evoke maximal responses. The subdivision is based on the potency of the agonists given alone.

responses it induces and how complete they are at maximally effective concentrations of the agonist. Although the subdivision shown in Table 7–1 applies to platelets from most species, there are some important exceptions: (a) epinephrine does not stimulate platelets from rat, rabbit, and dog; and (b) serotonin is more powerful with platelets from the cat than with those from most other species. As a rule, one platelet agonist can potentiate the effect of others. This is important during hemostasis *in vivo* where several agonists usually are made available to the platelets simultaneously. It is also important for the action of a single agonist *in vitro*, because if it triggers the secretion of dense granules and the liberation of arachidonate, additional agonists [adenosine diphosphate (ADP), serotonin, prostaglandins, thromboxane A_2] are automatically provided.

Activation of Platelet Responses

Each of the platelet agonists (Table 7–1) interacts with specific receptors on the platelet surface (for references, see 489). The interaction between an agonist and its receptor initiates a rapid increase in the metabolism of phosphatidylinositol and its polyphosphates, and causes an increase in the cytoplasmic concentration of Ca^{2+}. In general, platelet responses can be initiated by an experimental increase in cytoplasmic Ca^{2+}, which suggests that this ion may act as a second messenger. Since two platelet responses, secretion and liberation of arachidonate, lead to the appearance of more stimuli in the extracellular fluid, the propagation of platelet responses is amplified powerfully by a positive feedback loop.

Polyphosphoinositide Metabolism

The mobilization of cytoplasmic Ca^{2+} in cells, including platelets, probably is coupled to an increase in the metabolism of phosphatidylinositol and its polyphosphates that follows the agonist-receptor interaction (734). This metabolism entails

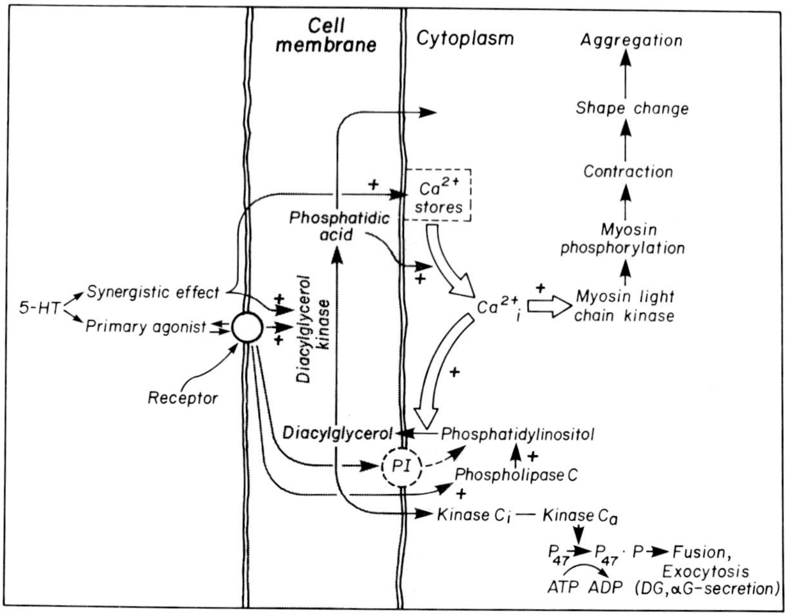

FIG. 7–3. Possible mechanisms for the activation of platelets by serotonin acting as a single agonist or as a synergistic partner. In general, when platelets are activated by an agonist interacting with its receptor the agonist-receptor complex affects the metabolism of phosphatidylinositol (PI) at three levels, coupling the formation of phosphatidic acid to the receptor occupancy: (a) direct activation of diacylglycerol kinase, a membrane-bound enzyme that phosphorylates diacylglycerol to phosphatidic acid with adenosine triphosphate (ATP); (b) activation of the non-membrane-bound phospholipase C leading to formation of more diacylglycerol; and (c) modulation of the topographical localization of phosphatidylinositol in the inner sheet of the plasma membrane so that it becomes available for cytoplasmic phospholipase C. A complication is the Ca^{2+} requirement of phospholipase C for activity (482); it is possible that the rapid but transient breakdown of phosphatidylinositol polyphosphates is connected with the mobilization of Ca^{2+} for this purpose. Little information is available concerning the metabolism of phosphatidylinositol polyphosphates except for its time course (6,93) when single agonists are used. In any event, phosphatidic acid directly converts bound Ca^{2+} to cytoplasmic free Ca^{2+} ($Ca^{2+}{}_i$), which activates phospholipase C to produce more diacylglycerol for the formation of phosphatidic acid and activates myosin light chain kinase which starts the reaction leading to shape changes and aggregation. The diacylglycerol formed activates protein kinase C, which in turn phosphorylates the 47-kd protein (P_{47}) which may be associated with fusion of the granular and plasma membranes, resulting in exocytosis of the dense granules (DG) and α-granules (αG). Serotonin can act as a single agonist. As a synergistic agonist it most likely enhances the mobilization of cytoplasmic Ca^{2+}. This is shown to occur by three mechanisms: (a) further activation of phospholipase C or modulation of the topography of phosphatidylinositol to ensure the availability of substrate; (b) further activation of diacylglycerol kinase; and (c) direct stimulation of the Ca^{2+} mobilization which is initiated by the phosphatidic acid produced by the occupancy of the receptors by the primary agonist.

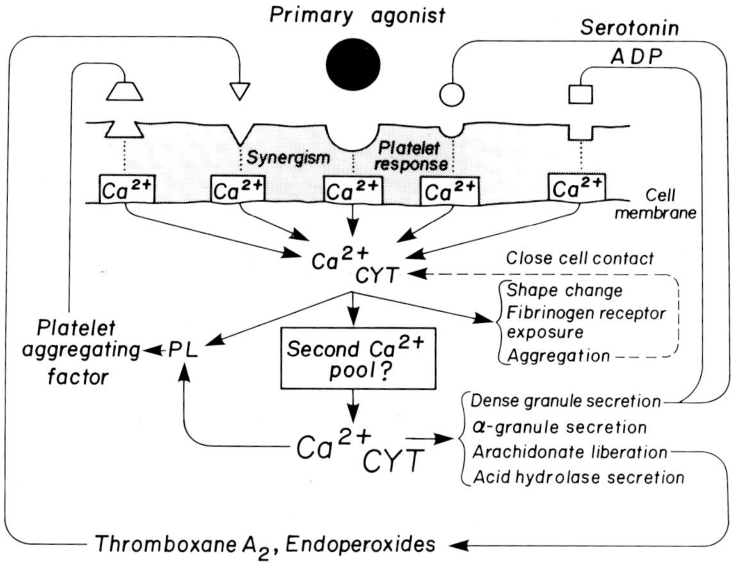

FIG. 7–4. Platelet activation mechanisms. A primary agonist interacts with its receptor in the plasma membrane which controls a certain amount of the second messenger, Ca^{2+}. As a result of occupancy of the receptor, the receptor's content of Ca^{2+} is released to the cytoplasm where it, probably via protein phosphorylation, initiates a number of responses. These seeding amounts of Ca^{2+} will also release Ca^{2+} from a larger store in quantities which are sufficient to stimulate the secretory responses and the liberation of arachidonate (by activation of phospholipase A_2). This in turn causes release of serotonin, adenosine diphosphate (ADP), thromboxane A_2, and prostaglandin endoperoxides, which will interact with their respective receptors on the plasma membrane of the platelets (positive feedback regulation).

a diesteratic breakdown of phosphatidylinositol and polyphosphates; this produces diacylglycerol, which is rapidly phosphorylated to phosphatidic acid, a divalent cationophore that could directly transport Ca^{2+} from internal or external stores to the cytoplasm (866). There is an apparent tight coupling between formation of phosphatidic acid and occupation of thrombin receptors (488), which suggests that the formation of phosphatidic acid may be directly controlled by the receptor and, hence, constitute the initial step in the "phosphatidylinositol response" (Fig. 7–3).

Cytoplasmic Ca^{2+}

The agonist-induced activation of platelet responses (Fig. 7–4) has been reviewed elsewhere (489). It is likely that the interaction of an agonist and its receptor leads to an increase in the cytoplasmic concentration of Ca^{2+} that may originate from both internal and external sources. When the Ca^{2+} concentration has reached about 5×10^{-7} M in the cytoplasm, phosphorylation of myosin light chains occurs and the platelets change shape (211). Whether shape change is sufficient or a further increase in the cytoplasmic Ca^{2+} concentration is required to cause exposure of the fibrinogen receptor, and hence aggregation, is not known. Distinctly higher concentrations of cytoplasmic Ca^{2+} must be reached to trigger the secretion of dense granules and α-granules (899), and a source of Ca^{2+} other than that providing the activation ions for shape change and aggregation may be involved. These secretory processes appear to be associated with phosphorylation of a cytoplasmic protein of 47 kd, which is a substrate for protein kinase C (1021). Even higher concentrations of Ca^{2+} must be mobilized to achieve maximal acid hydrolases and activation of phospholipase A_2 resulting in liberation of arachidonate. Thus, cytoplasmic Ca^{2+} serves as an important second messenger to transmit the signal initiated by the agonist-receptor interaction at the plasma membrane to the interior of the cell.

Other Possible Second Messengers

Other second messengers may be involved in the platelet responses. The most likely are diacylglycerol, which stimulates protein kinase C (1021), and cyclic guanosine monophosphate (GMP) (459). Inositol polyphosphates are formed very early in the stimulus-response sequence (6) and may function as second messengers. Cyclic adenosine monophosphate (AMP) is definitely not involved in signal processing during activation of the platelets. The elevation of this cyclic nucleotide causes powerful inhibition of all platelet responses, suggesting that it acts on common steps in the activation process (459).

Positive Feedback Regulation

When platelets are activated to the stage of secreting dense granules, ADP and serotonin appear extracellularly. These agonists cause further activation of the platelets in synergism with the primary stimulus. Likewise, if arachidonate is liberated, it is converted by cyclo-oxygenase to prostaglandins G_2 and H_2, which are further converted to thromboxane A_2. These prostanoids will interact with receptors on the platelet membrane and may enhance responses by providing more Ca^{2+} to the cytoplasm. The augmentation of the primary response is synergistic rather than additive; how this can be explained in terms of Ca^{2+} mobilization is unknown. Platelet aggregation by itself exerts a physical positive feedback, apparently because of close cell contacts (487). The release of 1-alkyl 2-acetylglycerophosphocholine (platelet-activating factor) could theoretically also contribute to the positive feedback mechanism, but its importance is doubtful (575) (Fig. 7–4).

SEROTONIN AND PLATELETS

The platelets bind, take up, and store serotonin (see ref. 559 and Chapter 4). When they release their dense granules, the liberated serotonin feeds back on the platelets to activate further platelet aggregation (Fig. 7–5). The serotonin stored in the dense granules of the platelet can be released by four different mechanisms: exocytosis, exchange, inhibition of transport at the granule membrane, and action of aminophores or protonophores.

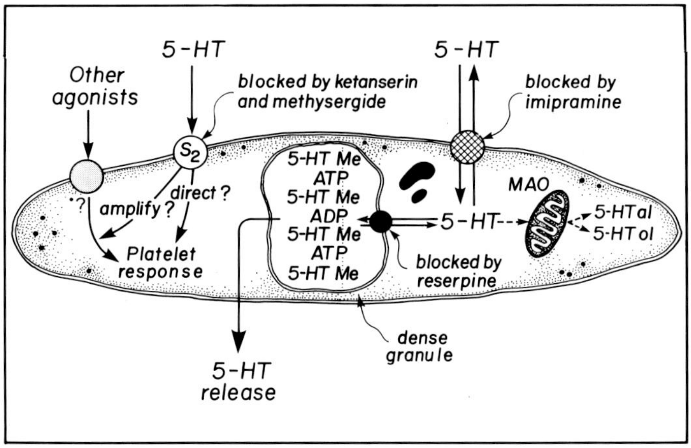

FIG. 7–5. Interactions of serotonin with platelets. Serotonin (5-HT) binds to two sites in the plasma membrane. One site *(hatched)* is that for active uptake and transport; the binding of serotonin to that site is blocked by imipramine and related compounds. The transport causes the influx of serotonin; the process is driven by the Na^+ gradient. Most of the serotonin that enters the platelet is transported into the dense granule by another transport system energized by the H^+ gradient. Inside the dense granule the amine is sequestered between stacked adenosine groups in large aggregates of adenosine diphosphate (ADP), adenosine triphosphate (ATP), and divalent metal ions (Me; Ca^{2+} in human and Mg^{2+} in porcine platelets). The second binding site in the plasma membrane is the S_2-serotonergic receptor (S_2), which controls intracellular Ca^{2+} (see Fig. 7–6) and thus platelet responses; binding to the receptor is specifically blocked by ketanserin. MAO = monoamine oxidase.

Exocytosis

This process involves a relative movement of the dense granule and the plasma membrane. The invaginations of the plasma membrane that constitute the open canalicular system and the granules most likely move toward each other. This is followed by a fusion of the granular and plasma membranes so that an opening is made which results in free access of the granular content to the extracellular space. Exocytosis is a rapid process. For example, when induced by thrombin, complete release of serotonin occurs within 15 to 25 sec at 37°C. The process is induced by a rise in the cytoplasmic Ca^{2+} concentration which is brought about by the agonist-receptor interaction. The exact mechanism of exocytosis is, however, not well understood. It is referred to as "dense granule secretion" (platelet release reaction) to distinguish it from those mechanisms that are connected with exocytosis of the contents of the α-granules and lysosomes, and to emphasize its similarity with the exocytotic step in secretion in general.

Exchange

Other aromatic amines (e.g., tyramine, tryptamine, or serotonin itself) can enter the dense granules and competitively displace serotonin from its binding sites within the matrix of nucleotide-metal aggregates (215). The exchange process is slow; for example, complete displacement of serotonin by tyramine takes 5 to 10 min. Unlike exocytosis, where serotonin is extruded together with the other constituents of the dense granules, only serotonin is displaced *(unpublished observations)*.

Inhibition of Transport at the Granule Membrane

A slow release of serotonin is observed from platelets treated with reserpine (146). Apparently, in the noninhibited state there is a constant, slow leakage of serotonin from the dense granules. However, the amine that reaches the extracellular space is efficiently recaptured by the transport system of the granule membranes. When the granular uptake is abolished (e.g., by reserpine), leakage predominates and causes a net efflux of serotonin out of the granule. Typical for this slow loss of serotonin is that only oxidized metabolites of serotonin leave the platelet (262). Thus, the storage in the dense granules protects serotonin from oxidation by mitochondrial monoamine oxidase.

Aminophores and Protonophores

These agents specifically release serotonin from platelets. The distinction between these two types of monovalent cationophores is that amine cations bind more tightly to aminophores and protons more tightly to protonophores. However, this selectivity is not absolute and in both cases the actual mechanism of release is mixed. Aminophores, for example lasalocid (X537A) (761) and lysocellin (698), release serotonin from platelets by forming a complex with the amine in its protonated form which then diffuses out of the cell to dissociate in the extracellular space. Increases in the intragranular pH, for example by addition of NH_3, cause serotonin release (1104) by reducing the H^+ gradient across the granular membrane; this is accompanied by conversion of the protonated serotonin to the free amine. Thus, an important gradient of free serotonin is established between the inside and outside of the dense granule, so that the amine diffuses rapidly to the cytoplasm and then to the extracellular space.

Serotonin in Platelet Activation

Direct Effects of Serotonin

Human platelets usually respond to serotonin with shape change and a small, reversible aggregation (1148). Cat platelets, on the other hand, respond with shape change, irreversible aggregation, secretion of dense granules, and liberation of arachidonate (731).

Synergistic Effects of Serotonin

In human platelets serotonin synergistically potentiates platelet aggregation induced by ADP, collagen, and epinephrine, and this effect is blocked by ketanserin (232). The secretion of α-granule constituents and the liberation of arachidonate

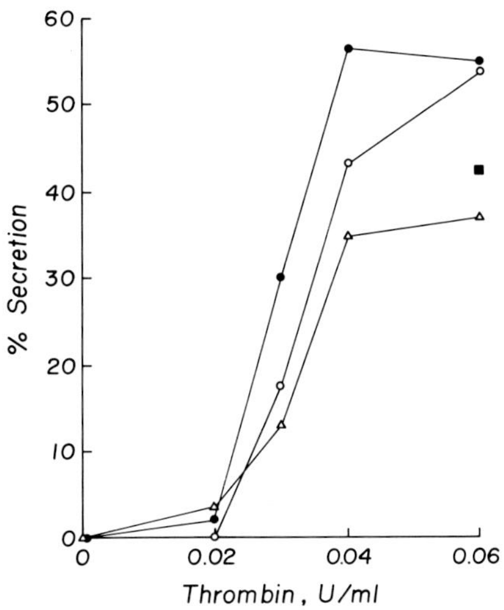

FIG. 7–6. Effect of serotonin and ketanserin on thrombin-induced dense granule secretion. Human platelets (in citrated platelet-rich plasma) were incubated with ^{14}C-serotonin and then gel filtered. Portions were incubated at 37°C for 60 sec with thrombin in the concentrations shown and analyzed for released ^{14}C-serotonin. Additions were: ○, control; ●, 5 μM serotonin; Δ, serotonin and 20 μM ketanserin; ■, 20 μM ketanserin (488).

TABLE 7–2. *Effect of serotonin and ketanserin on thrombin-evoked formation of phosphatidate*

Primary agonist	Synergistic partner: None	Serotonin	Epinephrine
Thrombin	4,476	5,822	6,554
Thrombin + ketanserin	4,286	4,260	6,442

Platelets (labeled with ^{32}P-orthophosphate and gel filtered) were incubated with 0.01 U/ml of thrombin at 37°C for 60 min in the presence and absence of various combinations of serotonin (5 μM), ketanserin (20 μM), and epinephrine (1 μM). The increment in the radioactivity of phosphatidic acid above controls (incubated without thrombin) was determined and shown in cpm/ml platelet suspension. Radioactivity in controls was 165/185 cpm/ml (see 488).

(estimated from the formation of thromboxane B_2) induced by collagen are also augmented by serotonin, and this effect is inhibited by ketanserin (233). The secretion of dense granules induced by a narrow concentration range (0.03–0.06 U/ml) of thrombin is also facilitated by serotonin; this potentiation is reduced by ketanserin (Fig. 7–6). The serotonin secreted by the platelets must interact synergistically with the primary agonist, thrombin (Fig. 7–4). The synergistic effects of serotonin on human platelets are, however, small as compared to those of epinephrine. They are much larger in cat platelets (243).

Although the mechanism(s) underlying this synergistic potentiation is far from clear, these observations suggest that (a) serotonin potentiates a step that is common to aggregation, secretion of dense granules, secretion of α-granules, and liberation of arachidonate induced by several primary agonists; and (b) serotonin effects this by binding to an S_2-serotonergic receptor. A likely candidate for such a common step would be the receptor-controlled mobilization of cytoplasmic Ca^{2+} (Fig. 7–3).

Synergism Between Serotonin and Thrombin and Formation of Phosphatidic Acid

The synergistic potentiation by epinephrine of platelet stimulation induced by thrombin causes a variable augmentation (50–250%) in the formation of phosphatidic acid *(unpublished observations)*. When measured during the interaction of platelets with thrombin, in the absence and presence of epinephrine, serotonin increased the thrombin-induced synthesis of phosphatidic acid by 30%, and this increase was abolished by ketanserin (Table 7–2). Under the same conditions, epinephrine increased the formation of phosphatidic acid by 50%, but this increase was insensitive to ketanserin.

— 8 —

Serotonin and the Heart: Pharmacological Evaluation with the S_2-Serotonergic Antagonist Ketanserin

F. T. Thandroyen, S. Saman, and L. H. Opie

Ischaemic Heart Disease Research Unit of the Medical Research Council of South Africa, University of Cape Town, Cape Town, South Africa

The hearts of several mammalian species, including humans, contain serotonin (64,80,672,989,1118). Serotonin is present not only in the platelets contained in the coronary blood vessels, but also in the cells of the myocardium, which can synthesize and metabolize the monoamine (753). This chapter reviews the possible effects of serotonin on myocardial function, with special emphasis on the cardiac effects of the S_2-serotonergic antagonist ketanserin (Table 8–1).

TABLE 8–1. *Possible beneficial effects of S_2-serotonergic blockers in heart disease*

Antiarrhythmic activity (high dose)
Prolongation of action potential duration
Inhibition of ischemic and reperfusion arrhythmias
Indirect antiarrhythmic activity by inhibition of coronary vasospasm
Myocardial contractility
Therapy of severe heart failure by causing arteriolar and venous dilatation
Prevention or reversal of carcinoid heart disease
Coronary vasculature
Prevent coronary vasospasm
Prevent platelet-induced initiation of vascular occlusion

HEART RATE

Physiology

Serotonin may play a role in regulation of the heart rate. Indeed, the monoamine (EC_{50} 0.1 μM) produces a positive chronotropic effect in atria of the kitten; this effect is not reversed by ketanserin (562). The receptor involved is different from S_1- and S_2-serotonergic binding sites. At the high concentrations of 10^{-5} M or 10^{-6} M, ketanserin has a negative chronotropic effect in the isolated heart of the rat (Fig. 8–1).

Pathology

The following observations support the hypothesis that serotonin could provoke or facilitate the occurrence of ventricular arrhythmias: (a) Serotonin can displace stored norepinephrine from sympathetic nerve terminals (338,351). An enhanced adrenergic activity may contribute to the development of ventricular arrhythmias (808). (b) Serotonin, infused into dogs 72 to 96 hr after acute myocardial infarction, provokes ventricular arrhythmias, and this is prevented by depletion of catechol-

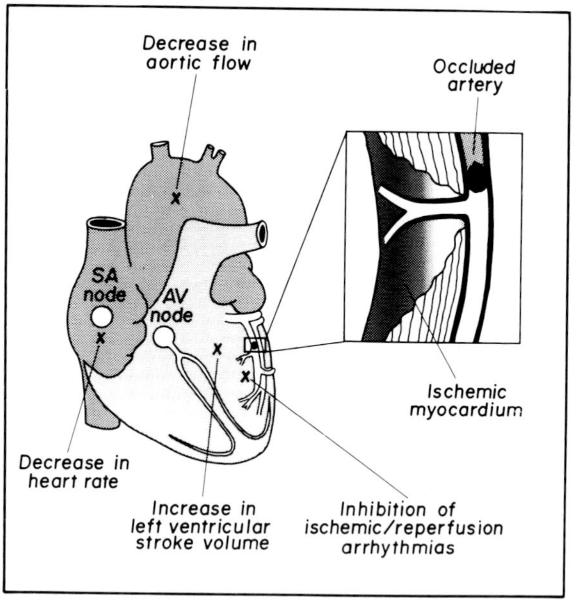

FIG. 8–1. Cardiac effects of ketanserin. Direct effects of ketanserin in high concentrations (10^{-5} M) on the isolated working rat heart include: (a) a bradycardia causing the heart rate to fall from 278 ± 12 to 180 ± 3 beats/min ($p<0.01$); (b) a fall in cardiac output from 64 ± 0.8 to 55 ± 2.6 ml/min ($p<0.05$); and (c) a rise in stroke volume from 0.24 ± 0.01 to 0.31 ± 0.01 ml/min ($p<0.01$). For details, see Saman, Thandroyen, and Opie (933). For antiarrhythmic effects, see Fig. 8–2 and Table 8–2.

amines (920). (c) Serotonin (10^{-5} M or more) induces aftercontractions in the papillary muscle of the kitten (562).

Activation of serotonin receptors may play a role in ventricular arrhythmogenesis during acute coronary artery occlusion or upon reintroduction of coronary flow. In these two situations, adrenergic activity may be enhanced.

Pharmacology

In the isolated heart of the rat perfused with artificial solutions, ketanserin (5×10^{-6} M or 10^{-5} M) prevents the development of spontaneous ventricular arrhythmias during acute regional myocardial ischemia, and on subsequent reperfusion of the ischemic myocardium (Table 8–2). This protection is not mediated by the preservation of coronary flow during ischemia or reperfusion, but is associated with a reduction in heart rate. However, in hearts paced at a rate similar to that observed in the control group, ketanserin (5×10^{-6} M) still prevents arrhythmias, which excludes the bradycardia caused by the drug as the reason for the protection. The antiarrhythmic effect of ketanserin does not imply the involvement of S_2-serotonergic receptors in the genesis of ventricular arrhythmias during ischemia

TABLE 8–2. *Effect of ketanserin on ventricular arrhythmias during ischemia and reperfusion of isolated rat heart*

	Ischemia		Reperfusion	
	VPB	VT	VT	VF
	Incidence %		Incidence %	
Control *N* = 18	100	89	100	100
Ketanserin 10^{-6} M *N* = 6	100	28[a]	57	71
Ketanserin 5×10^{-6} M *N* = 6	50[a]	0[a]	33[a]	16[a]
Ketanserin 10^{-5} M *N* = 7	33[a]	0[a]	16[a]	0[a]

[a] $p < 0.02$ vs control.

VPB = Ventricular premature beats; VT = ventricular tachycardia; VF = ventricular fibrillation; *N* = number of hearts. The model was the isolated non-blood-perfused rat heart (Langendorff technique). The perfusate was Krebs-Henseleit buffer, substrate 11 mM glucose, 95% O_2, 5% CO_2 (Po_2 of the perfusate was more than 500 mm Hg). The protocol was, in the control series: (a) 15 min preperfusion; then (b) left main coronary artery ligation producing acute regional myocardial ischemia of the left ventricle; and (c) reperfusion of the ischemic myocardium for 10 min. In the ketanserin series, the appropriate concentration of ketanserin was perfused for 10 min prior to coronary ligation and for the rest of the experiment.

and/or reperfusion, since the concentrations of ketanserin required to prevent arrhythmias in the isolated heart of the rat are severalfold higher than those required to cause S_2-serotonergic blockade (637). Since ketanserin in higher concentrations binds to α_1-adrenoceptors (1093,1098), the antiarrhythmic activity on the ventricle of 5×10^{-6} to 5×10^{-5} M concentrations of the drug may therefore be due to α_1-adrenergic blockade. Electrophysiological studies in the superfused papillary muscle of the guinea pig show that ketanserin (10^{-6}–10^{-5} M) prolongs the action potential duration at 90% of repolarization, but does not alter V_{max}, an index of sodium conductance (Fig. 8–2). Acute myocardial ischemia shortens the duration of the action potential (808). This electrophysiological alteration may predispose to ventricular arrhythmogenesis by (a) decreasing the refractory period or (b) causing dispersion of refractoriness between ischemic and nonischemic myocardium. Thus, one possible mechanism whereby ketanserin may prevent ventricular arrhythmias is by maintaining the duration of the action potential in the ischemic myocardium. In the dog another potent serotonergic antagonist, cyproheptadine (0.4–0.7 mg/kg), prevents ventricular arrhythmias of late onset during myocardial infarction. However, these doses of cyproheptadine are insufficient to antagonize the serotonergic receptors (920).

If S_2-serotonergic receptors on myocardial cells do not appear to play a major role in ventricular arrhythmogenesis during acute ischemia or upon subsequent reperfusion, in the intact organism, S_2-serotonergic antagonists may still inhibit ventricular arrhythmias indirectly by preventing the facilitating effect of serotonin on platelet aggregation (Table 8–1; see Chapter 15).

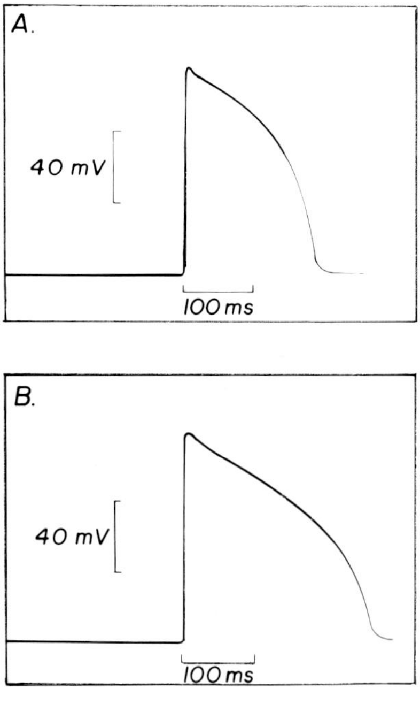

FIG. 8–2. Effect of ketanserin (10^{-6} M) on the cardiac action potential. Right ventricular papillary muscle was isolated from guinea pigs and superfused with oxygenated Krebs-Henseleit buffer (37°C). The preparation was stimulated at 0.5 Hz. Action potentials were recorded by conventional microelectrode techniques. The action potentials and the electronically differentiated maximum rate of phase 0 depolarization were displayed at two sweep speeds on two channels of an oscilloscope screen and photographed. After stabilization for 30 min, ketanserin was added to the superfusate to obtain a final concentration of 5×10^{-6} M. Observations were made for 40 min. Shown are control **(A)** and 40 min after ketanserin **(B)**. Ketanserin increased the action potential duration at 90% repolarization from a control value of 177 ms to 257 ms.

MYOCARDIAL FUNCTION

Physiology

Serotonin has a positive inotropic effect on the mammalian heart by increasing the concentration of norepinephrine in the vicinity of the myocardial cells (134,882). Several mechanisms may be involved in this effect: (a) stimulation of serotonergic receptors on nerve terminals, which triggers the release of the adrenergic neurotransmitter (351); (b) inhibition of the reuptake of norepinephrine by sympathetic nerve terminals (140,496); and (c) a tyramine-like indirect sympathomimetic action (338). The inotropic effect is not mediated by cyclic AMP (71), which probably is the intracellular second messenger of β-adrenergic stimulation. However, serotonin does not regulate cardiac contractility under basal conditions (989).

Pathology

Serotonin may provide inotropic support in the setting of impaired ventricular function (989). Indeed, alteration in the metabolism of tryptophan, the amino acid which is the critical factor in the synthesis of serotonin (337; see Chapter 1) may contribute to the pathogenesis of endocardial and myocardial damage and the development of heart failure. Thus, tryptophan deficiency causes cardiomyopathy in both the rat and the guinea pig (883,989). In the rat, an inverse correlation exists between the plasma level of tryptophan and the severity of the myocardial damage (883). In the guinea pig, the acute cardiomyopathy evoked by tryptophan deficiency can be prevented by the administration of serotonin (994). The histological profile of myocardial damage in the rat closely resembles that in patients with "African cardiomyopathy" (883). This syndrome is associated with low plasma levels of tryptophan and serotonin (884). Tryptophan supplementation prevents the progression of the disease in subclinical cases of "African cardiomyopathy" (882). The possibility that serotonin may function as an endogenous modulator of contractile failure is supported by the observation that the content and synthesis of the monoamine are decreased in the heart of the myopathic hamster and in that of patients with heart failure (989); speculatively, these alterations may be the consequence of the reflex increase in sympathetic tone which follows cardiac decompensation. Furthermore, excessive levels of circulating serotonin, as seen in carcinoid syndrome, produce endocardial and myocardial damage and predispose to heart failure (see Chapter 20).

Pharmacology

Ketanserin (10^{-7}–10^{-6} M) causes no alteration in cardiac output, peak aortic pressure, left ventricular power development, or myocardial oxygen consumption in the isolated working rat heart perfused with an artificial solution (Fig. 8–1; 933). At 10^{-5} M, however, the compound causes a decreased cardiac output without alteration in the myocardial contents of cyclic AMP or of high energy phosphates

(933). Since power production of the left ventricle is not altered, the fall in cardiac output is most probably due to the marked reduction in heart rate (Fig. 8–1; 933). These findings do not imply that ketanserin may have a deleterious effect on the mechanical function of the human myocardium, since the concentration of ketanserin reducing the cardiac output is 50 to 100 times higher than the therapeutic concentrations achieved in man (476,527). Furthermore, the stroke volume actually increased as the heart rate fell, presumably due to better ventricular filling. In this system where the preload and afterload are fixed, an increased stroke volume indicates either better filling or a positive inotropic effect. It should be noted that the preparation used is devoid of external autonomic nervous control and works against a fixed hydrostatic load, unlike the situation *in vivo* where the heart interacts dynamically with the peripheral circulation. Preliminary evidence indicates that ketanserin is beneficial in the acute treatment of severe heart failure refractory to digitalis glycosides and diuretics (265). Ketanserin, by causing arteriolar dilatation, decreases peripheral vascular resistance, unloads the left ventricle, and increases cardiac index and stroke work index (297). Because of its venodilator effect, the drug decreases the right atrial, pulmonary arterial, and pulmonary wedge pressures (297,1143). The dilator action of the drug is probably due to the prevention of the amplifying effects of serotonin (528). Thus, ketanserin is potentially a useful vasodilator agent in the therapy of severe chronic heart failure (Table 8–1). Its inhibitory effect on platelet aggregation may also improve focal areas of obstruction caused by platelet aggregates in the peripheral microcirculation.

CORONARY ARTERIES

Serotonin evokes contraction in isolated coronary arteries in several species, including humans (76,837,884), and also amplifies the vasoconstrictor effects of catecholamines (257,258). The receptors on the coronary smooth muscle belong to the S_2-serotonergic subtype (562). In the intact canine heart studied *in situ*, serotonin is a potent vasoconstrictor of epicardial coronary arteries and causes a reduction in the cross-sectional area of up to 42%; however, the monoamine does not produce a significant increase in coronary resistance (121). Serotonin is a more potent vasoconstrictor of the epicardial arteries than the α_1-adrenergic agonist phenylephrine, which suggests that the large-conductance coronary arteries may respond more vigorously to vasoconstrictors released from platelets than to neurohumoral stimulation. In particular, work with isolated coronary arteries and experiments in the intact dog suggest that in areas of endothelial damage, serotonin is implicated in the genesis of coronary vasospasm (Fig. 8–3) and may even initiate the acute thrombotic occlusion in the coronary artery (121,313,528,1067; see Chapters 9 and 10). If this were the case, S_2-serotonergic antagonists such as ketanserin should have a beneficial effect in patients with coronary vasospasm and/or obstruction (Table 8–1). In an evaluation of the effect of ketanserin on coronary artery vasospasm, the drug was compared with placebo in 5 patients with rest chest pain and evidence of ST segment elevation. Ketanserin proved ineffective

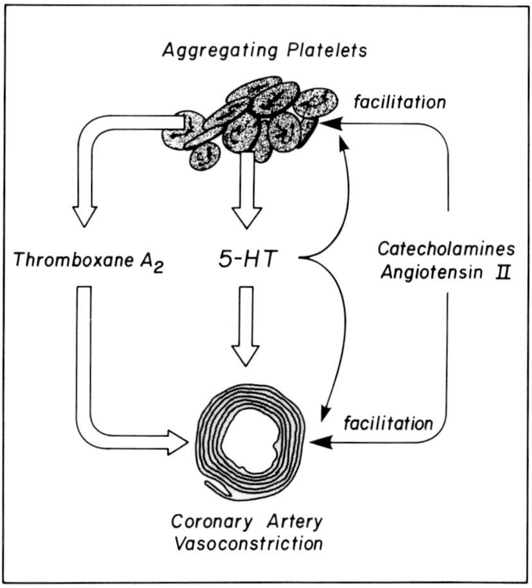

FIG. 8–3. Possible interaction between serotonin, catecholamines, and angiotensin II on platelet aggregation and coronary artery vasoconstriction.

in preventing vasospastic ischemic episodes. Thus, in the limited patient sample studied, serotonin did not appear to play a role in the pathogenesis of coronary artery vasospasm associated with chest pain at rest.

ACKNOWLEDGMENT

We thank the Medical Research Council of South Africa and the Chris Barnard Fund for support. Dr. S. Saman was supported by the Guy Elliott Fellowship. We are also grateful to June Chambers for secretarial assistance and to Errol Yon for electrophysiological experiments.

9

Serotonin and Vascular Smooth Muscle

*J. M. Van Nueten, *W. J. Janssens, and **P. M. Vanhoutte

**Department of Pharmacodynamics, Janssen Pharmaceutica, B-2340 Beerse, Belgium; and **Department of Physiology and Biophysics, Mayo Clinic and Mayo Foundation, Rochester, Minnesota 55905*

Serotonin has complex and multiple actions on cardiovascular function (818). In the intact animal, it can cause either increases or decreases in blood pressure, depending on the species studied and the dose used. The local effect of the monoamine on blood flow is similarly confusing: depending on the species, the vascular bed, the experimental conditions, the degree of sympathetic tone, and the route of administration, serotonin augments or reduces blood flow. At the microcirculatory level, serotonin induces edema; in certain species (e.g., the rat) it is more, but in others (e.g., the guinea pig and the rabbit) less potent than histamine in that regard (318). The changes in blood flow and in blood pressure caused by the monoamine reflect the changes in vascular tone that it can cause at both the pre- and the postcapillary levels. Indeed serotonin *in vivo* and *in vitro* has a number of direct and indirect actions on the blood vessel wall (Fig. 9–1). The balance between these various effects determines whether the monoamine will cause contraction or relaxation of the smooth muscle cells and hence constriction or dilatation of the blood vessel (1072,1073,1088,1103). This chapter reviews the major effects of serotonin leading to changes in contractile activity of vascular smooth muscle, and how local environmental changes or serotonergic inhibitors can affect these. The effects of both exogenously added and endogenously released serotonin are considered.

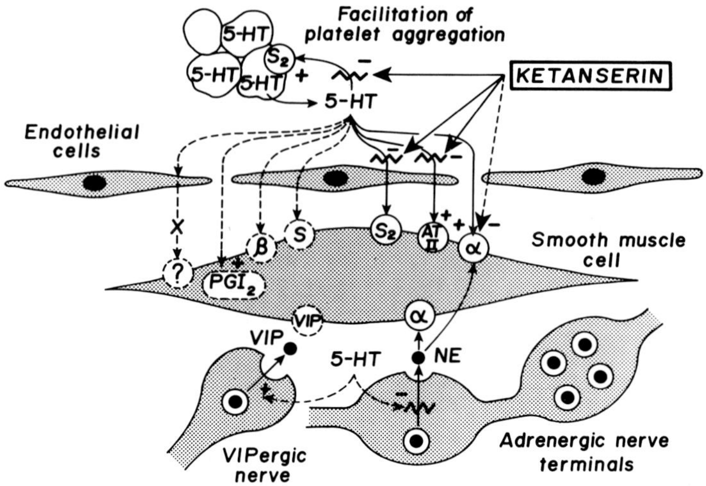

FIG. 9–1. Multiple effects of serotonin (5-HT) on the blood vessel wall. Vasoconstriction could be caused by (a) activation of S_2-serotonergic receptors on the smooth muscle cells; (b) amplification of the vasoconstrictor responses of other neurohumoral mediators (e.g., norepinephrine, NE; angiotensin II, AT II); (c) activation of postjunctional α-adrenoceptors (α); or (d) an indirect sympathomimetic effect, by displacement of stored norepinephrine from adrenergic nerve terminals. Free serotonin also activates S_2-serotonergic receptors on platelets causing enhanced release of the stored amine. Vasodilatation could be caused by (a) prejunctional inhibition of adrenergic neurotransmission; (b) activation of inhibitory autonomic nerves (presumably causing release of vasoactive intestinal polypeptide, VIP); (c) activation of inhibitory serotonergic receptors (S) on the smooth muscle cells; (d) activation of the production of prostacyclin (PGI_2); (e) stimulation of β-adrenoceptors (β); or (f) triggering of endothelium-dependent relaxation. Not all these effects have been demonstrated in all blood vessels. + = Excitatory effect, amplification (potentiation); − = inhibitory effect. (From ref. 1078, with permission.)

VASOCONSTRICTOR EFFECTS OF SEROTONIN

Direct Effects

Receptors

Serotonin causes contraction of most isolated large arteries. These contractions can be inhibited by ketanserin, spiperone, methysergide, and a number of other serotonergic antagonists (176,179,180,1092,1093,1095). The affinity of these serotonergic antagonists for S_2-serotonergic binding sites correlates well with their potency to antagonize contractions of isolated blood vessels to serotonin (see Chapter 5). Thus, serotonin activates S_2-serotonergic receptors on the vascular smooth muscle (643,1093,1095). However, large differences in sensitivity to the monoamine exist, depending on the anatomical origin of the blood vessels (Fig. 9–2). Serotonin is more potent than norepinephrine in coronary, cerebral, umbilical, and pulmonary arteries (179,180,301,1070,1088). Cutaneous and subcutaneous arteries contract strongly to serotonin, but in a number of cutaneous arteries these contractions are inhibited by α-adrenergic antagonists (34,673,1070,1159). Thus,

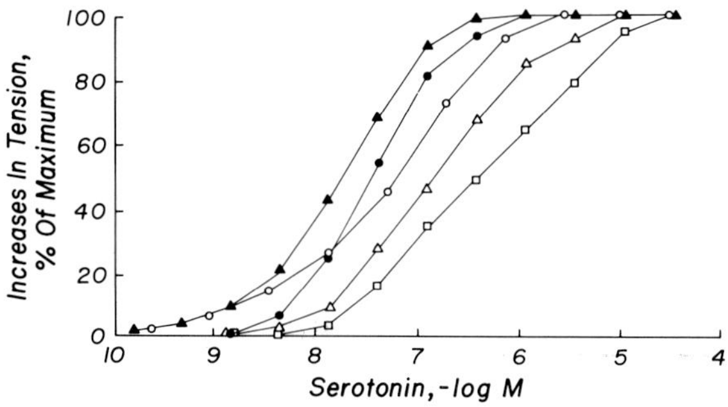

FIG. 9–2. Contractile responses of canine blood vessels to serotonin. ▲, Basilar artery (100% = 1.40 ± 0.16 g); ●, internal carotid artery (100% = 3.10 ± 0.71 g); ○, coronary artery (100% = 3.48 ± 0.81 g); Δ, gastrosplenic artery (100% = 7.23 ± 1.89 g); □, gastrosplenic vein (100% = 2.54 ± 0.65 g). (Modified from ref. 1102.)

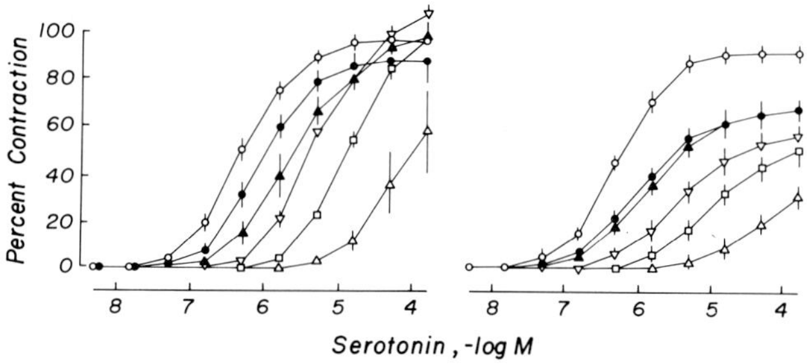

FIG. 9–3. Inhibition by ketanserin (*left*, ○, control; ●, 4×10^{-10}; ▲, 1.6×10^{-9}; ▽, 6.3×10^{-9}; □, 2.5×10^{-9}; △, 10^{-7}) and methysergide (*right*, ○, control; ●, 4.5×10^{-9}; ▲, 1.8×10^{-9}; ▽, 7×10^{-9}; □, 2.8×10^{-8}; △, 1.1×10^{-7}) of the contractile responses to serotonin in jugular veins of the rat. Data shown as means ± SEM.

serotonin may interact to some extent with α-adrenergic receptors. In certain pathological states in animals and in humans the vascular smooth muscle cells become hypersensitive to serotonin (184,1159,1161).

Most large veins constrict when exposed to serotonin, but their individual sensitivity differs (1070). In canine cutaneous veins, which contract to the same degree when activated with serotonin or norepinephrine, serotonergic contractions can be inhibited with α-adrenergic inhibitors (172,205,1070). In the canine gastrosplenic and saphenous vein ketanserin inhibits contractions induced by serotonin but not in a competitive manner (1093). In the jugular vein of the rat, which is devoid of α-adrenoceptors, the response to serotonin is inhibited competitively by ketanserin, spiperone, and methysergide (Fig. 9–3; 176). Thus, contractile responses of venous smooth muscle to serotonin can be mediated predominantly by S_2-serotonergic receptors, although in certain veins important similarities may exist between S_2-serotonergic and postjunctional α-adrenoceptors.

To judge from data in intact dogs, the smooth muscle of arteriovenous anastomoses contains serotonergic receptors, different from the S_2-subtype, which can activate the contractile process (939). In cerebral blood vessels, serotonergic receptors that do not belong to the S_2-subtype may contribute to the contractile response to serotonin (1095).

Activator Calcium

The contractions of vascular smooth muscle ultimately depend on increases in sarcoplasmic Ca^{2+} concentration, which can be due to either the influx of extracellular Ca^{2+} or its release from intracellular stores (535). In the caudal artery of the rat, most of the contraction to serotonin is inhibited by Ca^{2+}-entry blockers, suggesting that the monoamine triggers mainly the influx of extracellular Ca^{2+} to activate the smooth muscle cells (1087,1101). By contrast, in the ear artery of the rabbit serotonin induces contractions both by increasing the influx of extracellular Ca^{2+} and by releasing intracellularly stored ions (673). Although part of the contraction of this blood vessel is due to binding of serotonin to α-adrenoceptors, a specific serotonergic component is involved (673). In the bovine coronary artery, Ca^{2+} released from intracellular stores also contributes to the activation caused by serotonin (875). Thus serotonin is able to mobilize both intracellular and extracellular Ca^{2+}, but the relative importance of the two sources varies from one blood vessel to another. For example, in cerebral blood vessels of the rabbit, the contractions induced by the monoamine depend more on extracellular Ca^{2+} than those of systemic blood vessels from the same species (86).

Amplifying Effects

Subthreshold or threshold concentrations of serotonin, whether exogenously added or released from activated platelets, markedly potentiate (amplify) contractions of isolated blood vessels induced by other vasoconstrictor agonists, including angiotensin II, norepinephrine (whether exogenously added or released from sympathetic nerves), degradation products of fibrin and fibrinogen, prostaglandin $F_{2\alpha}$, thromboxane A_2, and the thromboxane A_2-mimetic U 44069 (238,349,599,695,759, 1087,1092–1094). The concentrations of serotonin causing amplification correspond to the concentrations of the monoamine in human plasma (197). In the femoral artery of the rabbit and the caudal artery of the rat, the amplifying effect of serotonin is inhibited by ketanserin, indicating that activation of an S_2-serotonergic receptor is required (1092,1093; Fig. 9–4). It is also prevented by the Ca^{2+}-entry blocker flunarizine, suggesting that an augmented Ca^{2+} entry is involved (1089).

Modulation by Local Factors

Hypoxia

In cerebral and coronary arteries hypoxia potentiates the contraction induced by serotonin (1090,1099,1100). The hypoxic augmentation of the response to serotonin is inhibited by ketanserin. Ultimately the augmentation involves an increased entry

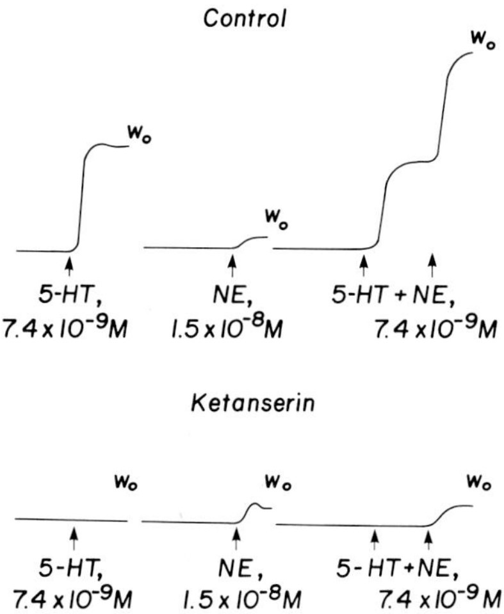

FIG. 9–4. Amplifying effect of serotonin (5-HT) on the contraction to a threshold concentration of norepinephrine (NE; **upper**). The amplification is abolished by ketanserin **(lower)**. W_o = Washout. (Modified from ref. 1093.)

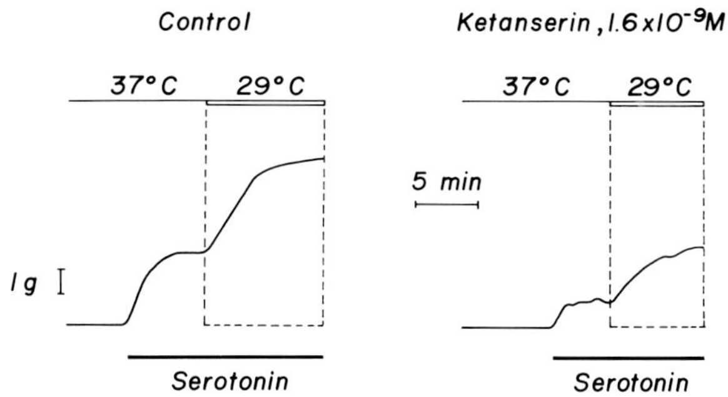

FIG. 9–5. Cooling from 37 to 29°C augments the contraction of the tibial artery of the rabbit induced by serotonin (7.4 nM). Both the contraction and its augmentation by cooling are inhibited by ketanserin. (Modified from ref. 1091.)

of Ca^{2+}, since it is attenuated by the Ca^{2+}-entry blocker lidoflazine. Lidoflazine inhibits normoxic contractions to both norepinephrine and serotonin, but not the hypoxic facilitation of the response to norepinephrine (1100). Thus, hypoxia probably does not enhance Ca^{2+} entry directly.

Cooling

Moderate cooling augments the contractions of cutaneous blood vessels induced by a variety of vasoconstrictor substances (1071). For instance, in cutaneous veins and arteries of the dog, cooling from 37 to 24°C markedly potentiates contractions induced by serotonin (649,1082). In the tibial artery of the rabbit, contractions evoked by serotonin also are augmented by moderate cooling, and ketanserin causes a concentration-dependent inhibition of this potentiation (Fig. 9–5; 1091). In the canine saphenous vein, the similar potentiation by cooling of the response to norepinephrine can be counteracted by α-adrenergic antagonists (529); it reflects an instantaneous increase in affinity of the postjunctional α-adrenoceptors (530). Likewise, moderate cooling may induce an increase of the affinity of S_2-serotonergic receptors of cutaneous vascular smooth muscle.

Aggregating Platelets

Isolated blood vessels contract when surrounded by aggregating platelets (179,180,238, 239,649,708,1088; Fig. 9–6). These platelets release thromboxane A_2 and serotonin. The relative contribution of these two substances in causing contraction may vary from one blood vessel to another, depending on its sensitivity for these agonists. For example, in the pulmonary artery of the dog thromboxane A_2 plays little, if any, role in the contraction (708). In the coronary and saphenous arteries of the same species and in the caudal artery of the rat, thromboxane A_2 contributes but is less important than serotonin (179,180,238,649). By contrast, in the basilar artery of the dog the release of thromboxane A_2 is the most important factor in causing platelet-induced contractions. The serotonergic component of the contraction can be inhibited by the S_2-serotonergic antagonists such as ketanserin and methysergide (179,180,238,1088). The platelet-induced contractions of the caudal artery of the rat are inhibited by the Ca^{2+}-entry blocker flunarizine, indicating that they are caused by an increased influx of Ca^{2+} into the smooth muscle cells (238,1089). The response to platelets is augmented by cooling in canine saphenous arteries (649) and by hypoxia in the internal carotid artery of the same species *(unpublished observations)*. Serotonin amplifies contractions induced by thromboxane A_2 (or U 44069) in the caudal artery of the rat and the coronary artery of the dog (238,239,759). Since both substances are released by platelets, mutual amplification may be an important determinant of the contraction of vascular smooth muscle cells exposed to aggregating platelets.

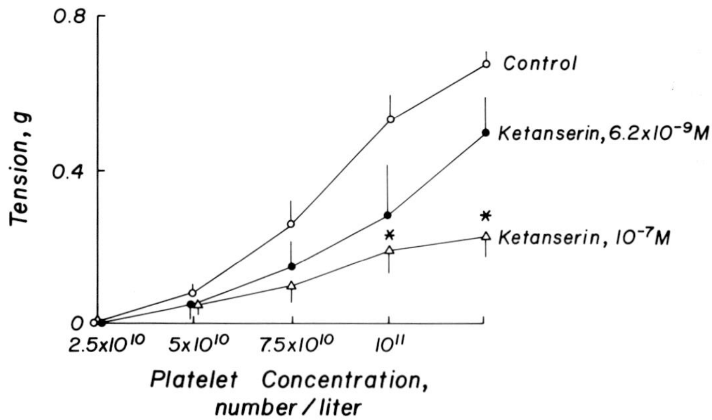

FIG. 9–6. Inhibitory effect of ketanserin on the contractions of the caudal arteries of the rat induced by increasing amounts of aggregating platelets. (Modified from ref. 238.)

VASODILATOR EFFECTS

Although the direct effect of serotonin on most vascular smooth muscle cells is to cause contraction, relaxations caused by the monoamine have been demonstrated in coronary arteries, cerebral arteries, saphenous veins, the canine nasal circulation, and the perfused vascular bed of the guinea pig stomach (179,180, 301,336,365,951,1093). In cerebral arteries serotonin may activate β-adrenoceptors to cause relaxation (301). However, in most other cases a serotonergic receptor appears to be involved, which differs from the S_2-serotonergic receptor of the vascular smooth muscle cells mediating the contractions caused by serotonin. Indeed, the relaxations of vascular smooth muscle obtained *in vitro* with serotonin can be inhibited by methiothepin, methysergide, pizotifen, and cyproheptadine, but not by ketanserin (179,180,336,951,1093,1095). In many instances, blockade of the contraction evoked by the monoamine even is a prerequisite to observe its dilator effect (1093; Fig. 9–7).

Although it cannot be excluded that serotonergic receptors mediating relaxation are located on the vascular smooth muscle cells themselves, the more potent dilator responses to serotonin most likely are indirect ones (Fig. 9–1) and involve other cells present in the blood vessel wall. These include (a) endothelial cells (see Chapter 10) and (b) adrenergic nerve endings. Indeed, serotonin can inhibit adrenergic neurotransmission with a resulting decrease in the amounts of norepinephrine liberated at the neuroeffector junction (503,599,710). The inhibition of adrenergic neurotransmission is not affected by ketanserin (Fig. 9–8) and thus is not mediated by S_2-serotonergic receptors (1093). It is, however, antagonized by S_1-serotonergic

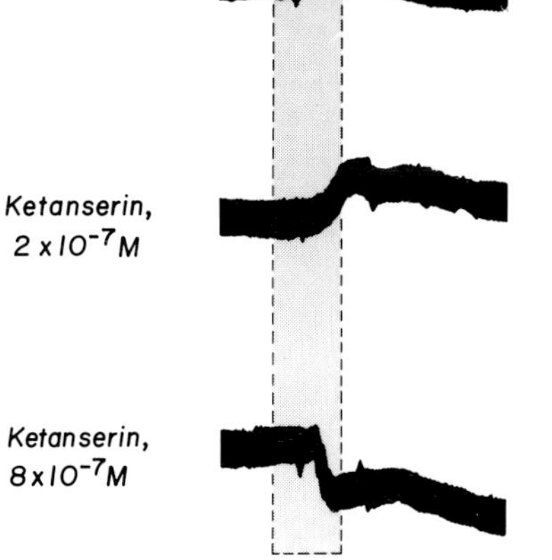

FIG. 9–7. In the isolated perfused stomach of the guinea pig ketanserin inhibits *(middle)* and even reverses *(bottom)* the vasoconstrictor response to serotonin. (Modified from ref. 1093.)

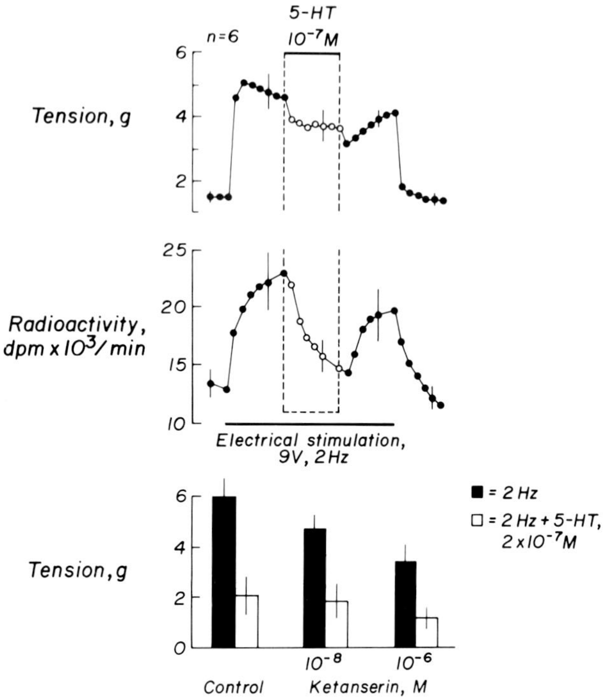

FIG. 9–8. Serotonin inhibits contractions induced by electrical stimulation in canine saphenous veins **(upper)** because of inhibition of the release of norepinephrine **(middle)**. The inhibitory effect of serotonin on contractions induced by electrical stimulation is not prevented by ketanserin **(lower)**. (Modified from refs. 709 and 1093.)

inhibitors such as methiothepin (308,660). Serotonin also induces the release of vasodilator substances [e.g., vasoactive intestinal polypeptide from peptidergic nerves (see 1103)]. It can stimulate the production of prostacyclin (see 1103).

PITFALLS IN ASSESSING SEROTONERGIC ANTAGONISM

The analysis of the effects of S_2-serotonergic antagonists on the responses to smooth muscle is complicated by several factors. First, a number of these drugs are partial agonists (1094,1095; Table 9–1). Their partial agonistic effects are due to activation of S_2-serotonergic receptors, since they can be blocked by ketanserin (1092). In cutaneous blood vessels, serotonergic antagonists may also activate α-adrenergic receptors (1070). Second, a proper pharmacological analysis of S_2-serotonergic antagonism should require previous inhibition of the receptor(s) mediating the relaxation induced by serotonin. Since no selective inhibitor of the latter receptor(s) is at present available, this problem can be overcome in part by the removal of the endothelium and the exclusion of neuronal effects of serotonin.

TABLE 9–1. *Properties of serotonergic blockers*

	Agonistic properties[a] rabbit femoral artery	Antagonism of serotonin-induced contraction in the caudal artery of the rat: IC_{50} (M)
LSD	+ + +	8.4×10^{-9}
Methysergide	+ +	2.7×10^{-9}
Bromo-LSD	+	6.7×10^{-9}
Mianserin	+	4.2×10^{-9}
Pizotifen	0	1.6×10^{-9}
Metergoline	+ +	1.7×10^{-9}
Cyproheptadine	0	2.3×10^{-9}
Metitepine	0	1.7×10^{-9}
Phenoxybenzamine	0	5.9×10^{-8}
Cinanserin	0	3.4×10^{-9}
Ketotifen	0	7.6×10^{-8}
Ketanserin	0	1.7×10^{-9}

[a]The agonistic properties are compared to those of LSD which was the most active. + + +, Strong effect; + +, moderate effect; +, weak effect; 0, no effect; highest concentration tested: 2×10^{-6} M. IC_{50} = Concentration causing 50% inhibition.
Modified from ref. 1094.

POTENTIAL IMPORTANCE IN THE INTACT ORGANISM

Although the plasma concentration of free serotonin is very low under normal conditions (197,510,695), it obviously will augment dramatically during platelet aggregation, as illustrated in embolized animals (1064; see Chapter 21). The monoamine can be taken up by adrenergic nerve endings and subsequently be released, which may explain why the blood vessel wall contains serotonin (695,1105). The effect of any amount of serotonin will depend on factors such as the integrity of the endothelium (see Chapter 10) and the degree of responsiveness of the vascular smooth muscle (see Chapters 11 and 12). Serotonin may be of particular importance in cerebral blood vessels, since their smooth muscle is very sensitive to the monoamine (Fig. 9–2). In the brain serotonin might diffuse from the neurons which release it to the interstitium, from which it can reach the vascular smooth muscle cells without passing the blood-brain barrier (301). Furthermore, cerebral vessels appear to be innervated or closely associated with serotonin-containing nerves (301,426).

The potentiation by hypoxia of vasoconstriction caused by serotonin released from aggregating platelets or from serotonergic nerves may contribute to or prolong vasospastic episodes in the cerebral and coronary arteries. The augmentation by cooling of the constriction of cutaneous blood vessels induced by platelets may well play a role in the vasospasm of Raynaud's phenomenon (see Chapter 19). Of the many actions of serotonin on the blood vessel wall (Fig. 9–1), it is tempting to assume that the amplifying effect, which occurs at very low concentrations, is probably the most relevant.

10

Platelets, Serotonin, and Endothelial Cells

*Richard A. Cohen and **Paul M. Vanhoutte

**Robert Dawson Evans Department of Clinical Research, Boston University School of Medicine, Boston, Massachusetts 02215; and **Department of Physiology and Biophysics, Mayo Clinic, Rochester, Minnesota 55905*

As the epithelial lining of the cardiovascular system, the endothelium acts as a barrier that excludes circulating cellular elements and harmful substances, or as a modulator of the function of the blood cells, the vessel wall, and the tissues that the blood vessel supplies. The endothelial cells participate in capillary transport, regulation of plasma lipids and proteins, and control of hemostasis. To accomplish their work, these cells possess a full complement of membrane receptors, ion exchange pumps, and synthetic and degradative enzymes (791,964,1081).

Platelets interact both physically and biochemically with the blood vessel wall. The endothelium plays an important role in these interactions. Where the endothelium is damaged, platelets aggregate on the subendothelial collagen and smooth muscle. As an inevitable consequence of their aggregation, platelets release serotonin, nucleotides, prostaglandins, catecholamines, and various proteins, all of which may have profound effects on other platelets, the surrounding endothelium, the smooth muscle, or the tissues beyond (743,966). The serotonin content of platelets (1–5 nmol/10^9 platelets) is more than sufficient to exert effects upon its release (209). When serotonin is released by platelets in numbers equivalent to those present in blood, the concentration reaches 0.6 μM (180). The local concentration reached at the vessel wall beneath a platelet thrombus may be much greater.

This chapter reviews interactions of serotonin and platelets with the blood vessel wall, with special emphasis on the role of endothelial cells.

ROLE OF THE ENDOTHELIUM

Anti-Aggregatory Action

The surface characteristics of endothelium aid in preventing platelet aggregation, but the primary mechanism by which endothelium inhibits platelet deposition is by the production of prostacyclin (743,966). This production is regulated in part by external influences, and possibly by substances released by platelets themselves. Thus, cyclic endoperoxides synthesized by platelets may provide a substrate for the formation of prostacyclin by the endothelium (686). Proteins secreted by the platelets also stimulate the production of prostacyclin (859). In addition, the synthesis of prostacyclin by the endothelial cells increases in response to acetylcholine, thrombin, and calcium ionophores (65,415,1140), suggesting that activation of receptors on the endothelial cell membrane stimulates the production of prostacyclin, possibly by initiating Ca^{2+} entry. Since serotonin, histamine, catecholamines, and intact platelets stimulate the influx of calcium into endothelial cells (208), this could serve to increase the production of prostacyclin and thereby inhibit further platelet aggregation.

Catabolic Role

Several enzymes contained within endothelial cells may inhibit platelet aggregation by destroying substances that facilitate it. Thus, the endothelium breaks down adenosine diphosphate released from platelets and thereby attenuates the pro-aggregatory influence of the nucleotide (139). Likewise, serotonin is metabolized by endothelial monoamine oxidase (see Chapter 3). Reduction of this enzymatic capability by disease could result in greater effective concentrations of serotonin in the vicinity of the platelets which may accelerate platelet aggregation (406).

Influence on Smooth Muscle Contractility

Though earlier evidence had pointed to the endothelium as a possible site where smooth muscle responses were initiated (826), the first evidence for a role of arterial endothelial cells in vascular responsiveness was provided by Dr. Robert Furchgott and his colleagues who demonstrated that arteries denuded of endothelium fail to relax when exposed to acetylcholine (371,372,386). The arterial endothelium actually contributes to or even plays an obligatory role in the relaxations due to arachidonic acid, adenosine di- and triphosphate, bradykinin, histamine, substance P, thrombin, and vasopressin (159,164,261,262,372,560,1068). These substances probably interact with specific receptors on the endothelial cells (Fig. 10–1). The relaxation due to calcium ionophores is also endothelium-dependent. Hence, the endothelium-dependent responses may be due to the influx of Ca^{2+} into the endothelial cells (974). The transmission of the signal between endothelium and smooth muscle requires direct contact between the two types of cell; it probably is a short-lived chemical agent (369). The relaxation due to arachidonic acid is

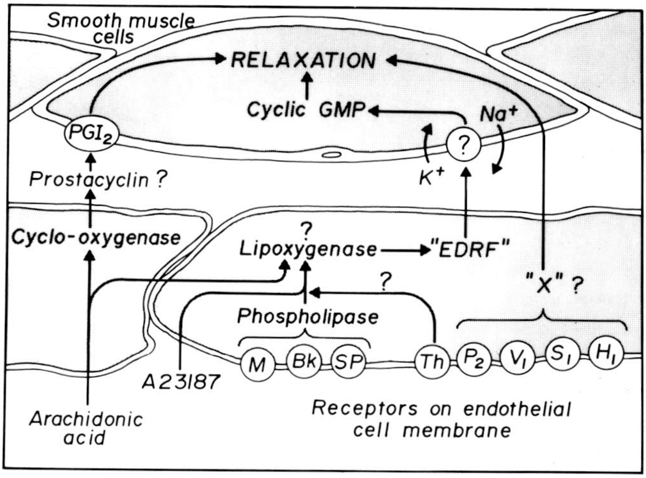

FIG. 10–1. Summary of endothelium-dependent inhibitory receptors. Bk = Bradykinin; H = histamine; M = muscarinic; P = purines; S = serotonin; SP = substance P; Th = thrombin; V = vasopressin; PGI_2 = prostacyclin; X = unknown factor(s).

probably caused by the formation of prostanoids by the endothelial cells (161,262). The cellular mechanisms underlying the other endothelium-dependent responses are not well understood. Since inhibitors of the enzymes phospholipase and lipoxygenase block the endothelium-mediated response to acetylcholine, an oxidation product of arachidonic acid metabolism could be involved; however, it is likely that the inhibitors interact chemically with substance(s) released by the endothelium. Apparently, endothelium-mediated responses to thrombin and adenine nucleotides do not depend on the metabolism of arachidonic acid (160,164,262,368,1081). Relaxations induced by the endothelium are associated with a rise in 3′,5′-cyclic guanosine monophosphate in the smooth muscle, as occurs during relaxation caused by nitrates (370,873). However, relaxation of vascular smooth muscle caused by nitrates or papaverine does not involve the endothelium (179,261,372).

VASCULAR RESPONSE TO PLATELETS

Smooth Muscle Contraction

Aggregating platelets cause contractions of isolated canine and porcine coronary arteries, canine saphenous arteries, pulmonary artery, femoral arteries and veins, saphenous vein, and rat caudal artery (180,238,304,497,649,708; see Chapters 9 and 15).

Mediators of Contraction

A major component of the contractile response to aggregating platelets is blocked by the serotonergic antagonists ketanserin, methysergide, and cyproheptadine

(179,180,238,708). The similarity in the action of the antagonists on the contractions caused by both platelets and serotonin supports the conclusion that serotonin released by the platelets mediates the contractions (179,180,708; Fig. 10–2). The action of the selective S_2-serotonergic antagonist ketanserin suggests that serotonin causes contraction by activating S_2-receptors on the smooth muscle (177,1093). However, thromboxane A_2 contributes also to the contractile action of platelets (239,304).

Inhibition of Contraction by the Endothelium

The contractions of isolated canine coronary arteries caused by platelets are larger in rings which have been mechanically denuded of endothelium. A similar difference is observed in the contractile responses to serotonin, but not in those to potassium chloride or prostaglandin $F_{2\alpha}$ (Fig. 10–3; 179,180). Thus, the endothelium attenuates contractions due to serotonin, as illustrated also by the fact that the sensitivity of perfused coronary segments to serotonin is less in segments with endothelium than in those without endothelium (181). By contrast, the contractions of canine pulmonary arteries caused by platelets and serotonin are unaffected by the removal of the endothelium (708).

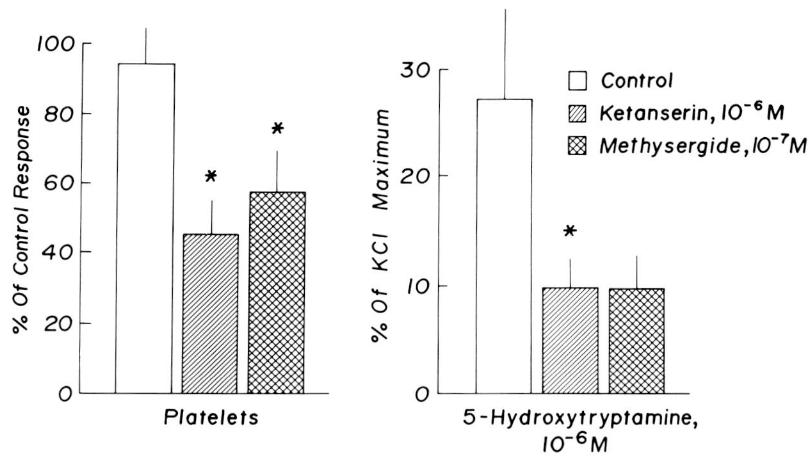

FIG. 10–2. Effect of the serotonergic antagonists ketanserin and methysergide on responses of isolated canine left circumflex coronary arteries to aggregating platelets **(left)** and serotonin **(right)**. Isometric tension was measured in coronary rings suspended in organ chambers filled with physiological salt solution. Autologous platelets were prepared in calcium-free citrate buffer to prevent aggregation; when added to the physiological salt solution in numbers equal to those present in normal plasma, the platelets aggregated due to the presence of calcium in the bath and collagen in the blood vessel wall. Responses to platelets in the presence or absence of antagonists are expressed as a percentage of a preceding control response in the absence of antagonist. The response to serotonin in the presence and absence of the antagonists is expressed as a percentage of the maximal contractile response to potassium chloride. *Asterisks* denote significant attenuation of response.

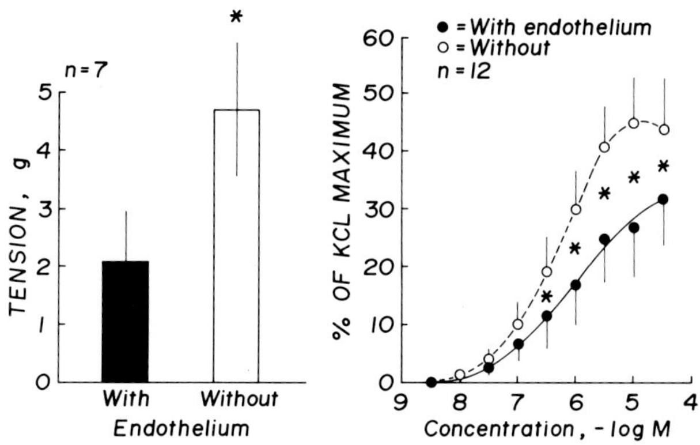

FIG. 10–3. Effect of mechanical removal of endothelium on contractile response of canine left circumflex artery to aggregating platelets **(left)** and to cumulative increases in serotonin concentration **(right)**. Responses to platelets and serotonin are of less magnitude in rings with intact endothelium. (From ref. 180, ©1983 by the American Association for the Advancement of Science; and ref. 179, with permission.)

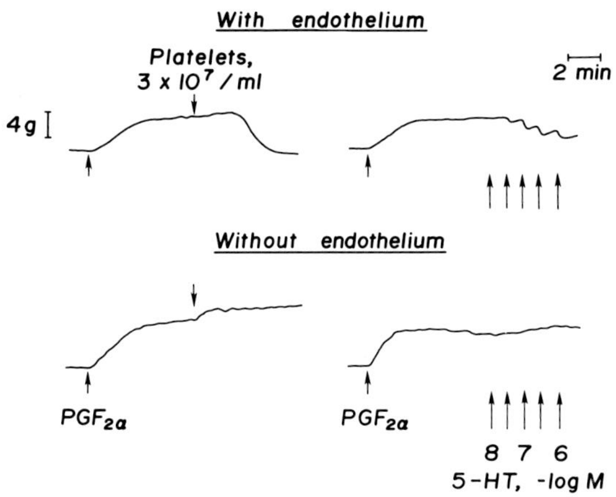

FIG. 10–4. Response to aggregating platelets and serotonin of two canine left circumflex artery rings contracted with prostaglandin $F_{2\alpha}$. In the ring with intact endothelium **(upper)**, platelets *(left)* cause a transient further increase in tension, followed by relaxation, while only further contraction occurs in the ring denuded of endothelium **(lower)**. The relaxation may be attributable to serotonin released by the platelets, since serotonin *(right)* also causes endothelium-dependent relaxation, and further contraction in the denuded ring. (From ref. 180, ©1983 by the American Association for the Advancement of Science; and ref. 179, with permission.)

Endothelium-Mediated Relaxation

In canine coronary arteries contracted with prostaglandin $F_{2\alpha}$, aggregating platelets cause opposite responses, depending on the presence or absence of endothelium. In rings with intact endothelium they induce relaxation, while in rings denuded of endothelium only a further contraction occurs (Fig. 10–4). These experiments suggest that the coronary endothelium not only protects the vessel wall from the effects of platelet aggregation, but also actively inhibits contractions evoked by serotonin and other vasoactive substances released from the platelets.

The contractile effect of thromboxane A_2 which is released by the platelets, as evidenced by the presence of its stable metabolite thromboxane B_2, is overcome by this inhibitory influence of the endothelium (179,180). All canine blood vessels do not react similarly to platelets: coronary arteries and femoral veins do, but femoral arteries and saphenous veins do not display endothelium-dependent relaxation in response to platelets (497).

Mediators of Relaxation

Since serotonin also causes endothelium-dependent relaxation of canine coronary artery (Fig. 10–4; 174,179), the relaxation caused by platelets may be attributable in part to the serotonin that they release. The relaxations caused by serotonin are attenuated by methysergide and methiothepin, suggesting that the response is initiated at serotonergic receptors on endothelial cells. Since the response is unaffected by ketanserin, these endothelial serotonergic receptors probably belong to the S_1-serotonergic subtype (Fig. 10–5; 179,180,497). Further evidence that serotonin released from aggregating platelets can activate endothelial receptors comes from studies in perfused segments of coronary artery. Aggregating platelets and serotonin cause smooth muscle relaxation only when given access to the intimal surface of the blood vessel, while extraluminal application causes only contraction (Fig. 10–6; 181).

Other agents released by platelets could have a role in the relaxation of coronary smooth muscle mediated by the endothelium. Since blockade of cyclo-oxygenase with indomethacin or of prostacyclin synthetase with tranylcypromine fails to inhibit the response, the synthesis of prostacyclin is not the primary mediator. Lipoxygenase products of the metabolism of arachidonic acid are not involved, since

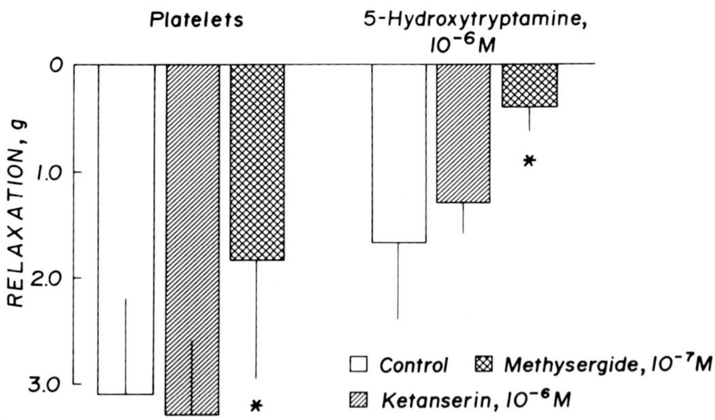

FIG. 10–5. Effects of the serotonergic antagonists ketanserin and methysergide on the relaxations (indicated in grams) caused by aggregating platelets and serotonin in coronary artery rings with endothelium contracted with prostaglandin $F_{2\alpha}$ (see Fig. 10–3). Methysergide, but not ketanserin, significantly attenuated the relaxations. The lack of effect of ketanserin suggests that endothelial serotonergic receptors are of a different subtype than the S_2-receptors which mediate smooth muscle contraction. (From ref. 179, with permission.)

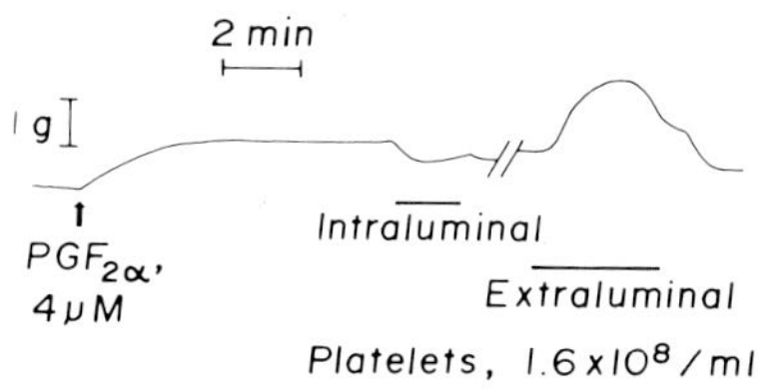

FIG. 10–6. Effect of intra- and extraluminal aggregating platelets in a perfused segment of canine coronary artery. Smooth muscle tension was measured by hooks within the vessel lumen. After contraction of the smooth muscle with prostaglandin $F_{2\alpha}$, addition of aggregating platelets to the perfusate caused relaxation. Following washout of the platelets, addition of the same number of platelets to the extraluminal bath caused contraction. This experiment indicates that the route of access to the blood vessel wall is important in determining the response of the smooth muscle, and that exposure of the intima to aggregating platelets can cause smooth muscle relaxation. (From ref. 181, with permission.)

inhibitors of that enzyme do not interfere with the endothelium-mediated response to platelets (180). Endothelium-dependent relaxations can be initiated by adenosine diphosphate (261). It is likely that adenine nucleotides released from the platelets indeed contribute to the platelet-induced relaxations (497).

POTENTIAL ROLE IN THE INTACT ORGANISM

Serotonin and Endothelium

Though the effect of serotonin on the cardiovascular system is multiphasic, a prolonged hypotensive action can occur following its intravenous infusion. This vasodilatation follows a brief depressor action, mediated by the Bezold-Jarisch reflex, and a pressor response, caused by activation of S_2-serotonergic receptors on smooth muscle (822,1093). It is possible that the endothelium mediates the hypotensive response. Like the serotonergic endothelial-dependent response of isolated coronary arteries, the vasodilator response to serotonin is not affected by S_2-serotonergic antagonists; it is mediated probably by serotonergic receptors of the S_1-subtype (551,1093).

Thrombosis and Hemostasis

The demonstration that aggregating platelets can cause relaxation of smooth muscle by activating endothelium implies that during thrombosis, serotonin as well as adenosine diphosphate may lead to vasodilatation (Fig. 10–7). Thrombin produced during coagulation could also contribute to endothelium-dependent vasodilatation (262). This vasodilator mechanism could be a key protective feature of the endothelium, tending to limit further aggregation by increasing blood flow (1078). The contractile action of serotonin and platelets may aid in hemostasis where the endothelium is interrupted or where blood extravasates and coagulates on the extraluminal surface of a blood vessel.

Vascular Diseases

Endothelial cell injury from hypercholesterolemia, hypertension, diabetes, or chronic cigarette smoking may be a key instigator of chronic vascular disease and

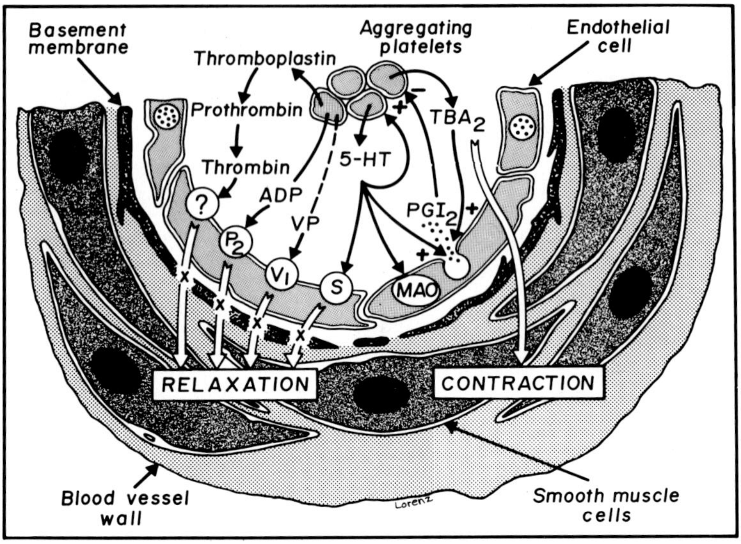

FIG. 10–7. Mechanisms of vasodilatation mediated by the endothelium during intraluminal thrombosis. Serotonin (5-HT) and adenosine diphosphate (ADP) as well as thrombin formed during coagulation can activate endothelium, which by an as yet poorly understood mechanism (X) causes smooth muscle relaxation. Monoamine oxidase (MAO) contained in the endothelium may contribute to its protective role by degrading serotonin. Prostacyclin (PGI_2), produced when stimulated by thromboxane (TBA_2), inhibits platelet aggregation. Human platelets also contain vasopressin (VP), which can cause endothelium-dependent relaxations. (From ref. 1078, with permission.)

in particular of atherosclerosis (916). Damage of endothelium may lead to accelerated platelet aggregation due to a decreased catabolic activity and the lack of prostacyclin synthesis. In a variety of immunological and thrombotic syndromes where the endothelium is damaged, depletion of platelet serotonin stores is observed, suggesting chronic platelet aggregation within the vascular system (193). Particularly in large arteries, smooth muscle contraction caused by vasoactive substances released by platelets at sites of endothelial damage may further limit blood flow which is already compromised by fixed atherosclerotic obstructions. Such vasospasm may contribute to angina pectoris, myocardial infarction, and Prinzmetal's variant angina (697,803) and could have a role in cerebrovascular disease and Raynaud's phenomenon (178). Since ischemia, which results from vascular obstruction, inhibits endothelial cell function (264,369,598), this pathological condition might exacerbate vasospasm by preventing the protective vasodilator function of the endothelium.

11

Responsiveness of the Hypertensive Blood Vessel Wall to Serotonergic Stimuli

*R. Clinton Webb and **Paul M. Vanhoutte

**Department of Physiology, University of Michigan, Ann Arbor, Michigan 48109; and **Department of Physiology and Biophysics, Mayo Medical School, Mayo Clinic, Rochester, Minnesota 55905*

The vascular responsiveness to many constrictor agents is increased in hypertension (1128). This is usually measured as a pressor response in the intact animal, or as a change in resistance to flow in a perfused vascular bed. Two factors contribute to the augmented vascular reactivity: (a) structural alterations, in particular an increase in wall thickness and wall-to-lumen ratio, provide a mechanical advantage for changes in lumen diameter and limit the minimal resistance to flow; and (b) functional changes in the components of the vascular wall (smooth muscle, endothelium, and nerve endings) cause a greater reduction in lumen diameter in response to a given level of sympathetic activation or to a given concentration of a vasoactive agonist. Folkow and co-workers have emphasized that the structural changes that occur in hypertension probably reflect adaptive responses of the vascular wall exposed to increased arterial pressures (342). This chapter describes experimental observations that characterize the abnormal responsiveness to serotonergic stimuli in the hypertensive blood vessel wall, with special attention to the smooth muscle cells and the adrenergic nerve endings. For an extensive overview of vascular changes in hypertension the reader is directed to more exhaustive publications (342,1128,1129).

Depending on the experimental situation, serotonin can cause either constriction or dilatation, and it has the unique property of interacting with other vasoactive agents (33,1073,1078,1087; see Chapter 10). The following paragraphs document these varied actions of serotonin on the vasculature of hypertensive animals.

AUGMENTED CONSTRICTOR RESPONSES TO SEROTONIN

Activation of Serotonergic Receptors

Serotonin

McGregor and Smirk (711) observed that the pressor response to serotonin in genetic and renal hypertensive rats was about twice that in normotensive controls. In the perfused mesenteric vasculature of both types of hypertensive animals the pressor responsiveness to norepinephrine and angiotensin II was also increased. However, the magnitude of the enhanced pressor responses to norepinephrine and angiotensin II (~50% above control) was greatly exceeded by the enhanced response that occurred when serotonin was administered (~400–700%; 711). Such a difference in responsiveness to different vasoactive agonists cannot be explained on the basis of a vascular structural change alone (711). Hence, the enhanced pressor responses to serotonin must reflect functional alterations in the components of the vascular wall.

Several other laboratories have confirmed that the pressor responsiveness to serotonin is greatly enhanced in hypertension (77,184,261). The sensitivity of the vasculature (threshold for constrictor effect) to serotonin is also augmented in hypertensive animals compared to normotensive animals, although the degree of increase varies from one vascular bed to another and from one type of hypertension to another (979,1127,1129,1161). Thus, in the renal vasculature of spontaneously hypertensive rats the constrictor responses to serotonin are potentiated to a greater degree than those to norepinephrine, angiotensin II, or barium (Fig. 11–1; 184). Furthermore, the relative changes in sensitivity (threshold response) observed in the hypertensive rats were greatest for the monoamine. In the perfused mesenteric artery of spontaneously, renal, and deoxycorticosterone hypertensive rats, the thresholds to serotonin are lower, the dose-response curves to the monoamine are steeper, and maximal constrictor responses to it are greater than in normotensive rats. In contrast, the threshold for constriction in hindquarter preparations of hypertensive animals does not differ from that in normotensive rats (Fig. 11–2; 434). The maximal pressor responses to serotonin are augmented in the hindquarters of renal and deoxycorticosterone hypertensive rats but not in the hindquarters or hindlimbs of spontaneously hypertensive rats (13,434). The aorta from spontaneously hypertensive rats is about 10 times more sensitive to serotonin, but the portal vein less than twice as sensitive as the corresponding blood vessels from normotensive rats (13). Thus, there is a marked variability in sensitivity to serotonin between different vascular preparations from hypertensive animals.

Although the specific increase in responsiveness to serotonin is a well-documented observation in hypertension, few studies have been performed to characterize the cellular mechanisms that may account for it. The cellular mechanisms

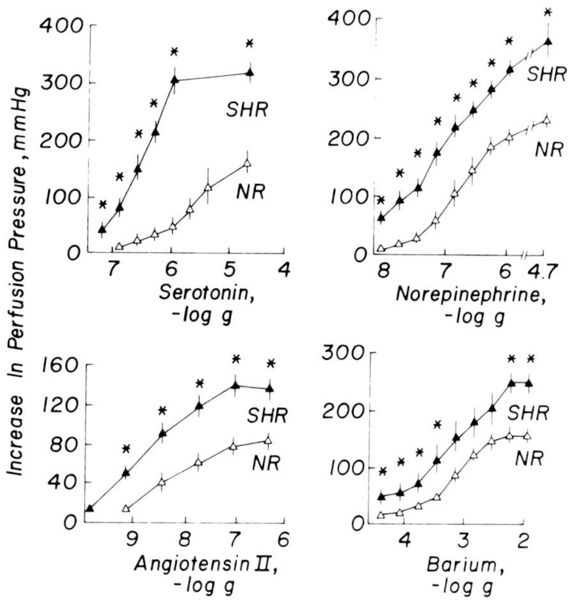

FIG. 11–1. Vascular reactivity to various constrictor agonists in the perfused renal vascular bed of spontaneously hypertensive rats (SHR) and normotensive controls (NR). The augmentation of the reactivity in response to serotonin (threshold) in the SHR is greater than that of the response to norepinephrine, angiotensin II, or barium. (From ref. 184, with permission of the American Heart Association.)

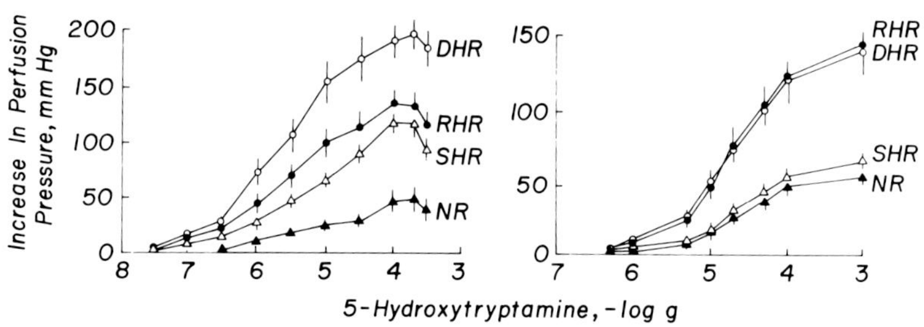

FIG. 11–2. Vascular reactivity to serotonin in three types of hypertension. **Left:** The responses to serotonin (5-hydroxytryptamine) of the perfused mesenteric artery from the spontaneously hypertensive rat (SHR), the renal hypertensive rat (RHR), and the deoxycorticosterone acetate hypertensive rat (DHR) are variably greater than those of the same artery from normotensive controls (NR). **Right:** Responses to serotonin of the vascular bed of the perfused hindquarter are elevated in the renal and deoxycorticosterone acetate hypertensive rat but not in the spontaneously hypertensive rat. (From ref. 434, with permission.)

that may contribute to the increased sensitivity to serotonin in hypertensive rats could include (a) an increased affinity of serotonergic receptors; (b) accelerated transmembrane movement of calcium following receptor activation; and (c) greater mobilization of activator calcium from calcium store(s) by serotonin (716). The competitive antagonism of serotonin-induced contractions by methysergide or ketanserin is similar in arteries from normotensive and deoxycorticosterone acetate hypertensive rats, suggesting that the increased sensitivity to the monoamine does not result from a change in receptor affinity. Following depletion of cellular calcium with ethyleneglycol bis(aminoethylether)tetraacetate (EGTA), dose-response curves to calcium in the presence of a maximal dose of serotonin are not significantly different in arteries from hypertensive rats compared to those from normotensive animals. In contrast, contractile responses to serotonin in calcium-free solution following loading of a cellular store(s) of calcium are greater by 50% in the arteries from the hypertensive rats (Fig. 11–3). Thus, the enhanced vascular sensitivity to serotonin in deoxycorticosterone acetate hypertension is not related to an alteration in the transmembrane movement of calcium following receptor activation, but rather to an altered mobilization of calcium from a cellular store(s).

A difference in the activation of receptor-operated ionic channels by serotonin may also account for the altered sensitivity to the monoamine in hypertension. Norepinephrine, angiotensin II, serotonin, and vasopressin induce an increased turnover of radiolabeled potassium in aortic segments from deoxycorticosterone acetate hypertensive and normotensive rats (543). With the exception of vasopressin, the dose of each agonist that causes a 50% change in potassium turnover rate is lower in arteries from hypertensive than in those from normotensive rats. The dose-response relationship for serotonin on potassium turnover is shifted more to the left of normotensive values than that for norepinephrine and angiotensin II (Fig. 11–4). Thus, the specific increase in vascular responsiveness to serotonin in hypertension may result from an altered relationship in the receptor activation of ionic channels important for membrane excitability.

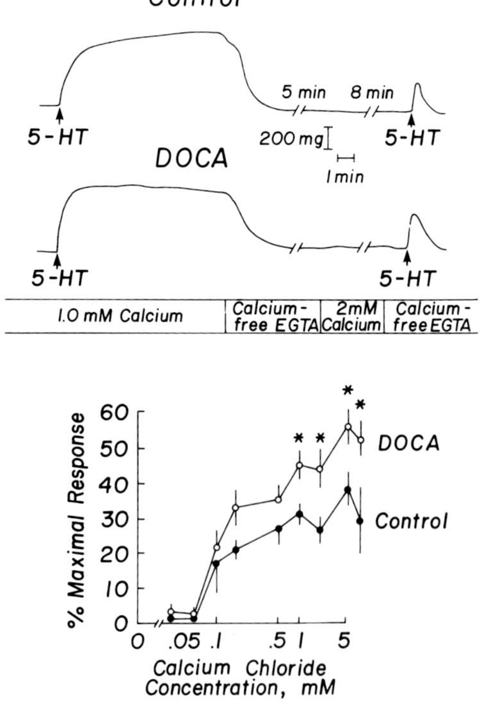

FIG. 11–3. Demonstration of a greater membrane store of calcium in femoral arteries of deoxycorticosterone acetate hypertensive (DOCA) rats. **Upper:** Femoral arteries were made to contract to serotonin (5-HT) (5.7×10^{-7} M) in solution containing 1.6 mM calcium; the strips were then depleted of calcium by placement in EGTA-calcium-free buffer for 10 min. Following this calcium depletion the strips were "loaded" for 5 min in buffer with a known concentration of calcium, placed in EGTA-calicum-free buffer, and then challenged with serotonin (5.7×10^{-7} M). **Lower:** Data are expressed as percent of contraction in 1.6 mM calcium buffer and represent the mean (± SEM) of responses from 6 femoral artery strips. *The contractile responses to serotonin in femoral arteries from deoxycorticosterone acetate hypertensive rats were significantly greater ($p < 0.05$) than responses from control rats at "loading" concentration of calcium above 0.5 mM. (From ref. 716, with permission.)

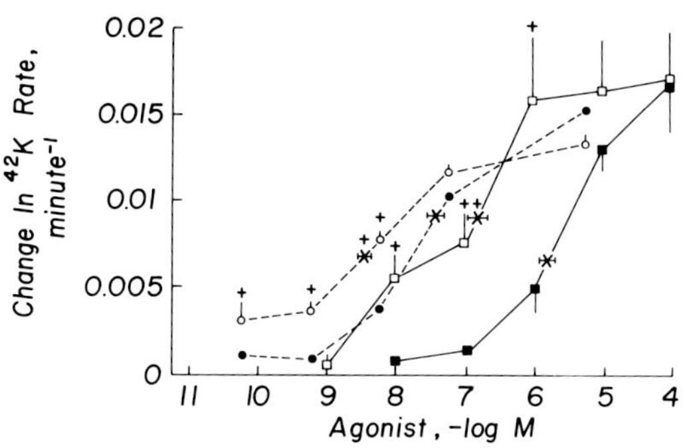

FIG. 11–4. Dose-dependent effects of norepinephrine (NE) and serotonin (5-HT) on the potassium efflux (^{42}K rate, min^{-1}) from aortic strips isolated from deoxycorticosterone acetate hypertensive (DOCA) and normotensive rats. The ED_{50} values are indicated by ×. The responses for hypertensive rats which are significantly different from the corresponding control are identified by (+). ○ = DOCA, NE; ● = control, NE; □ = DOCA, 5-HT; ■ = control, NE. Values are the mean (± SEM). (From ref. 543, with permission.)

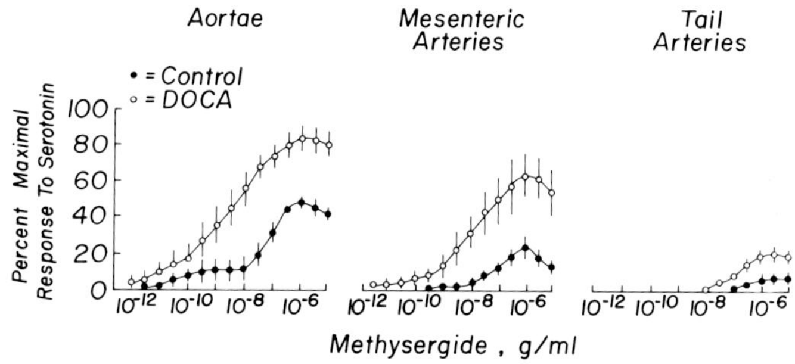

FIG. 11–5. Dose-response curves to methysergide in arteries from deoxycorticosterone acetate hypertensive (DOCA) and normotensive rats. The cumulative addition of methysergide to the muscle bath produced contraction. The contractile responses to methysergide are expressed as percent of the maximal response to serotonin since the agonistic properties of the serotonergic antagonist are mediated by serotonergic receptors. Arterial preparations from deoxycorticosterone acetate hypertensive rats were more sensitive to methysergide than those from control rats. Values are the mean (± SEM) for 6 control rats and 6 hypertensive rats. (From ref. 1129, with permission.)

Partial Serotonergic Agonists

In hypertensive animals the vascular responsiveness to the partial agonistic properties of methysergide is augmented (716). Thus, isolated arterial segments from mineralocorticoid hypertensive rats develop greater force and are more sensitive to methysergide than those from normotensive rats (Fig. 11–5). The relative change in responsiveness to methysergide is greatest in the aorta and least in the tail artery; mesenteric arteries are intermediate. Interestingly, this sequence is the reverse of that describing the density of innervation of these three blood vessels.

The altered vascular responsiveness to methysergide parallels the development of the increase in arterial blood pressure in deoxycorticosterone acetate hypertensive rats (1127). Placement of these hypertensive rats on a low-sodium diet reverses both the elevated blood pressure and the enhanced sensitivity to methysergide to near normal levels. An increased responsiveness to the partial agonistic properties of methysergide is also greater in genetic and renal models of hypertension (1127). Thus, the hypertensive process may be associated with an alteration in the sensitivity of the receptors that mediate the partial agonistic properties of methysergide. Hence, it is not surprising that attempts to lower blood pressure with methysergide have failed in hypertensive animals (434).

Potentiation and Attenuation of Constrictor Responses to Other Stimuli by Serotonin

In 1962 Bohr and Elliot (107) observed that serotonin potentiates contractile responses to epinephrine in isolated mesenteric arteries from the dog (Fig. 11–6). Following removal of serotonin, the contractile responses to epinephrine were attenuated (1093). The potentiating effect of serotonin on adrenergic responsiveness is inhibited by the S_2-serotonergic antagonist ketanserin (1093; see Chapter 9). The attenuation of the contractions induced by catecholamines following exposure to serotonin is blocked by methysergide and probably involves activation of the electrogenic sodium pump and depression of calcium influx during adrenergic activation (748).

Both the potentiating (amplifying) and attenuating effects of serotonin may be altered in hypertension. In the mesenteric vasculature of the spontaneously hypertensive rat the amplifying effect of serotonin is more pronounced than in that of normotensive rats (599). Likewise, contractile responses to norepinephrine in isolated femoral arteries from renal hypertensive rabbits are augmented by lower doses of serotonin than those in arteries from normotensive controls (862). The absolute magnitude of potentiation by the monoamine does not differ between arteries from hypertensive and normotensive rabbits. The attenuation of norepinephrine responsiveness that follows the exposure to serotonin is less in arteries from hypertensive than in those from normotensive rabbits. These changes induced by serotonin may be related to a greater influx of calcium into the smooth muscle cells from hypertensive rabbits during exposure to the adrenergic agonist (862).

Tachyphylaxis

The responsiveness of certain vascular preparations to repeated doses of serotonin decreases with time (184). This tachyphylactic response to serotonin is decreased in the renal vasculature of genetically hypertensive rats (Fig. 11–7; 184,261). The

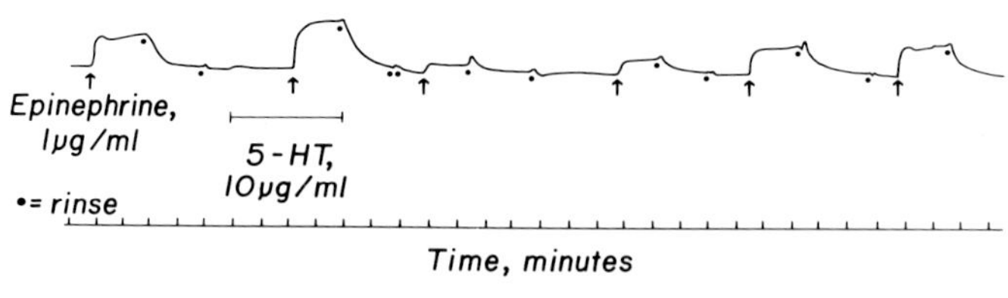

FIG. 11–6. Response of smooth muscle from resistance vessel (mesentery) to epinephrine with and following serotonin (5-HT) in the bath. When serotonin is added to the bath there is a potentiation of the response to epinephrine; after the serotonin has been washed out, repetition of the epinephrine stimulus elicits only a depressed response. After about 20 min recovery is complete. (From ref. 107, with permission.)

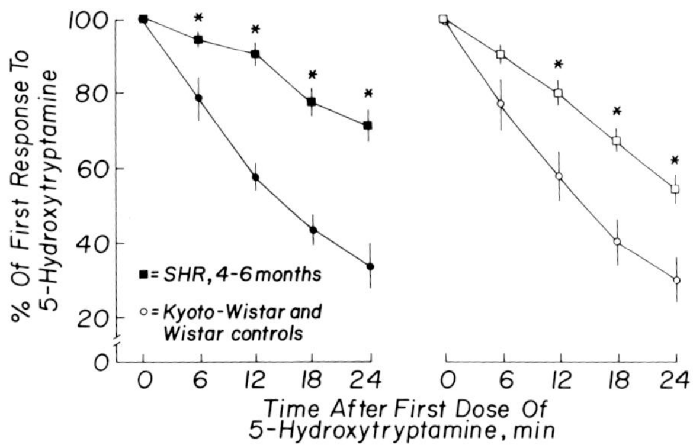

FIG. 11–7. Rate of development of tachyphylaxis to serotonin (5-hydroxytryptamine) in kidneys from male and female spontaneously hypertensive and sex-matched control rats. The graph shows mean percent decrease in response amplitude to repeated doses (2×10^{-5} g) of serotonin in kidneys from female control (○) and spontaneously hypertensive (□) rats **(right)**, and from male control (●) and spontaneously hypertensive (■) rats **(left)**. Significant differences between kidneys from spontaneously hypertensive and control rats are indicated by asterisks. (From ref. 184, with permission of the American Heart Association.)

effect is more pronounced in the male than in the female spontaneously hypertensive rat. The tachyphylaxis to serotonin is less in older than in younger spontaneously hypertensive rats (261). The cellular mechanism underlying the delayed tachyphylaxis in the renal vasculature of the spontaneously hypertensive rat is unknown; it is unrelated to the presence of activator calcium in the extracellular fluid (261).

Indirect Sympathomimetic Effect of Serotonin

Serotonin can be actively transported into adrenergic varicosities by the neuronal uptake mechanism in the blood vessel wall (1085,1105). In the adrenergic nerve endings serotonin can displace norepinephrine, causing vasoconstriction via an indirect sympathomimetic effect (e.g., 183). The monoamine can also be stored and released together with norepinephrine upon activation of the nerve endings (1105).

In innervated kidneys of young spontaneously hypertensive rats, the vasoconstrictor responses to serotonin are increased compared to normotensive values; this difference between the two animal groups is absent following adrenergic denervation with 6-hydroxydopamine (Fig. 11–8; 183). Thus, serotonin has a more pronounced indirect sympathomimetic effect in the hypertensive rats.

VASCULAR RELAXATION IN RESPONSE TO SEROTONIN

Serotonin can produce vasodilatation by either a direct action on the blood vessel wall (715,1072,1076) or an indirect effect resulting from inhibition of norepinephrine release from adrenergic nerve endings (503,709,1125). The inhibitory effect

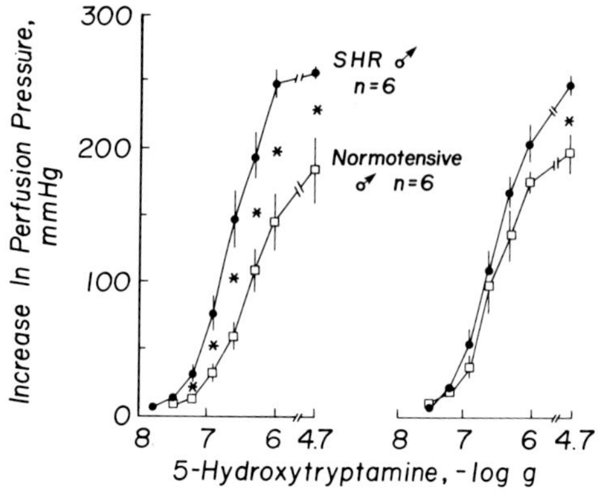

FIG. 11–8. Left: Dose-response curves to serotonin (5-hydroxytryptamine) in kidneys from 6- to 7-week-old spontaneously hypertensive (SHR) and control rats, before treatment with 6-hydroxydopamine. **Right:** After treatment. Mean increases in perfusion pressure of kidneys from spontaneously hypertensive (●) and control (□) rats. Significant differences between hypertensive and control values are marked by an asterisk. These experiments demonstrate that the indirect sympathomimetic effect of the monoamine is more pronounced in the young spontaneously hypertensive rat than in the control animals. (From ref. 183, with permission of the American Heart Association.)

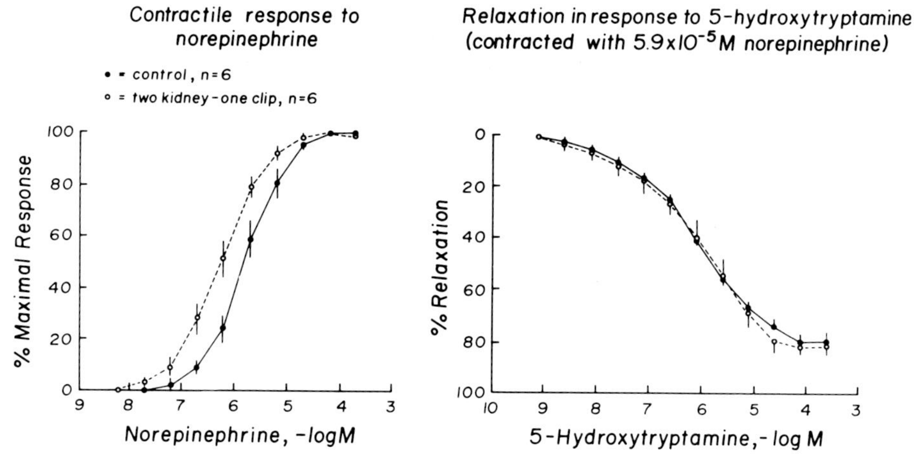

FIG. 11–9. Dose-response effects of norepinephrine and serotonin (5-hydroxytryptamine) in femoral veins from renal hypertensive rabbits. (From ref. 1128, with permission.)

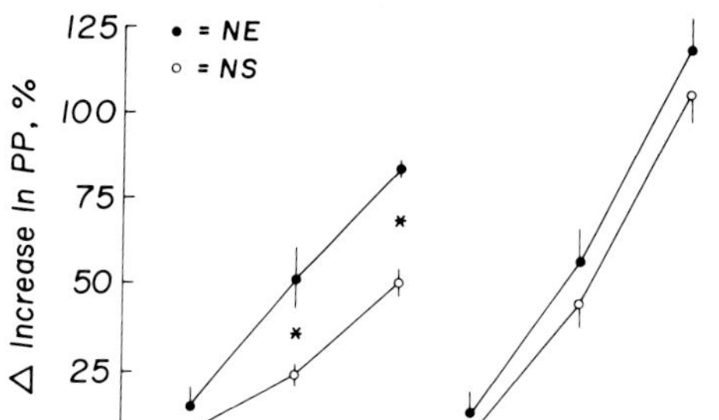

FIG. 11–10. Dose-response curves for the effects of serotonin (5-hydroxytryptamine) on the pressor response to nerve stimulation (NS) and norepinephrine (NE) infusion in perfused mesenteric arteries from normotensive (WKY) **(left)** and spontaneously hypertensive (SHR) rats **(right)**. Each point is the mean value of 5 experiments and vertical bars represent SEM. The asterisks indicate statistically significant differences between responses to norepinephrine and nerve stimulation. (From ref. 599, with permission.)

on the blood vessel wall may involve smooth muscle (180,715) and/or endothelial components (179; see Chapters 9 and 10).

Femoral veins from renal hypertensive rabbits are more sensitive to the contractile effects of norepinephrine than those from normotensive rabbits (Fig. 11–9). Serotonin causes relaxation of norepinephrine-induced contractions and this relaxant effect is not altered in femoral veins from the hypertensive animals (Fig. 11–9). Thus, the direct inhibitory effect of serotonin on the blood vessel wall is not altered in this model of hypertension. In contrast, the inhibitory effect of serotonin on the release of norepinephrine from adrenergic nerve endings may be decreased in hypertension. Indeed, constrictor responses to sympathetic nerve stimulation in the isolated mesenteric vasculature of spontaneously hypertensive rats are diminished to a lesser extent by serotonin than those in normotensive rats (Fig. 11–10; 599).

— 12 —

Peripheral Serotonergic Receptors and Hypertension

Paul M. Vanhoutte

Department of Physiology and Biophysics, Mayo Clinic and Mayo Foundation, Rochester, Minnesota 55905

In most patients with chronic hypertension, the etiology of the increase in blood pressure remains obscure. As the cardiac output of these patients is basically normal, the chronic increase in arterial blood pressure is due primarily to an augmented peripheral vascular resistance, owing to an abnormal narrowing of the systemic arterioles. Except perhaps in patients with renovascular hypertension or pheochromocytoma, no single derangement can be held responsible for the sustained increase in peripheral resistance. Factors such as genetic predisposition, increased activity of the sympathetic nervous system, altered function of the adrenergic neuroeffector junction, augmented activity of the renin-angiotensin-aldosterone axis, abnormal secretion of "natriuretic hormone(s)" or of vasopressin, and adaptive changes in the blood vessel wall all may contribute (Fig. 12–1; e.g., 964,1074). This chapter entertains the idea that the peripheral vasoconstrictor effects of 5-hydroxytryptamine (serotonin) may help to explain the sustained increase in peripheral resistance, at least in certain patients with chronic hypertension; it updates earlier, similar attempts (1015,1072,1073,1078) to which the reader is referred for

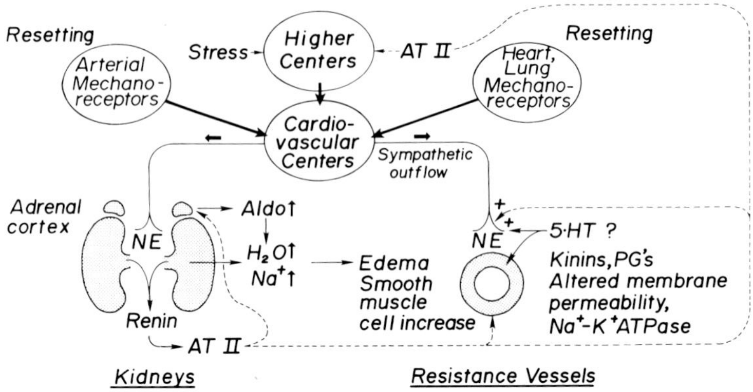

FIG. 12–1. Factors that may be involved in the genesis of essential hypertension. In addition, for each level of sympathetic nerve activity, the adrenergic nerve endings may liberate more norepinephrine than normal, and the α-adrenoceptors and S_2-serotonergic receptors of the vascular smooth muscle may be hyperresponsive. AT II = Angiotensin II; aldo = aldosterone; ATP = adenosine triphosphate; 5-HT = serotonin; NE = norepinephrine; PG = prostaglandin; ↑ = increase; + = facilitation. (Modified from ref. 964.)

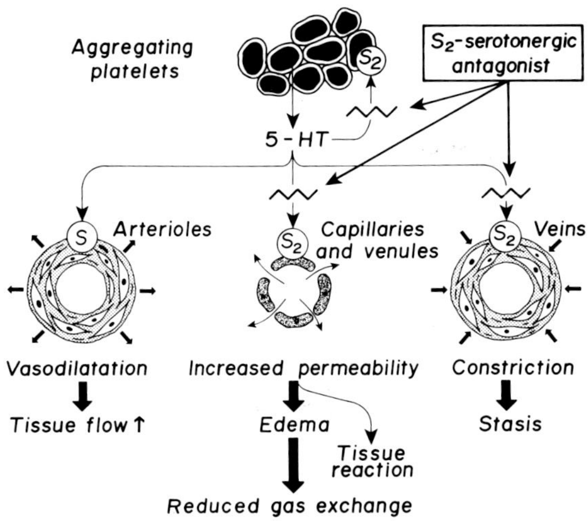

FIG. 12–2. Potential role of serotonin at the microcirculatory level. The release of serotonin from aggregating platelets would tend to cause vasodilatation at the arteriolar level **(left)**; increase capillary and venular permeability **(middle)**; and cause venoconstriction **(right)**. The combination of these effects favors local edema and stasis. The increased permeability and venoconstriction are mediated by S_2-serotonergic receptors. Selective antagonists acting at these sites could prevent the response. (From ref. 1078, by permission.)

details and original references not mentioned here. A shorter version of this text will be published (1077).

When serotonin is injected (or infused) intravenously or intraarterially in intact animals, it can cause either dilatation or constriction (see 820). As a general rule, the former is observed at the arteriolar level, while the latter is seen in large arteries and in postcapillary vessels (Fig. 12–2; see 1070, 1072, 1073; Chapter 9). In a given vascular bed, the response to serotonin may depend on the sympathetic vasoconstrictor tone, with vasodilatation being more prominent if the adrenergic nerves are active. Thus, it is not surprising that serotonin can cause either increases, decreases, or biphasic changes in arterial blood pressure (see 820). As in isolated blood vessels the monoamine can cause either contraction or relaxation of the vascular smooth muscle, the complex effects of serotonin observed in the intact organism can be reproduced *in vitro*, for example by varying the dose of serotonin or the route of its administration (Fig. 12–3), by altering sympathetic nerve activity, or by removing the endothelium (see 1075,1076,1079,1103; Chapters 9 and 10). The contractions evoked by serotonin can be attributed to (a) activation of S_2-serotonergic receptors on the smooth muscle (see 760; Chapters 9 and 11); (b) amplification of the contractile response to other vasoconstrictor neurohumoral mediators (see 1088; Chapter 9); (c) activation of α-adrenoceptors on the vascular smooth muscle; and (d) displacement of other endogenous vasoconstrictors, mainly norepinephrine from adrenergic nerves (Fig. 12–4). The relaxing effects of serotonin include (a) endothelium-dependent inhibition of vascular smooth muscle activity,

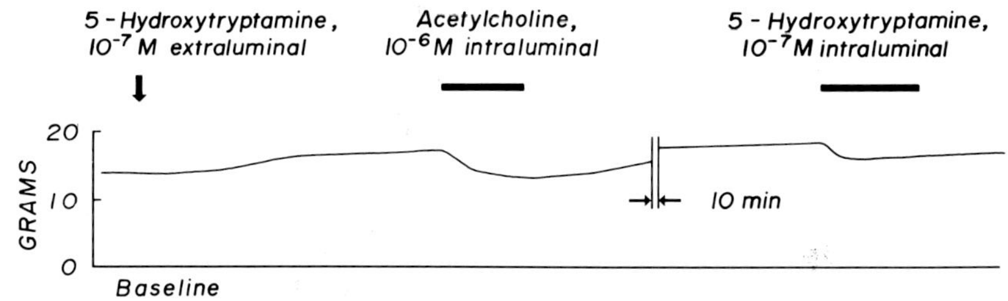

FIG. 12–3. In an isolated, perfused canine coronary artery with intact endothelium, serotonin (5-hydroxytryptamine) causes contraction if given extraluminally **(left)**, but relaxation if given intraluminally **(right)**. The inhibitory response to acetylcholine **(middle)**, demonstrates the presence of endothelial cells. (Data from ref. 181.)

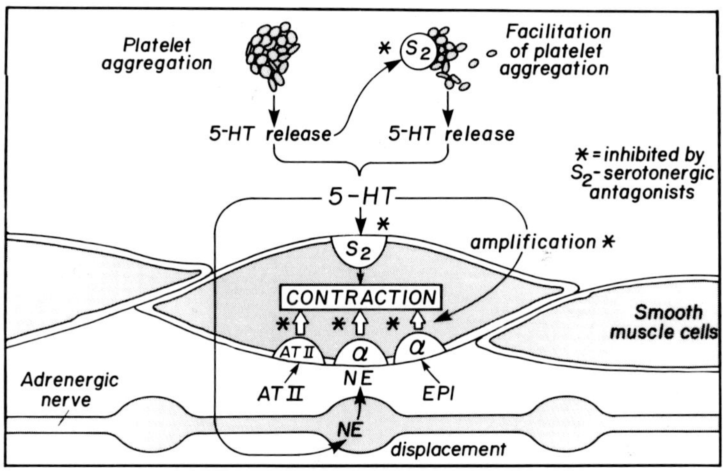

FIG. 12–4. Direct and indirect vasoconstrictor effects of serotonin. α = α-Adrenoceptor; AT II = angiotensin II; EPI = epinephrine; 5-HT = serotonin; NE = norepinephrine; S_2 = S_2-serotonergic receptor.

presumably by an action on S_1-serotonergic endothelial receptors (179–181,497); this effect may contribute to the overall protective role played by the endothelium (see Chapter 10; 1078); (b) activation of prejunctional S_1-serotonergic receptors with a resulting reduction of the evoked release of norepinephrine (308,660); this effect, in particular, helps to explain why dilator responses to serotonin are more pronounced when the sympathetic nervous system is activated; (c) release of inhibitory transmitters from peptidergic nerves; (d) direct inhibitory effects on vascular smooth muscle (e.g., β-adrenoceptor activation, stimulation of prostacyclin synthesis) (Fig. 12–5).

Because under certain circumstances serotonin is a powerful pressor agent, since its discovery by Page and his colleagues its potential involvement in the etiology of hypertension has been explored (see 820). However, the demonstration of the importance of serotonergic mechanisms in the hypertensive process has been complicated by the extreme variability of the vascular effects of serotonin, the lack of evidence that its plasma levels are elevated in patients with chronic hypertension, and the fact that most available serotonergic antagonists have no overt antihypertensive effects in humans. More recent observations may alleviate certain of these concerns.

VASCULAR RESPONSIVENESS TO SEROTONIN IN HYPERTENSION

Normal levels of circulating serotonin are not incompatible with the involvement of the monoamine in an increased peripheral resistance. This could be the case if the vasoconstrictor effects of the "normal" amount of serotonin were either augmented or prolonged, or if the vasodilator responses to it were blunted. To judge from data obtained mainly in hypertensive animals, both can be the case (Fig. 12–6).

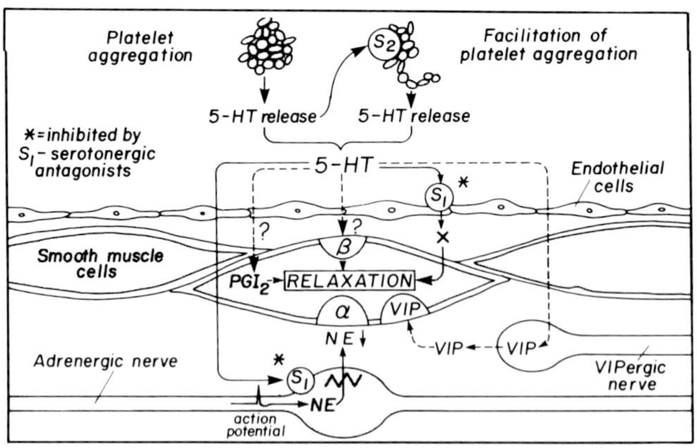

FIG. 12–5. Direct and indirect vasodilator effects of serotonin. α = α-Adrenoceptor; β = β-adrenoceptor; 5-HT = serotonin; NE = norepinephrine; PGI_2 = prostacyclin production; VIP = vasoactive intestinal peptide; X = unknown mediator; ? = uncertain; jagged line = inhibition of adrenergic neurotransmission.

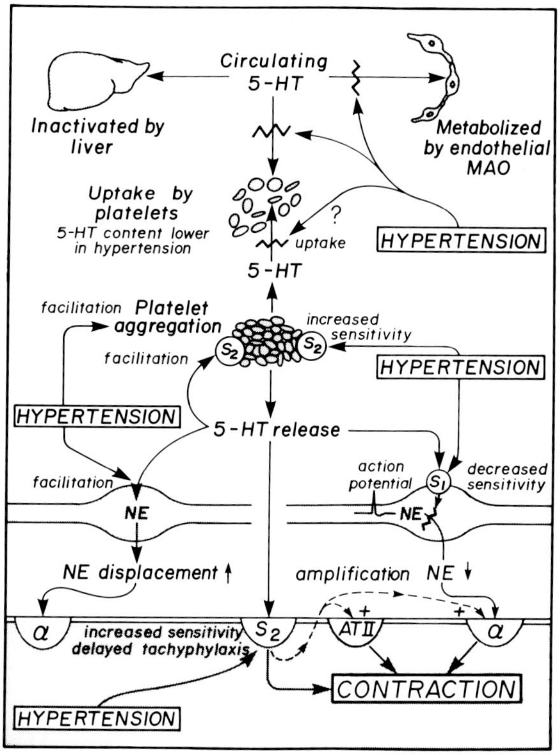

FIG. 12–6. Effects of high arterial blood pressure on the handling of serotonin (5-HT) by platelets and the blood vessel wall. $\alpha = \alpha$-Adrenoceptor; AT II = angiotensin II; MAO = monoamine oxidase; NE = norepinephrine; $S_2 = S_2$-serotonergic receptor.

Vasoconstrictor Effect

Indirect Effects

In very young spontaneously hypertensive rats, the indirect sympathomimetic effect of serotonin is exaggerated (e.g., 183; see Chapter 11). This may reflect the abnormal excitability of the adrenergic nerves at the early stage of the disease in this animal model (1085,1086). A modulatory role of serotonin at the adrenergic neuroeffector junction may be more important at the postjunctional level. Indeed, as serotonin can amplify the response to norepinephrine or angiotensin II (see 1088; Chapter 9), the vasoconstrictor impact of abnormally high levels of these potential mediators of the increased peripheral resistance (Fig. 12–1) may be considerably augmented by serotonin, even if only normal amounts of the monoamine were to reach the vascular smooth muscle at the same time. These indirect effects of serotonin may be of particular significance in early hypertension, when augmented levels of catecholamines or angiotensin II set the disease process in motion.

Direct Effect

In adult spontaneously hypertensive, renal-hypertensive, and corticosteroid-hypertensive animals, the vascular wall exhibits a profound hypersensitivity to the constrictor effect of serotonin (see Chapter 11). The augmentation of the vasoconstrictor response to serotonin is markedly more pronounced than that of those to angiotensin II or norepinephrine. Another important feature in spontaneously hypertensive animals is that the rate of decay of the vasoconstrictor response to repeated exposure to high doses of serotonin (tachyphylaxis) is considerably reduced (184,185,263). This is not seen for the tachyphylaxis for angiotensin II. Thus, it seems reasonable to conclude that an augmented vascular responsiveness to serotonin, due to both an increased sensitivity to and a prolonged effect of the monoamine, is an important characteristic of the hypertensive blood vessel wall. This does not necessarily imply that the exaggerated vasoconstrictor response to serotonin causes the hypertension. Indeed, augmented responsiveness and delayed tachyphylaxis to serotonin also progressively occur in aging, normotensive animals (Fig. 12–7; 263). In a population of normotensive and hypertensive rats of different ages, the maximal responsiveness to the vasoconstrictor effects of serotonin is linearly related to the systolic blood pressure (Fig. 12–8; 263). Hence, the exaggerated sensitivity and

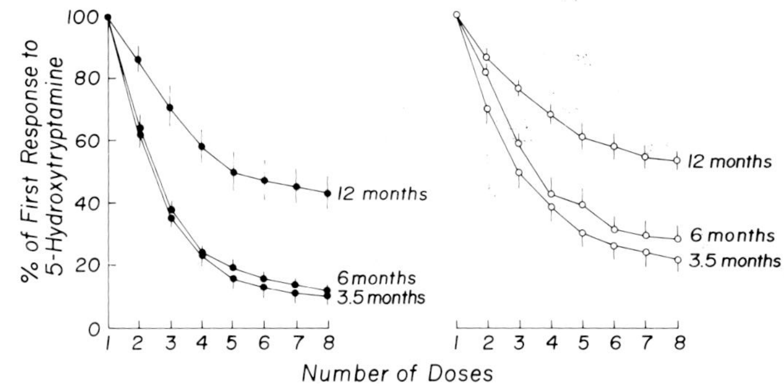

FIG. 12–7. Demonstration of the tachyphylaxis to maximal doses of serotonin (5-hydroxytryptamine) in isolated perfused kidneys of rats. In normotensive animals aging causes a reduced tachyphylaxis **(left)**. This is exaggerated in hypertensive animals, in which the tachyphylaxis is already abnormal in the youngest group studied **(right)**. (Data from ref. 263.)

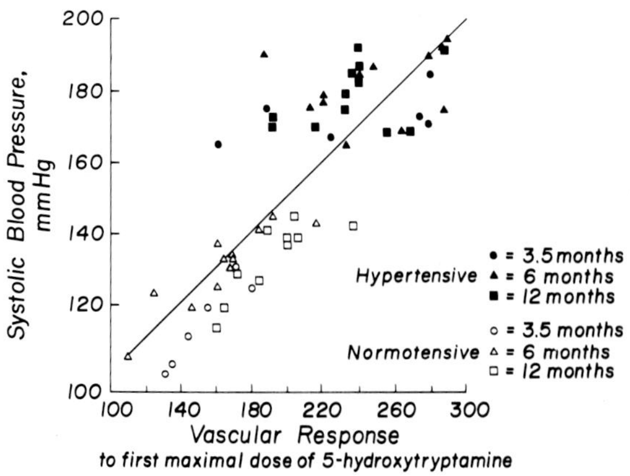

FIG. 12–8. In a population of normotensive *(open symbols)* and spontaneously hypertensive *(full symbols)* rats of varying ages, a linear correlation exists between the systolic blood pressure and the vascular responsiveness to serotonin (5-hydroxytryptamine). (Modified from ref. 263.)

the delayed tachyphylaxis to serotonin observed in hypertension probably reflect the premature aging of the blood vessel wall exposed to a chronically increased distending pressure (1072). However, even if the exaggerated vasoconstrictor effect of serotonin is a consequence of the hypertensive process, it obviously still could play a role in the maintenance of the increased peripheral resistance, provided serotonin can reach the vascular smooth muscle cells of the arterioles and does not trigger vasodilator responses.

Vasodilator Effect

The vasodilator effect of serotonin that could be of importance in hypertension relates to its endothelial (see Chapter 10) and prejunctional (see 1085; Chapter 11) actions. The latter is reduced in spontaneously hypertensive animals (599). Thus if normal or increased amounts of serotonin were to reach the hypertensive blood vessel wall, the vasoconstrictor potential of the monoamine would be offset less by the vasodilatation it normally causes.

LEVELS OF SEROTONIN

Although certain blood vessels may be innervated by serotonergic neurons (426), most appear not to be. Hence, the storage granules of the platelets (see Chapter 7) are the major potential source of serotonin to affect the precapillary resistance vessels, as platelets can reach all parts of the body. Under normal circumstances, the free serotonin present in the plasma is removed by either uptake and storage in nonaggregating platelets, or uptake and metabolic degradation in endothelial cells (see Chapters 3,4, and 7). The efficiency of those two "sinks" for serotonin is such that levels of free serotonin sufficient to activate vascular smooth muscle are reached probably only at sites of endothelial lesions where platelet aggregation occurs (1078), for example after traumatic interruption of the continuity of the blood vessel. Thus, it can be predicted that determination of the circulatory plasma levels of serotonin will provide little useful information as long as the sampling is not performed at the interface between the aggregating platelets and the immediately surrounding structures, as long as the normal mechanisms of removal remain operative, and as long as platelet aggregation is not abolished in the sample taken (1078). Furthermore, the implication of exaggerated levels of serotonin in any pathological conditions almost requires the demonstration that either the "sinks" for the monoamine are less avid or that it is released in quantities which can saturate the uptake in nonaggregating platelets and/or endothelial cells. Although foolproof evidence that this is the case in hypertension is not available, many observations concur to strongly suggest so (Fig. 12–6; 1015,1072,1073,1078).

Release of Serotonin

The platelets of hypertensive patients are more responsive to various aggregating agents (5). These include serotonin, which implies that whenever liberated the monoamine will exert a greater amplifying effect on the aggregating platelets than it does in normotensive subjects (see 233,1015). This, of course, would favor exaggerated platelet aggregation, causing further release of serotonin. To judge from parameters such as platelet survival time and circulating levels of β-thromboglobulin, the turnover of platelets is accelerated in patients with hypertension (see 1015,1072). This interpretation is supported by the finding that the urinary excretion of 5-hydroxindoleacetic acid, the principal metabolite of serotonin (see Chapter 1), is augmented in hypertensive patients.

Disposition of Serotonin

To judge from the clearance of serotonin by the lung, the uptake and/or metabolism by the endothelial cells is abnormally low in spontaneously hypertensive animals (923; see Chapter 3). The uptake of serotonin by platelets is reduced in hypertensive patients, resulting in a lower serotonin content (see 1015,1074).

ANTIHYPERTENSIVE EFFECTS OF SEROTONERGIC BLOCKERS

If serotonin contributes to the augmented peripheral resistance in chronic hypertension by its actions on blood vessels, serotonergic antagonists should lower arterial blood pressure in hypertensive patients because they antagonize the vascular effects of the monoamine. Of the serotonergic antagonists available for study in humans, the only one so far that appears to possess convincing antihypertensive properties is the S_2-serotonergic antagonist ketanserin (see 1015,1084; Chapter 13). However, ketanserin also interacts with other cell membrane receptors, in particular with α_1-adrenoceptors, which when activated cause vasoconstriction. Thus, several questions have to be answered before suggesting that the antihypertensive effect of ketanserin supports the involvement of the peripheral actions of serotonin in hypertension. Most of these questions have received at least a partial answer. In analyzing the answers, one should keep in mind that conclusive evidence in this matter only comes from studies in hypertensive humans, since animal models such as the spontaneously hypertensive rat may not allow better prediction of the efficacy of serotonergic blockers to lower blood pressure than they did years ago for β-adrenergic blockers.

Does Ketanserin Lower Blood Pressure?

The answer is yes. The drug lowers arterial blood pressure in hypertensive patients when given acutely or chronically (Fig. 12–9; see 26,30,249, 807,845,1015,1084,1143,1182; Chapter 13). It may be particularly effective in older patients (1015), which would fit the hypothesis that the hyperresponsiveness of vascular smooth muscle (and platelets) to serotonin in hypertension reflects premature aging of structures chronically exposed to an augmented pressure.

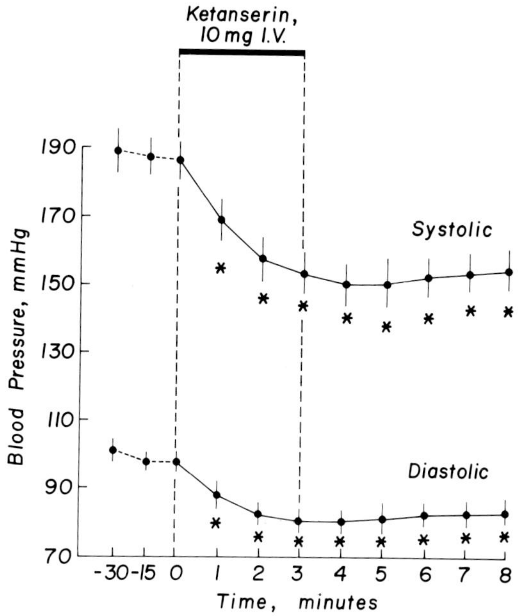

FIG. 12–9. In a group of elderly hypertensive patients, the intravenous (I.V.) injection of ketanserin causes a rapid and sustained reduction of both systolic and diastolic blood pressure. *The effect of ketanserin is statistically significant. (Data from ref. 249.)

Does Ketanserin, in Therapeutic Doses, Antagonize Responses to Serotonin?

The answer is yes. Both the vascular effects of serotonin in intact people and the effects of the monoamine on human platelets (*in vivo* and *ex vivo*) are prevented by acute and chronic treatment with ketanserin (see 1015).

Does Ketanserin, in Therapeutic Doses, Antagonize α_1-Adrenergic Activation?

Under acute conditions the answer is no, as the intravenous administration of ketanserin does not shift the pressor dose-response curves to α_1-adrenergic agonists to the right (56,941,1143,1144,1182). Under chronic treatment the answer is uncertain. Prolonged infusion of, or chronic oral treatment with, ketanserin causes displacement of the pressor dose-response curves to methoxamine and phenylephrine. However, unlike the selective α_1-adrenergic blocker prazosin, ketanserin does not shift the curves in a parallel manner but depresses the response to higher concentrations of the α_1-adrenergic agonists more than that to lower doses (Fig. 12–10; 56,324,885). In pharmacological terms, this "noncompetitive" behavior is hard to reconcile with the available data indicating that ketanserin, albeit in higher concentrations than those required to antagonize S_2-serotonergic receptors, is a competitive antagonist at α_1-adrenoceptors (e.g., 695,1093). The inhibitory effect of ketanserin on the response to α_1-adrenergic agonists could be explained by a nonspecific decrease in vascular responsiveness, resulting from the altered wall-to-lumen ratio following a reduction in arterial blood pressure (Fig. 12–11; 342,964). This issue

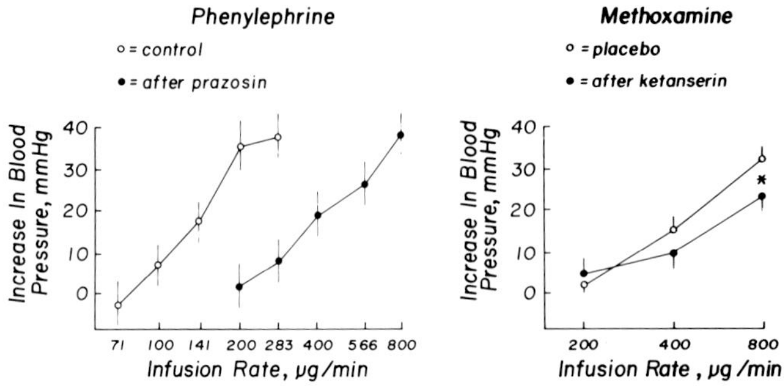

FIG. 12–10. In humans the selective α_1-adrenergic antagonist prazosin causes a parallel shift to the right of the dose-response curve to the pressor effect of the α_1-adrenoceptor agonist phenylephrine, which illustrates competitive inhibition (**left**; data from ref. 975). By contrast, ketanserin causes a nonparallel depression of the dose-response curve to another α_1-adrenoceptor agonist, methoxamine (**right**; data from ref. 324.)

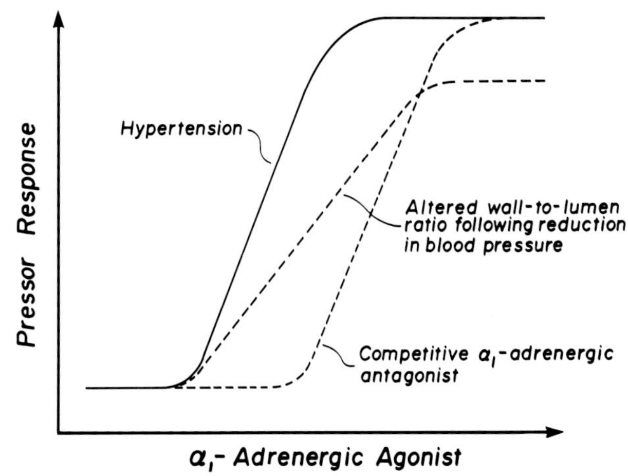

FIG. 12–11. Comparison of the theoretical effects of a competitive α_1-adrenergic antagonist and of nonspecific changes in wall-to-lumen ratio following successful antihypertensive treatment. (Redrawn from ref. 964.)

will be settled when the effect of chronic treatment with ketanserin on the pressor response to nonadrenergic vasoconstrictor agents is known.

The importance of α_1-adrenergic blocking properties of ketanserin for its antihypertensive effect differs in people and in hypertensive animals. In the latter, the acute decreases in blood pressure caused by the compound obviously are due primarily to α_1-adrenergic blockade (175,350,552,695,1168). However, with chronic treatment of rats with ketanserin, the α_1-adrenergic blockade may fade progressively, and this tolerance may then reveal a serotonergic component of the effect of the drug (200,849). By contrast, as discussed previously, in hypertensive patients α_1-adrenergic blockade does not appear to contribute to the acute antihypertensive effect of ketanserin, but may be involved with prolonged exposure to the drug.

Thus, the relevance of investigating the importance of the serotonergic component of the action of ketanserin (and other serotonergic antagonists) in hypertensive rats (and rabbits) may be questioned.

Can Ketanserin Lower Blood Pressure Without the Involvement of α_1-Adrenoceptors?

The answer is yes. Ketanserin lowers arterial blood pressure in patients with autonomic dysfunction, in whom α_1-adrenergic blockers have no effect (1144).

Do the α_1-Adrenoceptor and Serotonergic Blocking Properties of Ketanserin Interact?

The answer is uncertain, albeit probably yes. Previous treatment with the selective α_1-adrenergic antagonist prazosin reduces the hypotensive effect of the acute administration of ketanserin (941). This would be explained if in humans with a normal or hyperactive sympathetic nervous system the major vasoconstrictor effects of serotonin were the indirect ones (Fig. 12–3). Alternatively, the occupancy of S_2-serotonergic receptors on smooth muscle cells may favor the α-adrenergic blocking properties of ketanserin.

Why Ketanserin, but Not Other Available Serotonergic Antagonists?

The answer is uncertain. Possible explanations include the following: (a) Ketanserin possesses a greater affinity for α-adrenergic receptors than other serotonergic inhibitors. However, very comparable S_2-serotonergic/α_1-adrenergic binding-affinity ratios are obtained for mianserin and ketanserin; methiothepin has a greater affinity for α-adrenergic than for S_2-serotonergic binding sites (e.g., 637). (b) Ketanserin is a more selective S_2-serotonergic antagonist. As a consequence, it inhibits the vasoconstrictor (mediated by S_2-serotonergic receptors) and simultaneously unmasks the vasodilator (mediated presumably by S_1-serotonergic receptors) effects of serotonin and probably those of other products (e.g., adenine nucleotides, platelet activating factor) released from aggregating platelets (180,181,497,498). For example, endothelium-dependent relaxations caused by serotonin are antagonized by methysergide and methiothepin, and prejunctional effects of the monoamine by methiothepin, but not by ketanserin (181,308,497,1093). (c) Ketanserin is devoid of agonistic properties. Many serotonergic antagonists, in particular methysergide, have partial agonistic properties (e.g., 1083,1127). In hypertension the vascular smooth muscle becomes more responsive not only to serotonin, but also to methysergide (1127). Thus the hyperreactivity of the hypertensive blood vessel wall to the agonistic properties of methysergide may counteract the hypotension due to the prevention of the vasoconstrictor effect of endogenous serotonin (see 1072,1083; Chapter 11). (d) The effect of ketanserin is explained by actions other than inhibition of S_2-serotonergic (and/or α_1-adrenergic) receptors in vascular muscle and platelets. Among the possibilities are interference with the production and release of aldosterone (e.g., 1182), central nervous depressor effects (e.g., 705), and increase in the deformability of red blood cells (e.g., 1178).

13

Serotonergic Blockade and Hypertension

M. A. D. H. Schalekamp

Department of Internal Medicine I, Erasmus University, Rotterdam, The Netherlands

Serotonin is a neurotransmitter involved in the central regulation of arterial blood pressure, although its precise role in that regard is still uncertain (see Chapter 6). Even less is known about the peripheral role of serotonin in blood pressure regulation. There are at least two subtypes of serotonergic binding sites in the cortex of the rat brain (839). The serotonergic receptors subserving vascular smooth muscle contraction are of the S_2-serotonergic subtype (176,526,637,1093,1096,1097). Hence, it is not surprising that renewed interest in the cardiovascular effects of serotonin has been stimulated by the introduction of the S_2-serotonergic antagonist ketanserin. Other serotonergic antagonists such as methysergide, cyproheptadine, and pizotifen have inconsistent effects on arterial blood pressure. This might be related to lack of selectivity and to partial agonistic activity (1083). Some serotonergic antagonists have marked central actions, which cause serious side effects. Ketanserin is devoid of agonistic activity and animal studies seem to indicate that the compound is more potent as a serotonergic antagonist in the periphery than in the central nervous system (526,846). However, ketanserin, in higher concentrations than required to inhibit S_2-serotonergic receptors, also binds to α_1-adrenergic sites and antagonizes α_1-adrenergically mediated contractions of isolated arteries and veins (1093). Ketanserin has antihypertensive properties both in animals and in humans, but it remains to be established whether this effect is caused by serotonergic blockade, by α-adrenoceptor blockade, by a combination of both, or by some other action (see Chapter 12).

This chapter is a survey of studies on the acute hemodynamic effects of ketanserin in hypertensive patients. The rather limited experience with the chronic use of this drug in such patients is also reviewed.

ACUTE STUDIES

Systemic Hemodynamics

The earliest report of an acute lowering of arterial blood pressure by ketanserin in humans came from De Cree and colleagues (245,252), who measured the effect of the drug (10 mg intravenously) in 23 hospitalized elderly patients with essential hypertension. These investigators noted an immediate fall in blood pressure (from systolic: 187/diastolic: 98 to 151/81 mm Hg after 5 min), an effect that persisted for at least half an hour. Ketanserin had no effect on either heart rate or cardiac output, as assessed indirectly by measurements of systolic time intervals.

In a study of 12 patients with essential hypertension, a fall in blood pressure (from 188/89 to 159/75 mm Hg) was observed within 5 min after the intravenous infusion of ketanserin (10 mg) (1142). This effect persisted throughout a 2-hr period. However, an increase in heart rate (from 62 to 69 beats per minute) and a transient increase in cardiac output were also noted. The decrease in blood pressure was caused by a reduction in total peripheral resistance. The small rise in heart rate accompanying the vasodilator action of ketanserin was due probably to baroreflex-mediated withdrawal of vagal tone and increased sympathetic stimulation of the heart. The augmented sympathetic activity was reflected by an increase in the level of norepinephrine in the plasma (Fig. 13–1).

A similar reduction in arterial blood pressure (from 192/90 to 146/68 mm Hg) was observed in an expanded study including 17 patients (1142). The fall in blood pressure peaked at 3 to 5 min and was observed for 2 hr. A fall in total peripheral resistance (by 23%) accompanied the decrease in mean arterial pressure (18%). A rise in heart rate (by 13%) was noted 10 min after the injection of ketanserin. Although cardiac output increased by 8%, it returned to baseline levels within 2 hr. A decrease in mean arterial pressure of 22% was observed in 30 further hypertensive patients who received an intravenous injection of ketanserin (10 mg) (1144). Immediately after administration of the drug, blood pressure decreased from 177/86 to 139/68 mm Hg; it was 147/72 mm Hg 15 min later and remained significantly below baseline for 2 hr. Although the fall in blood pressure was accompanied by an initial rise in heart rate (from 68 to 78 beats per minute), heart rate returned to 73 beats per minute within 15 min and to control levels within 1 hr, despite the persistence of the decrease in arterial blood pressure. A small but significant increase in cardiac output (from 4.35 to 4.68 liters/min) was observed 5 min after the injection of ketanserin. This was followed by a reduction to 4.08 liters/min within 90 min. Stroke volume decreased from 68 to 61 ml 90 min after ketanserin administration. Thus the decrease in arterial blood pressure was mainly a consequence of a decrease in total peripheral resistance (Fig. 13–2).

In a double-blind, placebo-controlled study of 9 patients with untreated essential hypertension, Zabludowski and colleagues (1174) observed a fall in blood pressure during a 2-hr period after the intravenous administration of ketanserin (10 mg). The drug decreased blood pressure from 150/94 to 137/88 mm Hg, while a slight increase in pressure was observed after placebo. The decrease in mean arterial

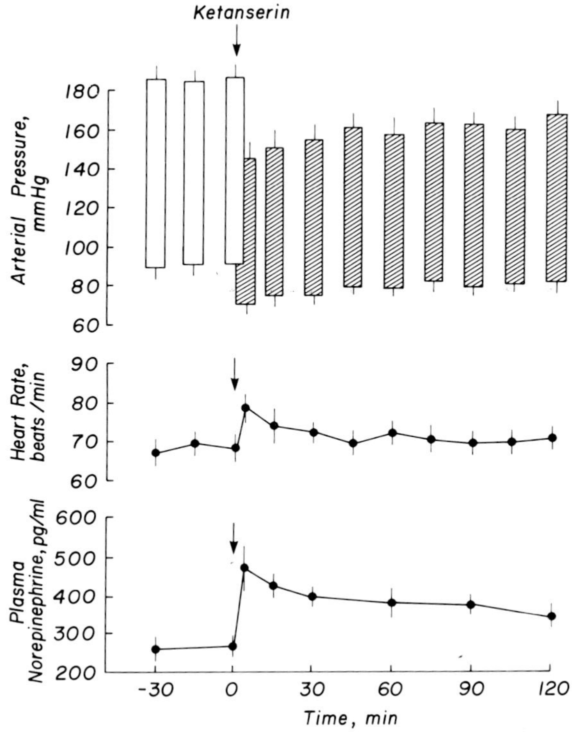

FIG. 13–1. Changes in systemic arterial blood pressure **(upper)**, heart rate **(middle)**, and plasma norepinephrine **(lower)** with a single dose of ketanserin (10 mg i.v.) in 20 patients with essential hypertension. Changes in arterial pressure and norepinephrine were significant throughout the observation period. Heart rate had returned to baseline after 45 min. (From ref. 1144, with permission of the American Heart Association, Inc.)

pressure in the 2 hr after the injection of ketanserin averaged 13.8 mm Hg, compared with placebo. Ketanserin also caused an immediate increase in heart rate, which peaked within 10 min of injection and remained elevated for 40 min. In response to head-up tilting (45°), both systolic and diastolic blood pressure fell after ketanserin from 144 to 123, and from 91 to 89 mm Hg, respectively; by contrast, on tilting after placebo, systolic pressure expectedly decreased from 155 to 146 mm Hg and diastolic pressure increased from 93 to 104 mm Hg. The increase in heart rate in response to tilting was similar after ketanserin and placebo.

Central Hemodynamics

In 12 patients with essential hypertension intravenous ketanserin (10 mg) decreased right atrial pressure, pulmonary artery pressure, and pulmonary capillary wedge pressure (1142). Pulmonary vascular resistance, however, was not significantly affected by the drug. An expanded study involving 17 essential hypertensives yielded similar hemodynamic findings (1144). Reductions in cardiac filling pres-

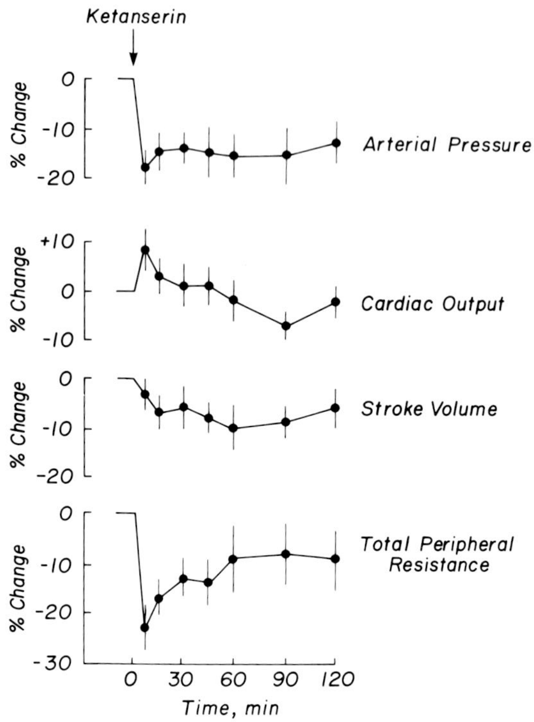

FIG. 13–2. Effects of ketanserin (10 mg i.v.) on systemic hemodynamics in 12 patients with essential hypertension. The decrease in arterial blood pressure **(upper)** was due to a decrease in total peripheral resistance **(lower)**. After 120 min arterial pressure and total peripheral resistance were still below baseline. (From ref. 1144, with permission of the American Heart Association, Inc.)

sures after ketanserin have also been observed in patients with cardiac failure (265). The hemodynamic profile of ketanserin is that of a vasodilator drug, which acts mainly on the arterioles but also has some effect on the venous side of the circulation. An action on the capacitance vessels might contribute to its blood-pressure-lowering effect. A combined effect on the arterioles and the veins might explain why, for a given effect on systemic arterial blood pressure, the increments in heart rate and circulating norepinephrine observed with ketanserin are less than with drugs such as hydralazine and diazoxide, which predominantly affect the resistance vessels (683,763). Theoretically, the arterial baroreflex could also be modified by ketanserin. However, intravenous injections of phenylephrine caused reductions in heart rate both before and during the infusion of therapeutic doses of ketanserin; the relationship between the changes in pressure and heart rate caused by phenylephrine was not altered by the S_2-serotonergic antagonist, which suggests that the drug does not alter the sensitivity of the baroreflex (Fig. 13–3).

CHRONIC STUDIES

Efficacy and Dose-Finding

The first double-blind, placebo-controlled, crossover trial of ketanserin given orally was conducted by De Cree et al. (245), who treated 14 elderly women with

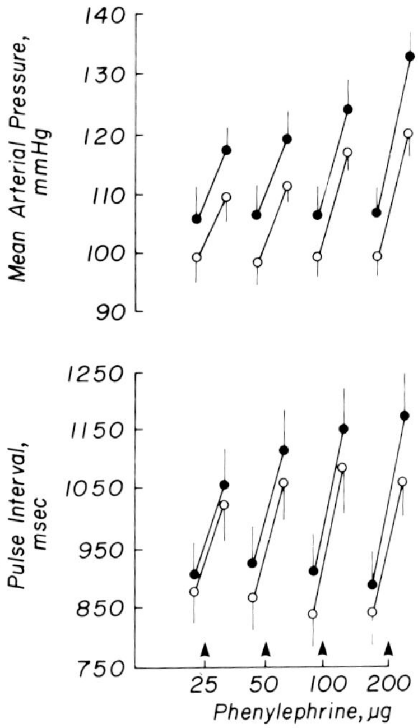

FIG. 13–3. Peak increases in mean arterial pressure **(upper)** and pulse interval **(lower)** caused by increasing doses of phenylephrine in 7 patients with essential hypertension before (●) and during (○) intravenous infusion of ketanserin (loading dose 10 mg in 2 or 3 min followed by a sustained infusion of 2 mg/hr). (From ref. 1144, with permission of the American Heart Association, Inc.)

predominantly systolic hypertension. Five of the patients received clonidine, and 7 diuretics, as concomitant therapy. Measurements of arterial blood pressure in the morning, at noon, and in the evening of the first, fourth, and eighth day of treatment with ketanserin (40 mg 3 times a day) revealed a significant decrease compared to placebo. The hypotensive effect of ketanserin was the most marked in patients with the highest baseline pressures. None of the 14 patients developed orthostatic hypotension, although 2 complained of dizziness and nausea on the first day of treatment. In the 10 patients who continued therapy with ketanserin for 3 weeks, blood pressure decreased further, which necessitated reduction of the dosage in 4 patients who experienced dizziness and tiredness.

These findings were extended to younger people in a double-blind, placebo-controlled, crossover study involving 10 patients with essential hypertension, 6 of whom had been unresponsive to previous antihypertensive therapy (245). All patients were weaned off other medications for at least 2 weeks before entering the trial. Arterial blood pressure fell from 173/105 to 160/98 mm Hg after 2 weeks of treatment with ketanserin (20 mg 3 times a day) and decreased further to 155/93 mm Hg after 4 weeks. No significant changes in arterial blood pressure were observed with placebo. Neither heart rate nor cardiac output (as assessed by measurements of systolic time intervals) changed during therapy with ketanserin. None of the 10 patients experienced side effects during treatment with the drug.

In another long-term study 16 essential hypertensives were treated orally with ketanserin (average daily dosage, 91 mg) for 9 months. The drug reduced the supine arterial blood pressure to 115/95 mm Hg (from a pressure of 167/102 mm

Hg observed during 2 weeks under placebo) (30). Long-term treatment with ketanserin caused a slight rise in heart rate to 73 beats per minute, as compared with 64 beats per minute with placebo. The arterial blood pressure and heart rate achieved under monotherapy with ketanserin were similar to those observed with the antihypertensive therapy used prior to the trial. There was no correlation between the plasma concentration of ketanserin and its antihypertensive effect. Ketanserin did not cause abnormal laboratory tests, changes in body weight, or side effects other than transient fatigue. That fatigue may be a dose-related phenomenon is suggested by a preliminary evaluation by the same investigators of the effects of three different dosage levels of ketanserin, given for 8 weeks; the highest incidence of fatigue occurred among patients receiving 60 mg of the drug 3 times a day *(unpublished observations)*.

In a single-blind study, 28 hypertensive patients were randomized to receive 40 mg of ketanserin orally, either once or twice daily for 8 weeks (468,845). The magnitude and timing of antihypertensive efficacy were similar at both dosage levels, with a similar reduction in supine mean arterial pressure relative to placebo in the groups receiving 40 mg either once (by 11%) or twice (7–10%) a day. Five patients in each study group failed to respond to ketanserin. No change in heart rate was seen in this study. Side effects of ketanserin were limited to mild and transient dizziness, fatigue, and lightheadedness. In contrast to earlier findings (245), there was no correlation between the magnitude of the hypotensive effect of ketanserin and the initial level of arterial blood pressure. Patients whose blood pressure had not been controlled with previous therapy (mainly diuretics and β-

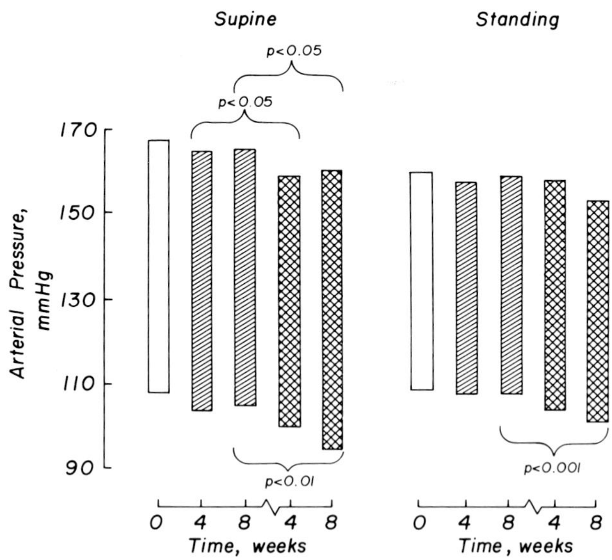

FIG. 13–4. Effects of ketanserin (40 mg twice a day) on supine and standing systolic and diastolic arterial pressure. *Open bars:* run-in period; *diagonal lines:* placebo period; *hatched bars:* ketanserin period.

blockers) did not respond well to ketanserin. The timing of maximal reductions in arterial blood pressure was correlated to the peak plasma levels of ketanserin. Although there was no correlation between peak plasma concentrations of ketanserin (which were similar for both dosages) and pressure reduction in individual patients, the steady-state plasma concentrations of the drug were significantly higher in the group receiving the two daily doses and were correlated to the decrease in systolic blood pressure observed after 8 weeks of therapy.

In an 8-week-long dose-finding study, only 2 of 17 hypertensive subjects achieved satisfactory blood pressure control with ketanserin at a dosage of 20 mg twice a day. To normalize blood pressure, the dose of the drug had to be increased to 40 mg in 8 and 60 mg in 3 patients (26). The same amounts of ketanserin taken once daily were not effective. From a supine mean arterial pressure of 186/100 mm Hg on placebo, ketanserin lowered pressure to 174/88 mm Hg. On cessation of therapy, the arterial blood pressure increased toward control levels within 24 hr in 12 patients, but in 3 patients it took more than 2 weeks to return to baseline. The investigators noted no significant changes in heart rate, body weight, or laboratory values. Side effects were infrequent (26). In a double-blind, placebo-controlled, parallel group study in 10 hypertensive patients, a reduction in supine arterial blood pressure (by 19 mm Hg systolic and 9 mm Hg diastolic) was noted after 6 weeks of treatment with ketanserin (142). In 3 patients the arterial blood pressure was controlled with 20 mg given twice a day, while 7 patients required 40 mg twice a day. The heart rate decreased by 6 beats per minute among ketanserin-treated subjects. Similar side effects were observed with placebo or active treatment.

Hemodynamics

The effect of ketanserin (40 mg 3 times a day) on the cardiovascular response to exercise in 8 patients with essential hypertension was investigated (468). A

TABLE 13–1. *Arterial pressure, heart rate, body weight, and hormonal changes during treatment with ketanserin for 8 weeks in 24 patients with essential hypertension*

	Placebo	Ketanserin	*p*
Mean arterial pressure (mm Hg)			
Supine	126 ± 3	119 ± 3	<0.01
Standing	128 ± 3	119 ± 3	<0.001
Heart rate (beats/min)			
Supine	71 ± 2	71 ± 2	NS
Standing	82 ± 3	81 ± 2	NS
Body weight (kg)	76.6 ± 2.4	78.0 ± 2.4	<0.001
Plasma norepinephrine concentration (pg/ml)	284 ± 25	314 ± 23	NS
Plasma renin concentration (μU/ml)	9.8 ± 1.6	9.2 ± 1.4	NS
Plasma aldosterone concentration (pg/ml)	110 ± 16	103 ± 13	NS

Data shown as means ± SEM. NS = The difference between patients treated with ketanserin and placebo is not statistically significant.

double-blind crossover study revealed that 8 weeks of treatment with ketanserin exerts little influence on the changes in arterial blood pressure and heart rate induced by dynamic exercise. However, the drug slightly attenuated the increase in arterial blood pressure associated with isometric exercise. Omvik and Lund-Johansen (807) performed a 9-month trial of ketanserin given orally. The reduction in arterial blood pressure that occurred in their 13 hypertensive patients was accompanied by a fall in heart rate and cardiac output and no change in stroke volume. They observed a decrease in total peripheral resistance during exercise but maintained that the drug does not exert a significant vasodilator effect; they also reported a high incidence of drowsiness and inability to concentrate.

An extensive assessment of the hemodynamic effects of chronic therapy with ketanserin was provided by Fagard and colleagues (324), who conducted a double-blind, placebo-controlled, crossover study involving 14 patients with essential hypertension. They found that ketanserin (40 mg 3 times a day) lowered supine arterial blood pressure from 150/84 to 141/77 mm Hg, a reduction that was not affected by exercise. They also observed decreases in total peripheral resistance (by 14%) and heart rate (5%), and increases in stroke volume (11%) and cardiac output (7%). Ketanserin did not affect pulmonary artery pressure or pulmonary capillary wedge pressure, but reduced pulmonary vascular resistance by 15%. Other effects of the drug included an increase in body weight (by 1 kg) and slight reductions in hemoglobin, hematocrit, red cell count, and creatinine clearance.

After a run-in period of 4 weeks on placebo, 24 patients with essential hypertension were randomized for allocation to either ketanserin (40 mg twice a day) or placebo, in a double-blind, crossover design of 8-week duration for each treatment period. Ketanserin reduced the supine systolic and diastolic arterial pressures by 4.21 ± 1.8 and 7.6 ± 1.7%, respectively (Fig. 13–4). Standing systolic and diastolic pressures were reduced by 3.1 ± 1.1 and 8.9 ± 1.2%, respectively. The decrease

TABLE 13–2. *Ambulatory intraarterial pressure in the morning, afternoon, evening, and night during treatment with ketanserin in 18 patients with essential hypertension*

	Placebo	Ketanserin	*p*
Systolic pressure (mm Hg)			
6–12 a.m.	149 ± 2	145 ± 3	<0.05
12–6 p.m.	153 ± 3	150 ± 2	NS
6–12 p.m.	155 ± 3	145 ± 3	<0.001
0–6 a.m.	132 ± 1	123 ± 2	<0.001
Diastolic pressure (mm Hg)			
6–12 a.m.	98 ± 2	91 ± 2	<0.001
12–6 p.m.	99 ± 2	94 ± 1	<0.01
6–12 p.m.	99 ± 3	90 ± 2	<0.001
0–6 a.m.	81 ± 1	75 ± 1	<0.01

Data shown as means (±SEM) over periods of 6 hr. NS = The difference between patients treated with ketanserin and placebo is not statistically significant.

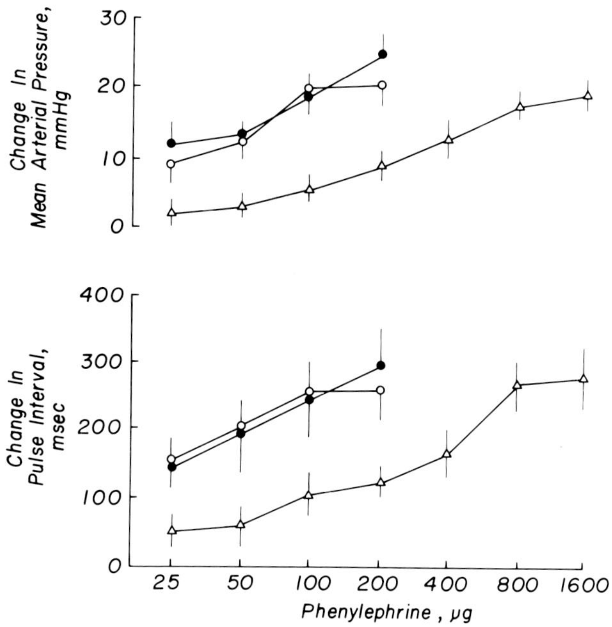

FIG. 13–5. Increases in mean arterial blood pressure **(upper)** and pulse interval **(lower)** caused by phenylephrine in 7 patients with essential hypertension before (●) and during (○) intravenous infusion of ketanserin (loading dose 10 mg in 2 or 3 min, followed by a sustained infusion of 2 mg/hr). In 5 patients (△) phenylephrine injections were also given after treatment with a dose of prazosin (4 mg), which caused a similar decrease in arterial blood pressure as ketanserin. (From ref. 1144, with permission of the American Heart Association, Inc.)

in blood pressure was associated with an increase in body weight. There were no significant changes in heart rate. Ketanserin did not affect the plasma concentrations of norepinephrine, renin, and aldosterone (Table 13–1). Hemoglobin, hematocrit, and plasma creatinine were unchanged. Apart from the effect on body weight, the side effects of ketanserin and placebo were not significantly different (Woittiez, Van den Meiracker, Wentling, Man in't Veld, Ritsema Van Eck, Boomsma, and Schalekamp, *unpublished observations*).

Twenty-Four-Hour Recordings

In 18 patients with essential hypertension, 24-hr blood pressure measurements were made. After a treatment-free period of at least 4 weeks, the patients were allocated at random to either ketanserin (40 mg twice a day) or placebo for a period of 4 weeks in a single-blind, crossover design. Ketanserin or placebo was taken at 8 a.m. and 6 p.m. At the end of each period the intraarterial ambulatory blood pressure was monitored for 24 hr (414). The intraarterial cannula was inserted into the brachial artery of the dominant arm and connected to a portable transducer-perfusion unit. Signals of blood pressure and electrocardiogram were recorded on magnetic tape using an analog type recorder (Medilog, Oxford Medical Systems).

The regimen of ketanserin tested was capable of lowering blood pressure for most of the 24-hr period (Table 13–2).

SEROTONERGIC ANTAGONISM VERSUS α_1-ADRENERGIC BLOCKADE

Ketanserin may lower arterial blood pressure because it possesses α-adrenergic antagonistic properties, albeit at higher concentrations than those required to inhibit S_2-serotonergic receptors. Thus in anesthetized normotensive rats and in conscious spontaneously hypertensive rats, the doses of ketanserin required to lower arterial blood pressure are 25 to 100 times higher than those required to inhibit responses to serotonin. At these high doses the pressor response to α_1-adrenoceptor agonists was also blocked. Hence, in the rat ketanserin may lower blood pressure primarily because of its α_1-adrenoceptor blocking properties (350,553).

It is unlikely, however, that intravenous injection of 10 mg of ketanserin causes α_1-adrenoceptor blockade in man, since the pressor responses to bolus injections of phenylephrine are not altered by the serotonergic blocker (1144). This contrasts with the shift of the dose-response curve to the α_1-adrenergic agonist observed after treatment with prazosin in doses that have a similar hypotensive effect as 10 mg of ketanserin (Fig. 13–5).

A slight but significant reduction of the pressor responses to the α_1-adrenoceptor agonist methoxamine has been observed during oral treatment of normotensive and hypertensive subjects with therapeutic doses of ketanserin (885). Methoxamine is more specific for vascular α_1-adrenoceptors than phenylephrine. Although the

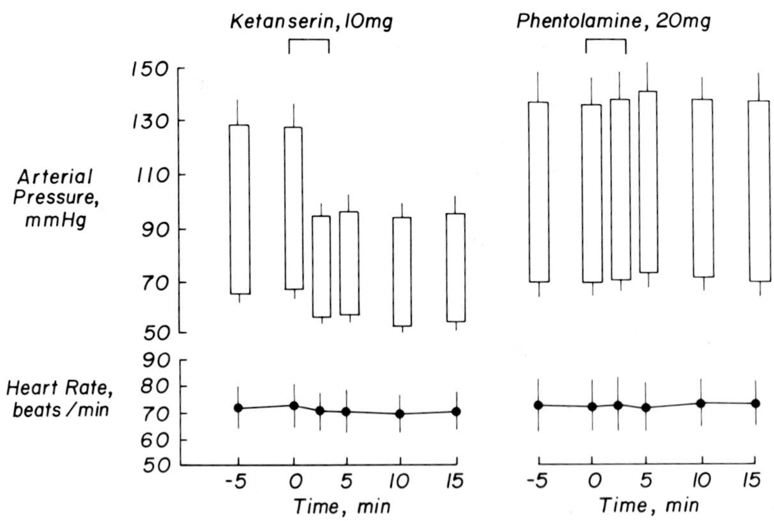

FIG. 13–6. Effects of ketanserin (10 mg intravenously) and phentolamine (20 mg intravenously) on arterial blood pressure **(upper)** and heart rate **(lower)** in patients with autonomic insufficiency. Mean values from observations in 4 patients are shown.

vasopressor action of phenylephrine is due mainly to direct activation of α_1-adrenoceptors on vascular smooth muscle, it also has a tyramine-like action resulting in the displacement of norepinephrine from adrenergic nerves and therefore can cause stimulation of β- and α_2-adrenoceptors as well. Thus methoxamine may be more suitable than phenylephrine for studying the importance of α_1-adrenoceptors in the action of ketanserin. In practice, however, the tyramine-like effect of phenylephrine appears to be of little importance. Indeed, the pressor responses to bolus injections of doses of phenylephrine that are used in human studies are completely blocked by clinically effective amounts of the selective α_1-adrenoceptor antagonist prazosin. Ketanserin, unlike prazosin, causes no parallel shift of the dose-response curve to methoxamine; as a matter of fact, it only affects the highest of the three doses of methoxamine studied (324,885). These data provide little evidence that ketanserin acts as an α_1-adrenoceptor antagonist in man (see Chapter 12).

That ketanserin can lower blood pressure independently of α_1-adrenoceptor blockade is substantiated in patients with autonomic insufficiency. The nonselective α_1-adrenoceptor antagonist phentolamine (20 mg intravenously) has no effect on blood pressure and heart rate in these patients, whereas it causes hypotension and tachycardia in normal individuals (Fig. 13–6). This demonstrates the presence of an efferent sympathetic lesion. However, ketanserin (10 mg intravenously) has a distinct hypotensive effect in these patients (1144).

— 14 —

Serotonin and the Preeclampsia-Eclampsia Syndrome

Carl P. Weiner

Department of Obstetrics and Gynecology, Division of Maternal Fetal Medicine, University of Iowa Hospitals and Clinics, Iowa City, Iowa 52242

Preeclampsia-eclampsia is a clinical syndrome defined as hypertension occurring after 20 weeks of pregnancy in association with either proteinuria or pathologic edema. The diagnosis of preeclampsia is made in 3 to 12% of deliveries (165). Preeclampsia is one of the three most common causes of maternal mortality. Maternal complications include seizure, hypertensive encephalopathy, heart failure, acute renal tubular necrosis, bilateral renal cortical necrosis, disseminated intravascular coagulation, and thromboembolic disease. The incidences of fetal growth retardation, asphyxia, and demise are greatly increased. Preeclampsia is the major identifiable factor associated with premature delivery and is indirectly the major known cause of perinatal mortality in the United States (358). The etiology and the pathophysiological mechanisms underlying preeclampsia are unknown. There is no accepted pharmacological therapy. The only effective treatment for preeclampsia-eclampsia is delivery.

No laboratory test provides a foolproof diagnosis of preeclampsia and there are other causes of hypertension and proteinuria during pregnancy. The finding of glomerular endotheliosis upon renal biopsy is probably the most reliable indicator of the disease (166). Based on biopsy studies, the clinical diagnosis of preeclampsia appears to be incorrect in 20 to 30% of nulliparous women and in 60 to 70% of multiparous women (706). This rate of diagnostic error results primarily from an inability to distinguish preeclampsia or chronic hypertension with superimposed preeclampsia from either previously unrecognized chronic hypertension, underlying renal disease, or latent hypertension.

The preeclampsia-eclampsia syndrome has been associated with factors such as seasonal variation, first pregnancy, genetic predisposition, abnormal placentation, multiple gestation, preexisting renal disease, and chronic hypertension. This picture suggests that preeclampsia is not a single disease, but rather a syndrome with multiple etiologies. The similarities of different clinical cases suggest a shared pathophysiological mechanism. This chapter reviews the evidence that serotonin plays a major role in causing the signs and symptoms of preeclampsia.

SEROTONIN DURING PREGNANCY

The search for a role of serotonin in the preeclampsia syndrome was prompted by the observation that the monoamine produced spasm of the placental vasculature 10 times greater than that obtained with norepinephrine (44,382). Progress was hampered by inaccurate assay techniques, varying definitions of preeclampsia, and, presumably, a high diagnostic error. Initially the pharmacological effect of serotonin was investigated. The parenteral administration of serotonin to the nonpregnant animal caused renal ischemia. The response of the pregnant animal to serotonin was somewhat different. Mortality occurred with smaller doses of the monoamine and renal lesions different from those in the nonpregnant animal but similar to glomerular endotheliosis were observed (863,864,1126).

Serotonin is difficult to assay accurately in either whole blood, plasma, or tissue (769; see Chapter 1). However, most investigators have reported an increase of serotonin concentration in whole blood during normal pregnancy independent of the assay method used to measure the monoamine (148,431,597,961). Higher levels of serotonin were reported in the blood, placenta, and amniotic fluid of women with preeclampsia compared to those of normal pregnant women (431,659, 864,936,961). Because the fetal content of serotonin is high, the excess serotonin might originate in the fetus (544,587). When placental monoamine oxidase activity was determined, the results were conflicting (271,574,961). However, determination of placental monoamine oxidase activity using serotonin as the substrate for the enzyme revealed no difference between preeclampsia and normal pregnancy (574). The older fluorescent and recent chromatographic serotonin assays correlate best in platelet samples (769). Whigham and co-workers observed that platelet serotonin concentration was reduced in women with preeclampsia (1147). These different observations indicate that such women have an excess of circulating serotonin, despite a reduced concentration of the monoamine in the major peripheral carrier, the platelets. The consistent findings of increased levels of serotonin and enhanced sensitivity to serotonin during normal and preeclamptic pregnancy suggest that the metabolism of serotonin is altered and that the amine may play a role in the pathophysiology of preeclampsia. Support for this interpretation comes from clinical investigations using the S_2-serotonergic inhibitor ketanserin.

KETANSERIN AND PREECLAMPSIA

In a randomized, double-blind crossover study, 20 women with preeclamptic hypertension received ketanserin intravenously, immediately postpartum (1135).

Preeclampsia was defined by the standard criteria of elevated blood pressure, proteinuria, thrombocytopenia, and central nervous symptoms (165). Nine women with either hypertension prior to pregnancy or a documented elevation of the mean arterial blood pressure during the first (90 mm Hg) or early second (95 mm Hg) trimesters were classified as chronically hypertensive with superimposed preeclampsia; only one of these women was primiparous. Eleven of the 20 patients were preeclamptic without evidence of chronic hypertension; 10 of the women comprising this group were primiparous. The study was completed within 3.5 hr to minimize the effect of spontaneous resolution of the disease. Ketanserin was given as a bolus (10 mg) followed by an infusion (initial rate 4 mg/hr). Supplemental boluses were administered after 5 min if necessary; the infusion rate was increased as required to maintain the diastolic pressure below 95 mm Hg.

All patients responded to ketanserin. The systolic, diastolic, and mean arterial blood pressures were significantly reduced (Fig. 14–1). Patients who received placebo first subsequently had a significant decline in blood pressure during the ketanserin infusion. The apparent decline in pressure during the placebo infusion was due to premature cessation of the placebo because of persistent hypertension and patient dropout. The diastolic pressure remained above 95 mm Hg in 2 women during the ketanserin infusion. Both women had chronic hypertension and belonged to the group with superimposed preeclampsia. The side effects of ketanserin were minimal.

In a second study ketanserin was administered to a woman with severe recurrent preeclampsia while a flow-directed catheter was in place in the pulmonary artery (Fig. 14–2; 1132). The fetus had died shortly prior to the woman's admission. Five milligrams of ketanserin were administered rapidly and an infusion was initiated (4 mg/hr). In response, the blood pressure and systemic vascular resistance declined rapidly. Cardiac output remained stable. Though a transient hypertensive episode occurred simultaneously with the induction of labor and the bolus infusion of fluid, the blood pressure was rapidly controlled by additional ketanserin injection. The

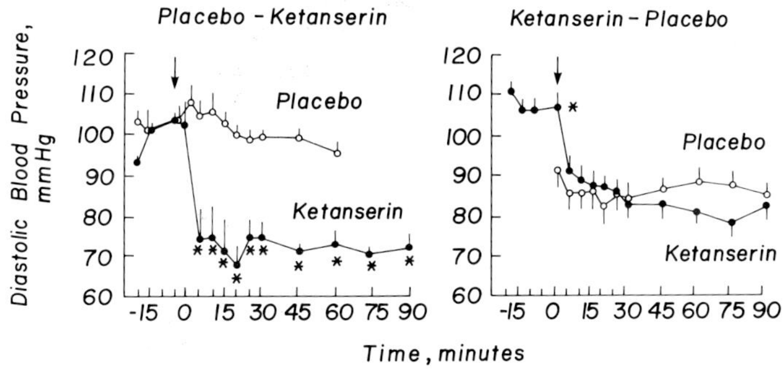

FIG. 14–1. Effect of ketanserin and placebo on diastolic arterial pressure: blood pressure in women with preeclampsia. Ketanserin, but not placebo, lowers arterial blood pressure **(left)**. If placebo is given following ketanserin, the carryover effect of the drug maintains a low diastolic blood pressure **(right)**. (From ref. 1135, with permission.)

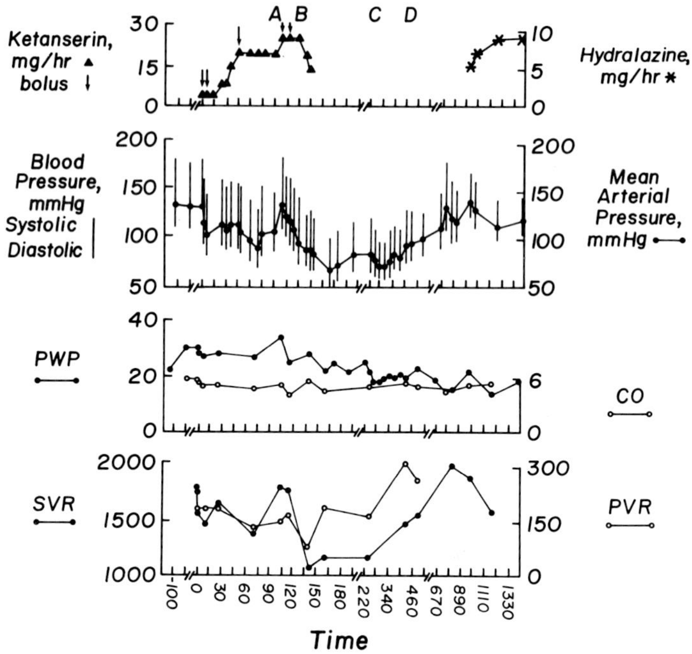

FIG. 14–2. The hemodynamic effect of ketanserin in a woman with preeclampsia. The drug is injected as a bolus *(arrow)* and then infused continuously at a rate adjusted for by blood pressure. The effects on arterial blood pressure, pulmonary wedge pressure (PWP), cardiac output (CO), systemic vascular resistance (SVR), and pulmonary vascular resistance (PVR) are shown. A = Induction of labor with prostaglandin E_2; B = injection of a fluid bolus; C, D = injections of isotonic solutions because of continued oliguria. At the end of the study, hydralazine was given to reduce pulmonary and peripheral vascular resistances. (From ref. 1132, with permission.)

ketanserin infusion was discontinued after 140 min when the blood pressure had stabilized. The arterial blood pressure remained stable for 4 hr, although the systemic and pulmonary vascular resistance began to rise.

Both studies demonstrate that ketanserin effectively controls the hypertension of preeclampsia. The reduction of blood pressure observed in both studies and the beneficial effect on cardiac function documented in the second study suggest a possible therapeutic role for ketanserin in preeclampsia.

THE ETIOLOGY OF HYPERTENSION IN PREECLAMPSIA—HYPOTHESIS

Numerous investigators have suggested that abnormal coagulation causes preeclampsia, despite their inability to consistently demonstrate alterations in common coagulation tests. This suggestion is reinforced by the fact that more sensitive parameters of clotting activity demonstrate that most women with preeclampsia exhibit hypercoagulability (1130,1131).

Histologically, fibrin deposition constitutes the major portion of glomerular endotheliosis, the renal lesion characteristic of preeclampsia (752). The liver and placenta are also sites of excess deposition of fibrin when the pregnancy has been complicated by preeclampsia (37,73). The levels of fibrinopeptide A, the first peptide cleaved from fibrinogen during thrombin-mediated formation of fibrin, increase gradually during normal pregnancy (1133). They are abnormally high in women with preeclampsia (285,1136). While fibrin generation is maximal, the fibrinolytic activity of whole blood normally decreases between 30 and 32 weeks of gestation; it declines further in women with preeclampsia (36). The first symptoms of preeclampsia often appear at this time.

Antithrombin III is the major *in vivo* inhibitor of the formation of thrombin; it may be consumed without fibrin generation. A significant decrease in functional antithrombin III activity occurs in women with preeclampsia (1130,1131,1134). The decline in antithrombin III precedes, in at least some women, the appearance of symptoms (1131). Similarly, consumption of factor VIII increases prior to the onset of preeclampsia (880,1147).

A decline in platelet number may be one of the earliest laboratory findings associated with the disease (879). Though thrombocytopenia is uncommon, platelets from women with preeclampsia are less responsive to some aggregating agents, contain less serotonin, and have a shortened lifespan (869,1121,1147). These observations suggest that exaggerated platelet aggregation has occurred. This premise is supported by the findings of other investigators who noted elevated levels of platelet factor 4 and β-thromboglobulin in women with preeclampsia (285,878,911,987). The degree of elevation correlates directly with the severity of preeclampsia. The level of β-thromboglobulin increases before the appearance of symptoms in at least some women (911).

The question remains as to the cause of these clotting abnormalities. Since they appear prior to the hypertension, they cannot be secondary to the increase in arterial blood pressure. As a matter of fact, women with chronic hypertension not complicated by preeclampsia have normal antithrombin III activity (1130,1131,1134). Both the abnormalities in platelet function and the clotting cascade could result from a common cause; for example, damage to the vascular intima would simultaneously activate both processes.

A number of prostaglandins have been implicated in the regulation of uterine blood flow. The synthesis of prostacyclin increases during normal pregnancy (632). This prostanoid reduces the tone and spontaneous motility in myometrium obtained from pregnant women (806). A high concentration of prostacyclin is present in uterine venous blood (384). Human placental and umbilical vessels have a high capacity for prostacyclin synthesis; a strong correlation exists between the umbilical blood flow and subsequent *in vitro* production of prostacyclin by the isolated human umbilical artery (678). During preeclampsia the peripheral concentrations of prostacyclin and the *in vitro* synthesis of 6-*keto*-prostaglandin $F_{1\alpha}$ are reduced in the fetal and maternal tissues (287,416,678,888,1117). In contrast, thromboxane synthesis is either unaltered or slightly increased over normal in women with preeclampsia (1117). There are at least two possible mechanisms by which abnormal prostaglandin metabolism might lead to excess coagulation activity. First, the reduction in prostacyclin synthesis may alter blood flow within the placenta or uterus causing excess turbulence and intimal damage. The generation of prostacyclin by the vascular intima is stimulated by both endothelial cell damage and platelet aggregation (693,745). Second, in women with preeclampsia, the abnormally low ratio of prostacyclin to thromboxane may interfere with the normal modulation of platelet aggregation and favor excess platelet aggregation and vasoconstriction.

The serotonin released from the platelets would not only facilitate further platelet aggregation, but also amplify the vasoconstrictor action of other neurohumoral mediators and eventually cause direct contraction of the vascular smooth muscle (1078; see Chapter 9). In addition, serotonin activates *in vitro* the synthesis of prostaglandins D and E_2 and prostacyclin (195,973). These three vasodilator substances are derived from a common precursor, prostaglandin H_2. In the absence of platelet-derived growth factor, the maximal response to serotonin is diminished (195). That serotonin may also stimulate the production of vasodilator prostanoids in the intact organism is suggested by the observation that ketanserin reduces the level of prostaglandin E_2 in patients with carcinoid syndrome (32). Deficient prostacyclin synthesis in women with preeclampsia might favor a build-up of serotonin at the microvascular level. The reduced production of prostacyclin would favor the vasoconstrictor effects of the monoamine.

On the basis of these observations a hypothesis can be advanced to explain hypertension in preeclampsia (Fig. 14–3). An abnormality of the metabolism of various prostanoids, reflected by a reduction of the prostacyclin-to-thromboxane ratio, occurs within the placental bed early in women who will become preeclamptic. This same decrease in prostacyclin production enhances the sensitivity of the maternal blood vessels to the vasopressor effects of angiotensin II. Whether secondary to the normal wear of the intimal lining, or abnormal turbulent blood flow secondary to faulty regulation, deficient prostacyclin synthesis leads to excessive aggregation and degranulation of platelets and activation of the coagulation cascade. The release of vasospastic amines such as serotonin from the platelets perpetuates and extends this cycle. Initially the generation of fibrin is prevented by endogenous inhibitors such as antithrombin III. The consumption of factor VIII increases. Because fibrin is not generated, the platelet plugs formed are unstable and are disrupted by the blood flow. The platelet number appears stable, but platelet lifespan is shortened because of the prior aggregation. Eventually platelet activation reaches a magnitude sufficient to cause excess generation of fibrin within the placenta. The deposition of fibrin and the associated vasospasm reduce the metabolic capacity of the placenta, and suboptimal fetal intrauterine growth results. Fibrin monomers eventually escape the placenta and are trapped in the microvasculature of such organs as the kidney and liver. The fibrin deposited within the renal and hepatic microvasculature has the same obvious effects as in the placenta—turbulent blood flow, vascular intimal damage, and platelet activation. Systemic hypertension follows because serotonin released from the platelets initiates either spasm of the

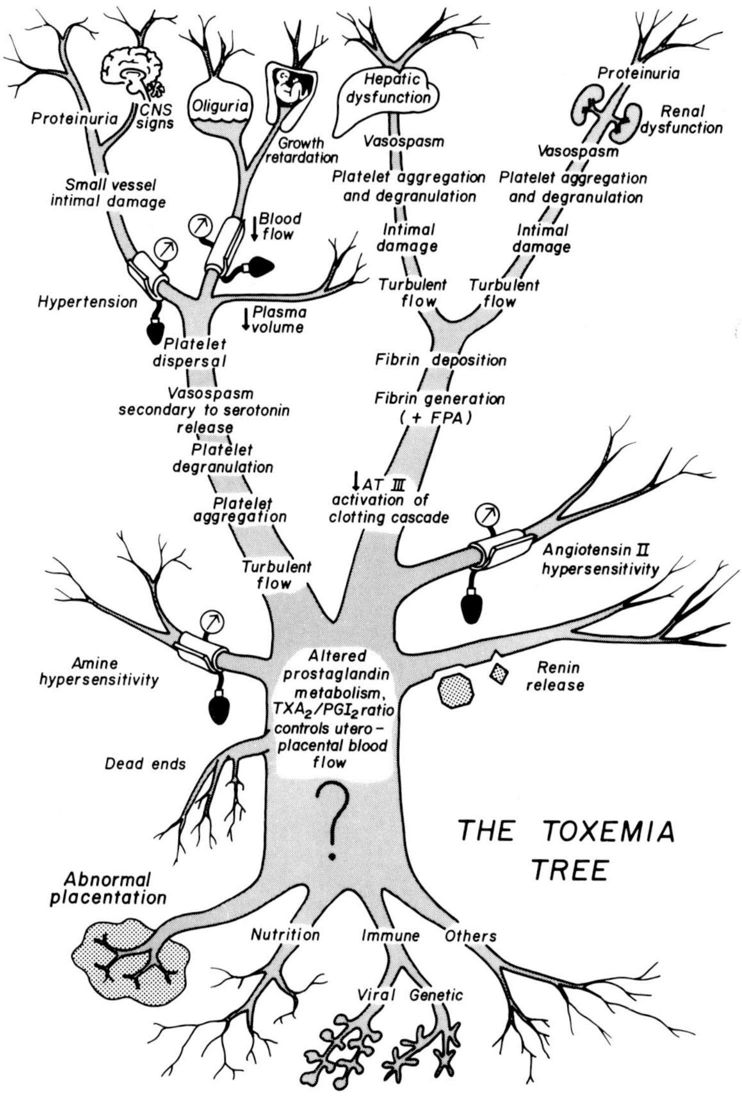

FIG. 14–3. Possible factors involved in preeclamptic hypertension. AT = Angiotensin III; CNS = central nervous system; FPA = fibrinopeptide A; PGI_2 = prostacyclin; TXA_2 = thromboxane.

juxtaglomerular arterioles and activation of the renin-angiotensin II systems, or generalized peripheral vasoconstriction. The progressive proteinuria may be secondary either to hypoxic destruction of the nephron or to permeability changes caused by substances released from the platelets. The deposition of fibrin within the hepatic sinusoids, if severe, produces anoxia and dysfunction of the hepatocyte. Hypertensive cerebral vascular damage in addition to exaggerated platelet activation leads to increased irritability of the central nervous system manifested as a low seizure threshold. The augmented levels of circulating serotonin sensitize the peripheral vasculature to vasoconstrictor neurohumoral mediators, in particular catecholamines and angiotensin II. The arterial blood pressure becomes characteristically labile.

This hypothesis seeks to describe the mechanism of hypertension in the preeclampsia syndrome. In the absence of the ability to identify and treat the causes of abnormal prostaglandin metabolism, treatment with serotonergic antagonists such as ketanserin may prevent both the clinical appearance and the sequelae of the preeclampsia syndrome.

— 15 —

Platelet-Derived Serotonin and Abnormal Tissue Perfusion

*Fred De Clerck and **Robert S. Reneman

**Laboratory of Haematology, Janssen Pharmaceutica Research Laboratories, B-2340 Beerse, Belgium; and **Department of Physiology, University of Limburg, Maastricht, The Netherlands*

Several blood-borne substances interact with the blood vessel wall and these interactions can have both functional and structural consequences. Thus thrombin, derivatives of fibrinogen, fibrinolytic enzymes, and kinins generated from plasma precursors (681,1009,1107), metabolites of arachidonic acid generated by platelets and white blood cells (206,1009), adenine nucleotides and amines released from red blood cells and platelets, respectively (231,238,239,946), and platelet-released β-thromboglobulin and platelet-derived growth factor(s) (195,492) can affect the barrier function of endothelial cells and contractility of vascular smooth muscle.

The interactions between platelets and the blood vessel wall can contribute to the pathogenesis of certain vascular disorders. Indeed, atherosclerosis and its complications are accompanied by an increased platelet sensitivity to aggregating agents, a shortened platelet lifespan, and raised plasma markers of platelet release (103,219). In this regard, interest has focused mainly on the contribution of the metabolites of arachidonic acid formed by the enzymes cyclo-oxygenase and lipoxygenase (934). This chapter suggests that serotonin released from platelets also can play a role under conditions of abnormal tissue perfusion.

PHARMACOLOGY

Effects of Serotonin

Platelets

Serotonin activates the platelets and produces shape changes, aggregation, and possibly the release of cellular products (see Chapter 7). Serotonin is taken up by the platelets by means of an energy-requiring transport system (see Chapter 4). These two processes are mediated by different serotonergic receptors (235,614,848). The receptor involved in the activation process resembles the S_2-serotonergic subtype (233,235,242,243,614,637,841,848). The activation of the platelets by the monoamine is (a) not affected by blockade of α-adrenergic, β-adrenergic, or histaminergic receptors; (b) not accompanied by detectable release of adenine nucleotides or formation of thromboxane B_2; and (c) not reduced by inhibition of cyclo-oxygenase or lipoxygenase enzymes (242). The serotonergic activation of human platelets is linked to the turnover of phosphatidylinositol which transduces the signal by the production of phosphatidic acid, a potential calcium translocator (*unpublished observations*; 314; see Chapter 7).

The consequences of the activation of the platelets by serotonin depend on local or pathological conditions. In normal human platelets, serotonin induces a shape change and a weak transient aggregation. However, the monoamine amplifies responses (cellular aggregation, release of intraplatelet products, and formation of prostaglandins) of the platelets to low concentrations of other agonists (Fig. 15–1; 233,235). In sensitized normal platelets exposed to subthreshold levels of other agonists and in platelets of patients with scleroderma or vascular disease, serotonin induces a full aggregation/release reaction by activation of S_2-serotonergic receptors (233,621; see Chapter 19). In platelets which are hyperreactive to S_2-serotonergic stimuli, the serotonin released during the initial platelet reaction contributes to the secondary wave of aggregation and to the further liberation of platelet products (87,243). Thus serotonin can behave as a potent activator of platelets and hence

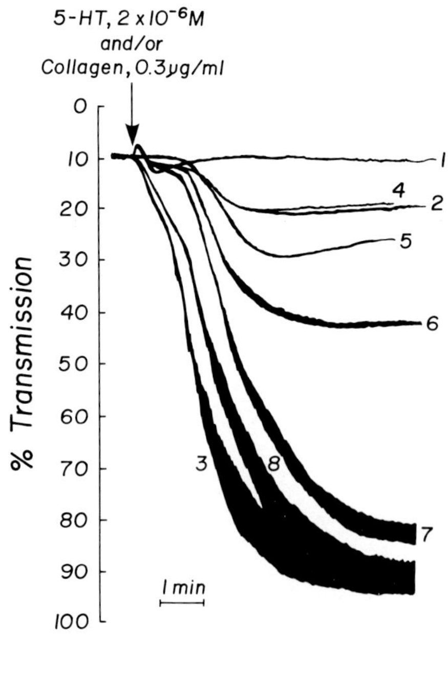

FIG. 15–1. Inhibition by ketanserin of the amplification by serotonin (5-HT) of human platelet aggregation induced by threshold concentration of collagen. (Reproduced from ref. 233, with permission.)

may favor the mechanical obstruction of blood vessels by formation of platelet aggregates.

Endothelial Cells

In cultured endothelial cells serotonin stimulates a transient influx of Ca^{2+} which may be involved in the activation of the actin-myosin-like proteins of these cells (234,965). Such activation may contribute to the increase in microvascular permeability induced by serotonin upon topical and/or intravascular administration in most vascular beds including skin, muscle, mesentery (Fig. 15–2), and the surface of the brain (236,377,674–676,789,802,1059). This increase in vascular permeability is restricted to the collecting venules (674–676,789) and results from the gap formation between endothelial cells and the combination of arteriolar dilatation and venous constriction which results in an increased capillary pressure (Fig. 15–3; 66,157,674–676,789).

Studies with ketanserin indicate that this process is mediated by S_2-serotonergic receptors of the venular wall in the skin (Fig. 15–4), the cremaster muscle, the mesentery of the rat (Fig. 15–2), and the frog brain (236,802; Verheyen, *unpublished observations*). Leakage of fluid into the interstitial space may increase the tissue pressure, especially in tissues surrounded by nonelastic structures (e.g., skeletal muscles) (889). The resulting decrease in transmural pressure limits blood flow (891). Since a change in tissue pressure of only a few millimeters of mercury determines whether or not capillaries in skeletal muscle are perfused (891), a limited amount of extravasation already might initiate an insufficient perfusion when the arterial circulation is compromised, as is the case in patients with intermittent claudication (890,891).

Vascular Smooth Muscle

At the microcirculatory level serotonin might induce constriction of the venules by a direct effect on the venular smooth muscle (20; see Chapter 9). In the skeletal muscle and mesentery of intact animals the monoamine probably causes dilatation of arterioles, especially those smaller than 70 μm in diameter (809,1101; Reneman, *unpublished results*). The mechanism underlying the vasodilator properties of serotonin is incompletely understood. It is probably not due to an enhanced synthesis of prostacyclin (2,195,196,492), since neither arteriolar dilatation *in vivo* nor endothelium-dependent relaxation of isolated arteries is inhibited by inhibitors of cyclo-oxygenase (708; Blackshear, *unpublished observations*). The vasodilatation in arterioles induced by serotonin *in vivo* (809; Reneman, *unpublished observations*) is inhibited by methysergide, a nonselective S_1- or S_2-serotonergic antagonist, but not by ketanserin, a selective S_2-serotonergic blocker (637). Thus the vascular receptors modulating this type of dilatation probably belong to the S_1-subtype, as do those mediating endothelium-dependent responses of large canine coronary arteries (see Chapter 10).

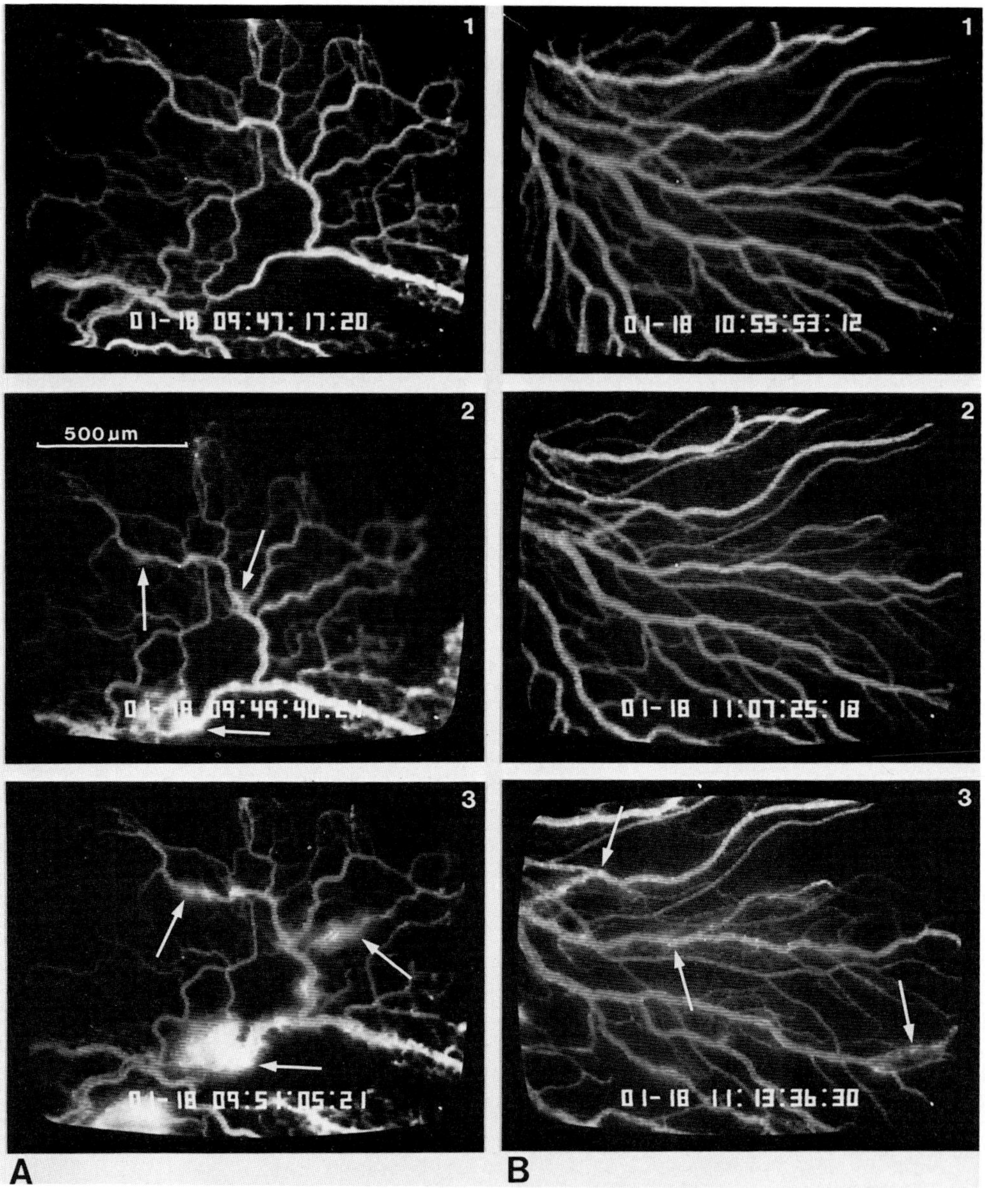

FIG. 15–2. Rat mesentery microvasculature as visualized with fluorescein isothiocyanate-bovine serum albumin, using fluorescence intravital microscopy. **A** and **B** are derived from different rats. **A:** (1) Control. (2) 1 min after a second topical application of 2 μg serotonin; 1 min interval between the applications. The *arrows* point to leaky sites in the venules. (3) 2.5 min after this application. More extensive leakage is observed *(arrows)*. **B:** (1) 2 min after an intravenous injection of ketanserin (0.16 mg/kg). No leakage is observed. (2) 6.5 min after a second topical application of 2 μg serotonin; 1 min interval between the applications. The second dose of serotonin was applied 7 min after the injection of ketanserin. No leakage could be detected over the 6.5 min period. (3) The same preparation 1 min after a second topical application of 10 μg histamine; 1 min interval between the applications. Diffuse leakage at various sites in the venules *(arrows)*, indicating that the preparation is still responsive to other amines. These experiments demonstrate that the increase in venular permeability induced by serotonin can be inhibited by ketanserin, indicating that this process is mediated by S_2-serotonergic receptors.

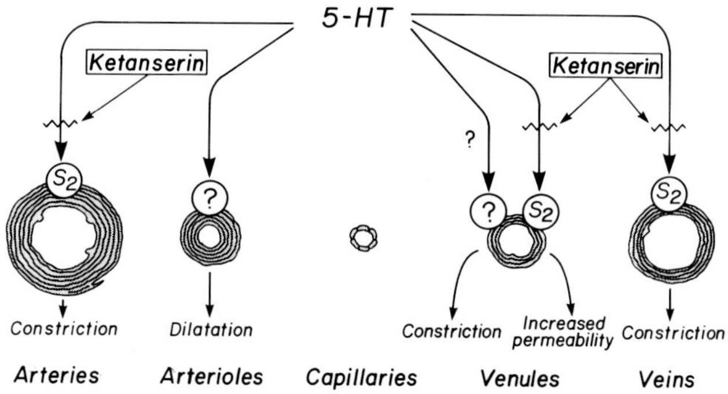

FIG. 15–3. The effects of serotonin (5-HT) that may interfere with tissue perfusion. + = Vasoconstriction; − = vasodilatation; S_2 = S_2-serotonergic receptor; ? = unknown receptor; jagged line = inhibitor effect.

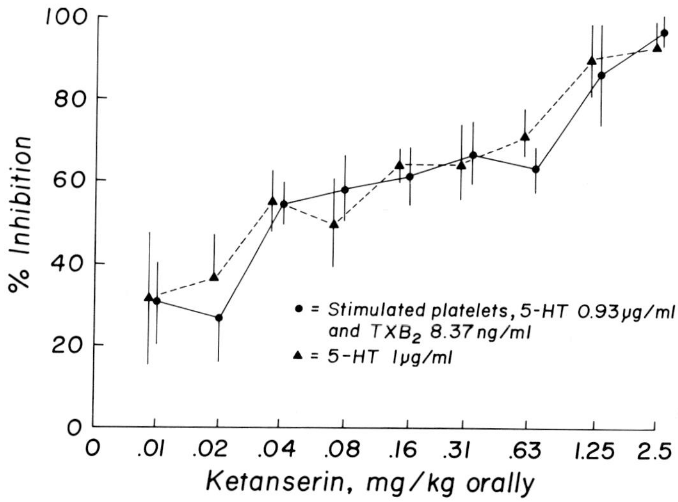

FIG. 15–4. Dose-related inhibition of the increase of cutaneous vascular permeability induced by thrombin-stimulated rat platelets [●; the figures show the release of thromboxane B_2 (TXB_2) and serotonin (5-HT)] or exogenous serotonin (▲) in rats treated with ketanserin. Data shown as mean ± SEM ($N = 6$).

Effects of Aggregating Platelets

Platelet-Derived Mediators

Besides causing mechanical obstruction by the formation of aggregates, the platelets contain and release a number of vasoactive products that can restrict tissue perfusion. Thus human blood platelets accumulate substantial amounts of serotonin and histamine and trace amounts of other amines from the surrounding plasma either by means of active carriers or by passive diffusion. In the case of serotonin the energy-dependent uptake outnumbers passive diffusion; it is a saturable process requiring Na^+ and Cl^- ions directly coupled to K^+ efflux (217). The captured serotonin is stored in the dense granules of the platelets as a complex with adenine nucleotides, calcium, and magnesium; in the granules it is protected from the mitochondrial monoamine oxidase activity (217; see Chapters 4 and 7). Platelets metabolize arachidonic acid and produce the vasoactive prostaglandins D_2, E_2, and $F_{2\alpha}$ as well as thromboxane A_2 (238,239,304,746,1009). Activated platelets also may generate vasoactive metabolites of arachidonic acid formed by lipoxygenase (971). Obviously, the final response of the blood vessel wall to metabolites of arachidonic acid released by platelets will depend on the production of prostacyclin formed from endogenous substrates or from platelet-derived endoperoxides by the endothelial and vascular smooth muscle cells (745,746,1009). Another modulating factor is the fact that proteins contained in the platelet granules may trigger the release of mediators from mast cells in the vicinity of blood vessels, thus affecting vascular tone and permeability indirectly (771,772,814).

When blood platelets are activated, they liberate a number of the vasoactive products which they store or synthetize, including serotonin. The nature and the amounts of released mediator(s) are determined by the stimulus causing the activation (485). Potent stimuli acting on specific receptors on the platelet membrane (see Chapter 7) include (a) thrombin generated from activated plasma (494,495); (b) subendothelial material such as collagen exposed in vessels with damaged endothelium (42,1002); and (c) combined multiple agonists leading to full aggregation with the release of a variety of products (235). High shear stresses in regions of turbulence distal to atherosclerotic lesions may induce the liberation of serotonin from the platelet either directly or indirectly by causing the release of adenine nucleotides from sheared red blood cells, which in turn activate the platelets (348,946). Ulcerated atherosclerotic plaques without substantial narrowing of the artery also must activate platelets; their presence is frequently associated with the occurrence of transient ischemic attacks (1036). The relevance of these processes in pathology is suggested by the increased levels of markers of platelet release [β-thromboglobulin, platelet factor 4, 5-hydroxyindoles, thromboxane B_2 (558)] in the plasma of patients with a variety of cardiovascular diseases (103,801,925,1177).

Effects on Vascular Smooth Muscle

Platelets activated *in vitro* with thrombin or after contact with an isolated, de-endothelialized vessel produce a substantial contraction of various large-sized blood vessels of different anatomical origin including the rat caudal artery, the canine coronary, femoral, and pulmonary artery, the rabbit pulmonary artery, and human digital arteries (238,239,304,708,755,1078). The released serotonin contributes to the platelet-induced contractions, but its relative importance varies with the type of blood vessel studied. For example, in canine pulmonary arteries serotonin is the predominant mediator (708), while in rat caudal artery (238,239) and human digital arteries (755) it contributes 60 and 50%, respectively, whereas the remaining contraction is caused by released prostanoids. Besides the direct activation of vascular smooth muscle, the released serotonin also amplifies the response to other platelet-derived vasoconstrictor agents, in particular thromboxane A_2 (239). Hence, depending on the type of blood vessel and the status of the endothelial lining (1078), serotonin can greatly contribute to the contraction of vascular smooth muscle induced by aggregating platelets. The serotonergic component of the contractile response to platelets is inhibited by ketanserin (238,239,1078), illustrating that it is mediated by activation of S_2-serotonergic receptors on the vascular smooth muscle (see Chapter 9).

Effects on Vascular Permeability

When injected intradermally in mice, rats, or rabbits, intact platelets of the rat or cell-free supernatants of platelets from rats, pigs, or humans, previously exposed to stimuli such as thrombin or collagen, cause an immediate and intense increase in vascular permeability as well as a protracted inflammatory reaction with infiltration of polymorphonuclear cells (125,767,814). The mediators involved in the increase in vascular permeability include histamine, serotonin, ATP, cationic proteins from lysosomes or platelet-aggregating-factor (60,125,767,771,772,814).

After depletion of serotonin with reserpine, platelets of the pig are as active as those from nontreated donors when injected into rabbit skin. The administration of antihistamines prior to the intradermal injection suppresses the increase in permeability (814). This then suggests that in this species cationic platelet proteins liberate mediators (in particular histamine) from cutaneous mast cells (771,772,814) which largely contribute to the platelet-induced response.

In the rat the vascular leakage caused by thrombin-stimulated platelets is not markedly affected by inhibitors of cyclo-oxygenase, thromboxane A_2 synthetase, or lipoxygenase, which indicates that the metabolites of arachidonic acid play no major role in this type of vascular defect. However, the reaction is strongly reduced or abolished by reserpine or ketanserin (Fig. 15–4). These findings demonstrate that under certain conditions serotonin released from platelets can play a major role in the formation of edema caused by aggregating platelets (236).

Hemostasis

The interaction between the platelets and the blood vessel wall is an essential step in primary hemostasis when platelets aggregate at the site of vascular damage where local vasoconstriction occurs. The contribution of platelet-derived serotonin to these processes is evidenced by the prolonged bleeding time and the increased blood loss in animals which are depleted of bioamines or which have been treated with ketanserin in concentrations that do not affect plasma coagulation, fibrinolysis or the platelet content of serotonin. Similar results are obtained with ritanserin and compound R 56 413, which are selective S_2-serotonergic antagonists, devoid of the α_1-adrenoceptor blocking effect of ketanserin (237). Thus platelet-derived serotonin contributes to the platelet aggregation during formation of the hemostatic plug and/

or to the local vasoconstriction induced by the activated platelets at the site of vessel injury.

Thromboembolism

Serotonin, which can activate both vascular and respiratory smooth muscle, may play a role in the cardiopulmonary dysfunction provoked by thromboembolism (see Chapter 21). Thus in dogs treatment with methysergide or ketanserin reverses the increases in mean pulmonary arterial pressure and pulmonary vascular resistance, and the associated disturbance in oxygenation caused by pulmonary embolization with a fresh thrombus (511,1045).

Collateral Circulation

In the cat hindleg an adequate collateral circulation usually is formed following mechanical obstruction of the abdominal aorta. However, proper collateralization is prevented by the presence of a stasis thrombus in the aorta, which provides a source of continuing platelet activation (778,942,943). This loss of collateral perfusion is reduced by depletion of serotonin or treatment with cinanserin, which suggests that platelet-derived serotonin is involved. This interpretation is strengthened by the finding that ketanserin improves the local circulation whether measured by objective (e.g., hindleg/foreleg systolic blood pressure ratio; blood flow determined by venous occlusion plethysmography; collateral vascular resistance) or subjective (scored hindleg function) means (Fig. 15–5). Exogenous serotonin constricts peripheral collateral vessels in the dog (809). Ketanserin corrects the impairment of collateral blood flow induced by exogenous applied serotonin in the rat (1109). Thus it appears likely that platelet-derived serotonin can impair tissue perfusion by constricting collateral blood vessels (778).

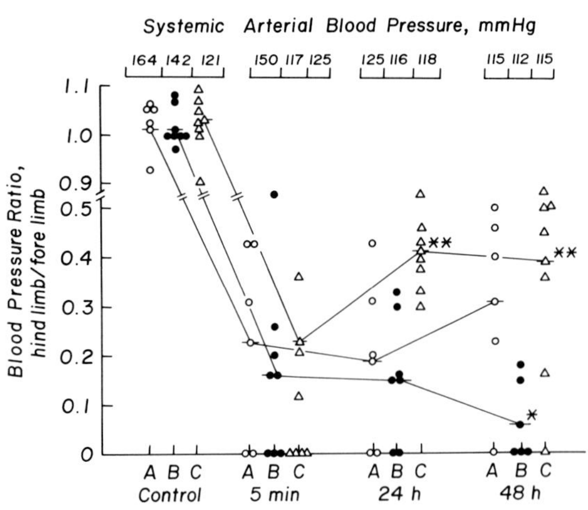

FIG. 15–5. Restoration by ketanserin of postthrombotic reduction of systolic blood pressure hindleg/foreleg ratio in cats as a function of time after aorta occlusion. Median (○, ●, △) and individual (A, ○, $N=5$) values of aorta ligation, ligation with a stasis thrombus (B, ●, $N=7$), and ligation with thrombus treated daily with ketanserin 1.25 mg/kg intraperitoneally (C, △, $N=7$). *$p \leq 0.05$ versus A; **$p \leq 0.05$ versus C. Mann-Whitney U test.

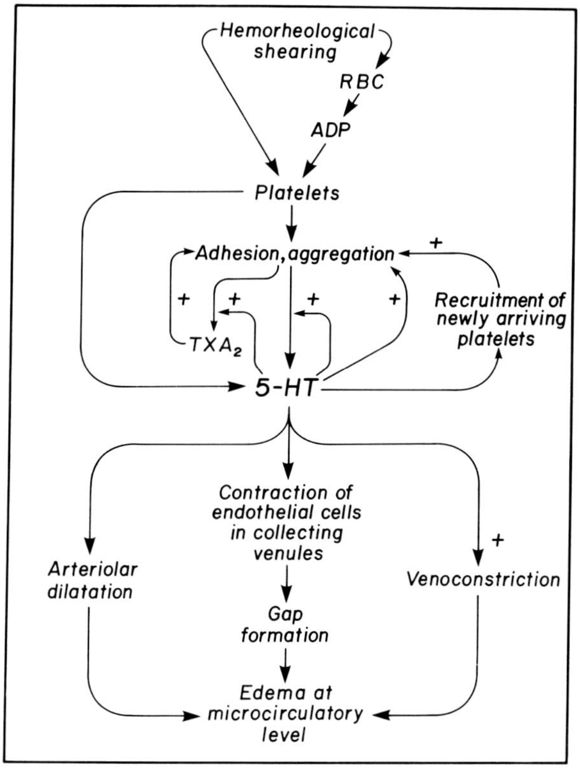

FIG. 15–6. Multiple interactions of platelet-derived serotonin (5-HT) with platelets, endothelial cells, and vascular smooth muscle cells. RBC = Red blood cell; + = amplification; TXA_2 = thromboxane A_2; ADP = adenosine-5′-diphosphate.

OBSERVATIONS IN HUMANS

Platelets display multiple abnormalities in a variety of cardiovascular diseases (103,219). Increased sensitivity to aggregating agents *in vitro* has been described in patients with myocardial infarction (294), diabetes mellitus (610), transient cerebral ischemia (18), cerebrovascular disease (700), hyper-β-lipoproteinemia (539), Raynaud's phenomenon (1175), and hypertension (173,700,861,1116). Hence, in these patients a facilitated release of platelet-bound mediators must occur (231), as documented by (a) the increase of markers of platelet release, in particular β-thromboglobulin (558,1117) in the plasma of some patients with venous thrombosis (666), myocardial infarction (266,801,925), peripheral arterial disease (51), Raynaud's phenomenon (1175), hypertension (571,721), preeclampsia (110,268,878), and diabetes mellitus (141,865); (b) the elevation of thromboxane B_2 [the stable degradation product of thromboxane A_2 formed from arachidonic acid bound to the platelet membrane (231)] in patients with myocardial ischemia (628, 723,724,902,1018); (c) the diminished serotonin content in platelets of patients with preeclampsia and hypertension (88,1146); and (d) the elevated plasma levels of 5-hydroxyindoles in patients with myocardial infarction (801,968).

The effects obtained with the S_2-serotonergic antagonist ketanserin in man probably illustrate the involvement of serotonin released from platelets in the pathogenesis of impaired tissue perfusion under certain pathological conditions (Fig. 15–6). In man, the intravenous injection of serotonin into a digital vein causes local constriction, while infusion of the monoamine into the brachial artery causes a rapid decrease in digital temperature and characteristic color changes (445,521). The former is inhibited by ketanserin (521). Threshold doses of serotonin enhance the reduction of forearm flow induced by the intraarterial infusion of norepinephrine (953). Cold-induced vasoconstriction in human volunteers (463), as well as the symptoms of Raynaud's phenomenon, can be corrected by ketanserin (519, 957,1006; see Chapter 19).

ACKNOWLEDGMENT

The authors are indebted to Dr. P. Janssen for his stimulating interest and criticism. This study was supported partly by a grant from the Instituut tot Aanmoediging van Wetenschappelijk Onderzoek in Nijverheid en Landbouw.

16

Segmental Vascular Responses to Serotonin

Norman K. Hollenberg

Departments of Radiology and Medicine, Harvard Medical School and Brigham and Women's Hospital, Boston, Massachusetts 02115

VASOCONSTRICTOR AND VASODILATOR EFFECTS OF SEROTONIN

The term "serotonin" implies a factor in serum that causes vasoconstriction: Page (815) first assayed the vasoconstrictor response to serotonin using the contractile response of an artery of substantial size, the isolated ear artery of the rabbit. Most schemata that describe a role for serotonin in physiology and pathophysiology have borrowed from the concept that constriction of the microvasculature induced by the monoamine is important.

Certainly, contractile responses to serotonin have since been observed consistently in a variety of isolated vascular preparations *in vitro* (35,97,176, 300,335,709,1044,1092,1093,1171). The contraction of large arteries induced by platelet activation is blocked by the serotonergic antagonist ketanserin (238), which demonstrates that serotonin is released from aggregating platelets in amounts sufficient to activate the arterial smooth muscle. The arterial preparations used for these *in vitro* studies share a common characteristic: all were prepared from large arteries.

By contrast, studies in which blood flow was measured in relatively intact animals have often revealed a surprising *vasodilator* response to serotonin (95,99, 305,433,649,707,726,895,1114,1119). Net arteriolar vasodilatation, with an increase in blood flow, has been documented even in vascular beds in which the conduit arterial vessels contract when exposed to serotonin *in vitro*. There is no simple explanation for the arteriolar dilatation and increase in blood flow caused by serotonin. In the dog limb, for example, inhibition of cyclo-oxygenase does not modify serotonin-induced vasodilatation, whereas in the dog kidney it converts serotonin from a vasodilator to a potent vasoconstrictor agent (98,99). The apparent consistency with which serotonin induces renal vasoconstriction in the rat (4,184,261) may reflect, at least in part, the unusual failure of the rat kidney to respond with vasodilatation to prostaglandins, which relax renal arteriolar smooth muscle in other species (52).

When *in vitro* and *in vivo* studies show divergent results, three major possible explanations exist. First, it is possible that serotonin exerts differential effects on large vessels and on the arterioles that determine blood flow. Alternatively, *in vitro* studies may be anomalous because of the artificial milieu in which the vessels are placed and the trauma involved in isolating and preparing them. Third, it is possible that removal of the vascular tissue from the body separates it from a system with which serotonin can interact: in this case, specifically, connections with the sympathetic nerves and with the central nervous system.

LARGE ARTERIES AND ARTERIOLES

In earlier studies *in vivo*, segmental vascular resistances in the limb were assessed by employing a pump-perfused, constant-flow system and measuring the segmental pressures. The results suggested that serotonin exerted a differential, directionally opposite influence at different levels (433,707). Thus *in situ*, serotonin enhanced the resistance to flow in proximal arteries but dilated the arterioles. As the latter determine the blood flow, the net supply of blood to the tissue was increased. The rather modest changes in segmental vascular resistance of the proximal large arteries are not easily interpretable in terms of either alterations in the dimensions of the large blood vessels or their hemodynamic impact.

The relative loss of interest in serotonin over several decades probably reflects the fact that such blood-flow measurements indicated a vasodilator action that belied its name. The various schemes demanded that serotonin induce vasoconstriction. Interest in the dynamic role played by large arteries, however, and in the vasoconstrictor action of serotonin at that level, has been awakened by observations that indicate a substantial contribution of spasm of these blood vessels to the genesis of occlusive or obliterative vascular disease. Indeed, it is now recognized that interactions between platelets and the blood vessel wall may play an important role in vascular spasm, in part due to the release of platelet stores of vasoconstrictor agents, which include serotonin and thromboxane (180,721). The increasing use of percutaneous transluminal angioplasty to open partially occluded major arteries has further demonstrated the importance of vasospasm (283). Moreover, the responsiveness of large arteries to serotonin is augmented under pathological circumstances, which may suggest that the monoamine plays a role in peripheral vascular disease. An additional substantial contribution to that interest has arisen from the development of the S_2-serotonergic antagonist ketanserin (637,1084).

More compelling evidence than that provided by measurement of segmental vascular resistance demonstrates that serotonin indeed exerts a dramatic influence on even normal large arteries. The assessment of the changes in arterial dimensions induced by serotonin *in vivo* has come from angiographic studies of the coronary circulation (121), the renal vascular bed (99), and the circulation of the limb (98). All of these studies show substantial constriction of major arteries in response to serotonin.

PATHOLOGICAL CONDITIONS

In the normal arterial tree the constriction of large arteries induced by serotonin is insufficient to dominate the hemodynamics. The net effect of serotonin is an increase in blood flow despite the substantial constriction of the large arteries. However, under pathological conditions, when larger arteries are affected by disease, the situation changes. First, an already compromised lumen requires a much smaller reduction in diameter before vasoconstriction becomes hemodynamically significant. Second, and perhaps more importantly, preliminary evidence suggests that the pathological area becomes supersensitive to the constrictor action of serotonin, so that smaller quantities of the monoamine induce vasoconstriction (98,649,1171).

The question of importance is obviously whether or not a sufficient quantity of the monoamine will be available *in vivo* to induce responses similar to those obtained with exogenous serotonin. The answer is unequivocally yes. In most of the studies reported, serotonin was infused into the arterial tree at doses of 1 to 10 μg/min, a quantity sufficient to increase the serotonin concentration of whole blood to several hundred nanograms per milliliter. Similar concentrations have been reported in the blood in several studies, and reflect the release of serotonin from a small fraction of the platelets traversing the vascular bed investigated (209,586). Most calculations indicate that secretion of the entire platelet content, as might occur with platelet disruption, would release substantially more serotonin. A stimulus for local platelet aggregation such as an embolus or an atherosclerotic plaque could easily liberate sufficient serotonin to produce local spasm, particularly if the activation or disruption of the platelets was accompanied by local stasis and platelet aggregation at the site of injury.

Atherosclerosis today is the leading cause of morbidity and mortality in Western society. It appears that the action of serotonin on large blood vessels may be relevant to that pathological process. The vascular sensitivity to serotonin released from aggregating platelets *in vitro* is enhanced when the endothelium is damaged or absent (180). Isolated aortas from atherosclerotic rabbits exhibit a striking increase in sensitivity to serotonin (1171). Strong support for a potential role of serotonin in this setting has come from the observations of Heistad and colleagues (649). They demonstrated that in monkeys with experimental atherosclerosis, the constriction of large arteries of the hindlimb in response to serotonin is enhanced; the enhancement is specific for serotonin, as responses to norepinephrine were not augmented. Whereas in normal monkeys serotonin causes a net vasodilatation and an increase in blood flow in the hindlimb, the response is changed to a net vasoconstriction and a fall in limb blood flow in atherosclerotic monkeys. This study demonstrates clearly that the large-artery responses become dominant in the diseased limb (649). This sequence of events is not specific for the hindlimb. Partial occlusion of the canine coronary arterial tree causes progressive total occlusion over a relatively short interval. Ketanserin, however, blocks the progression to complete occlusion in over 90% of the experiments (504).

The enhanced sensitivity of damaged large arteries to serotonin probably does not depend entirely on the absence of endothelium for its expression (98). A relatively modest lesion in the adventitia of the femoral artery, induced by the placement of a flowmeter-probe for a few hours, strikingly enhances the local constrictor response to serotonin after several weeks, but not acutely. In both cases the endothelium is disrupted occasionally, but not routinely. It is easy to understand how removal of the intima, thus exposing the smooth muscle in the media to serotonin, would enhance vascular responses (see Chapter 10). However, understanding how an adventitial lesion would lead to hyperresponsiveness is less straightforward. The vasa vasorum are increased strikingly in the area of arterial plaques and may contribute to the pathogenesis of the atherosclerotic lesion (59,1162). The delayed enhancement of responses of the femoral artery to serotonin when the adventitia is damaged could reflect an abnormal function of the vasa vasorum.

COLLATERAL CIRCULATION

In the presence of a hemodynamically important obstruction or of a complete occlusion of a major artery, the tissue supplied with blood by that artery lives or dies depending on the effectiveness of the collateral circulation. In the cat the collateral arterial vessels supplying the limb are very sensitive to serotonin (943). The same is true in the dog (Orlandi, Blackshear, and Hollenberg, *unpublished observations*). Thus, in the animal, 2 to 3 weeks after the superficial femoral artery has been occluded, vasodilatation normally induced by serotonin becomes vasoconstriction, with a reduction in calf blood flow (Fig. 16–1). Angiography reveals virtually complete occlusion of the collateral arterial supply to the occluded region in response to low concentrations of serotonin (Fig. 16–2). Whether this exquisite sensitivity reflects a special response of a new and growing arterial network, or is due to the critical size of the collateral channels, is uncertain. The augmented response may not be specific for serotonin, as a potentiated response of collateral vessels to catecholamines has also been described (1153). Further, it is unknown whether "mature" arterial collateral vessels retain this extreme sensitivity to serotonin. The answers to each of these questions may have important clinical implications. In the dog spasms of collateral arterial vessels in response to serotonin are prevented and reversed by ketanserin (Orlandi, Blackshear, and Hollenberg, *unpublished observations*). Essentially similar findings have been obtained in the rat, where serotonin induces intense vasoconstriction of collateral vessels in the limb; ketanserin blocks that response (1110). Nevelsteen and colleagues (778) have demonstrated that ketanserin helps to restore the limb circulation of the cat with collaterals in a model in which thrombosis and platelet activation

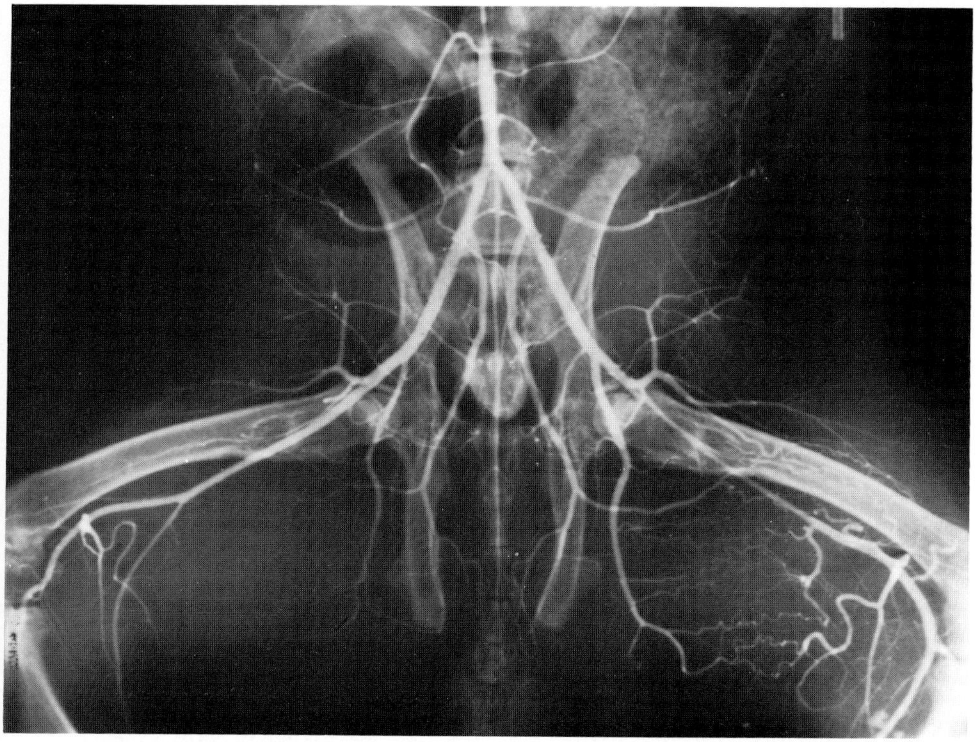

FIG. 16–1. Arteriogram in a rabbit 2 weeks after ligation of superficial femoral artery on the left. Note the striking collateral arterial supply.

occur, an observation that is consistent with a critical role for serotonin in this setting.

THERAPEUTIC IMPLICATIONS

The observations on the influence of ketanserin on collateral arterial vessels may have clinical relevance. In accord with the observations in the cat that ketanserin allows a better circulation to the limb after aortic occlusion (778), preliminary clinical trials with ketanserin in patients with peripheral vascular disease have indicated, at least in some patients, a striking improvement (250). In these patients ketanserin may induce a preferential increase in blood flow to the ischemic tissue. This interpretation is supported by the fact that following ketanserin, the systolic blood pressure measured at the calf level by plethysmography appears higher than that measured at the ankle by a doppler technique.

The use of "nonspecific" vasodilator agents in patients with peripheral vascular disease has earned well-deserved disrepute. The better controlled the clinical trial, the less evidence of efficacy these agents have had. To date, no vasodilator agent has proved to be more effective in dilating arterioles in the ischemic region than

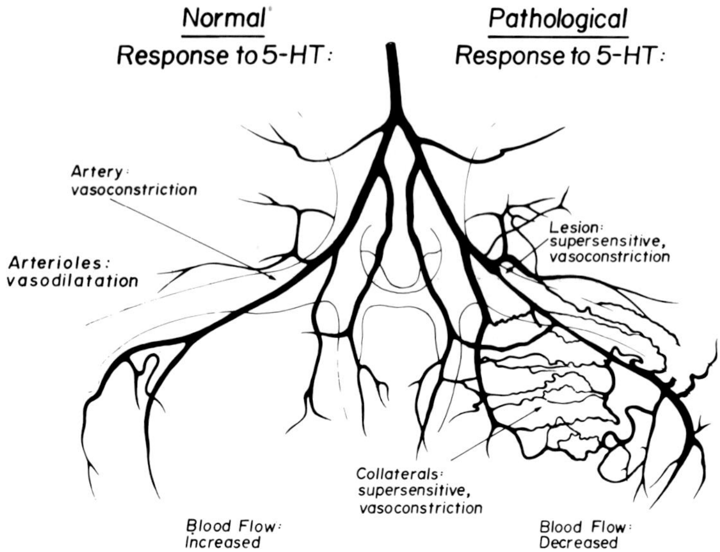

FIG. 16–2. Tracing reconstituting the distal blood supply of the arteriogram shown in Fig. 16–1. The normal response to serotonin (5-HT) and the responses to serotonin in the presence of arterial pathology are summarized.

the metabolic effects of ischemia *per se*. However, the studies in animal models of peripheral vascular disease have raised the possibility that spasm of larger proximal collateral arteries outside the ischemic zone may become "rate-limiting" under some circumstances. It is reasonable to presume that these collateral arterial vessels are not under the direct metabolic influence of the ischemia. Moreover, spasm at this level clearly can be induced by serotonin. Future studies in humans may document an important role of serotonin-induced spasm of the collateral vessels to the limb as limiting blood flow under some circumstances, and improve our understanding of the pathogenesis of peripheral vascular disease, which affects more than 75,000 people each year in the United States (283). If ketanserin were shown, beyond any doubt, to be effective in the treatment of atherosclerotic disease, one of its major therapeutic advantages might be that it has antihypertensive properties (1014). The control of an increased arterial blood pressure in patients with advanced peripheral vascular disease often exacerbates symptoms due to the peripheral vascular disease when currently available hypertensive agents are employed.

ACKNOWLEDGMENT

Personal research described in this chapter was supported by grants from the National Institutes of Health (CA 32849, HL 05832, HL 14944), Janssen Pharmaceuticals, and the National Aeronautics and Space Administration (NSG 8078).

17

Platelet Activation in Patients with Atherosclerosis of the Arteries of the Limbs

M. Verstraete

Center for Thrombosis and Vascular Research, Department of Medical Research, University of Leuven, Belgium

In the vast majority of people with intermittent claudication, atherosclerosis of the leg arteries is the underlying disease. Thus it seems logical to prevent the occurrence or retard the progression of atherosclerosis in these patients. The harsh debate around the recommendations of the British Coronary Prevention Group (129) demonstrates that there is no consensus on how to prevent atherosclerosis. The management of patients with hyperlipidemias is even more controversial. Indeed, new drugs, which are potentially important in this regard, risk submergence by the criticism with which clofibrate has been met, in particular if they are excessively and inappropriately prescribed. Although ignorance still prevails on the primary causes of atherosclerosis, smoking, hypertension, exercise, and diet are recognized risk factors in individuals with a high prevalence of the disease. This chapter discusses why treatment with antithrombotic drugs may be beneficial in atherosclerotic patients with intermittent claudication.

PLATELETS, COAGULATION, AND ATHEROSCLEROSIS

Platelet aggregation and blood clotting play a crucial role at three stages of atherosclerosis. The three stages of development of atherosclerotic lesions (atheroma) are:

1. Fatty streaks.
2. Fibrous plaque.
3. Complicated advanced atheroma:
   a. Ulceration.
   b. Hemorrhage.
   c. Calcification.
   d. Cell necrosis and mural thrombi.

Initial Events

Activated platelets release a growth factor that enters the arterial wall, where it stimulates mitosis and migration of smooth muscle cells (915–918). The growth factor also enhances the synthesis of collagen and the incorporation of cholesterol in the vessel wall. Thus platelets contribute directly to the development of the atherosclerotic plaque. Several animal studies support this conclusion. In chronically homocystinemic baboons the endothelial desquamation and the decreased platelet survival time are accompanied by marked smooth muscle proliferation. After 3 months of homocystinemia the animals show lesions identical in appearance to the fibromuscular lesions and the early fibrous plaques in man (453). By contrast, pigs with von Willebrand disease do not become atheromatous when fed a diet rich in cholesterol, because of the reduced adhesiveness of their platelets (376). Similarly, rabbits fed a high-cholesterol diet do not develop atherosclerotic plaques if they are made thrombocytopenic (355). Ideally, a drug that would prevent the synthesis, the release, and/or the action of platelet-derived growth factor might be able to prevent atherosclerosis.

Intermediate Stage

When atherosclerotic plaques are formed, platelets tend to adhere and aggregate on the sites of endothelial lesions of these plaques where no prostacyclin is formed (Fig. 17–1). The aggregated platelets may narrow the artery either mechanically or by the production of thromboxane A_2, which can induce vasospasm. Alternatively, larger thrombi adhering to a plaque can fragment in a shower of platelet

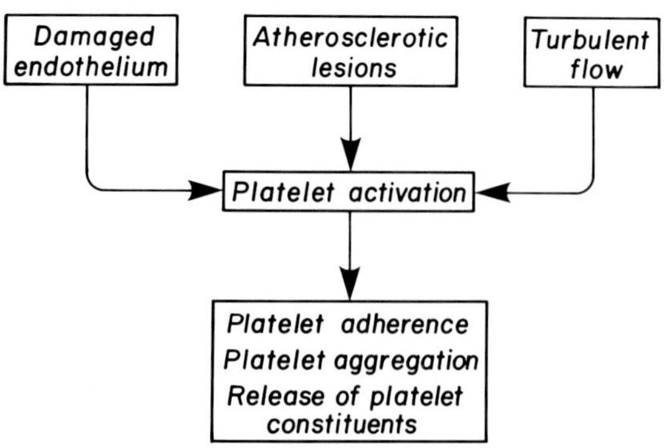

FIG. 17–1. The three main causes of platelet activation.

emboli, obstructing the microcirculation downstream. A typical example is the "fugace blue toe" syndrome. Theoretically, drugs impeding the adhesion and aggregation of platelets and the formation or the action of thromboxane A_2 may prevent additional lumen narrowing and microplatelet emboli.

Late Stage

A frequent complication is that a large thrombus totally occludes the already restricted lumen of the atherosclerotic artery (Fig. 17–2). One of the major functions of platelets in this regard is to provide a catalytic surface for the formation of intrinsic factor X-activating and prothrombin complexes. Stimulation of platelets by collagen or thrombin changes the phospholipid composition of the activated platelet membrane by a transbilayer movement of lipids or "platelet flip-flop" (1184). This increases the number of prothrombinase-binding sites with a marked enhancement of thrombin formation and release of lipid-protein particles (platelet factor 3). Hence, it seems logical to investigate inhibition of the formation of coagulation factors bound to platelets (antithrombin III, heparin, α_1-antitrypsin, α_2-macroglobulin), and of the interaction between these factors and the platelets. Similarly, more information is required on (a) the alteration of the composition of the platelet membrane, its fluidity, and its charge; (b) the possibility of preventing release of platelet factor 3 by selective inhibitors of microtubules and microfilaments; and (c) the effect of thrombomodulin on the release of protein C and the means to prevent the inactivation of this inhibitor.

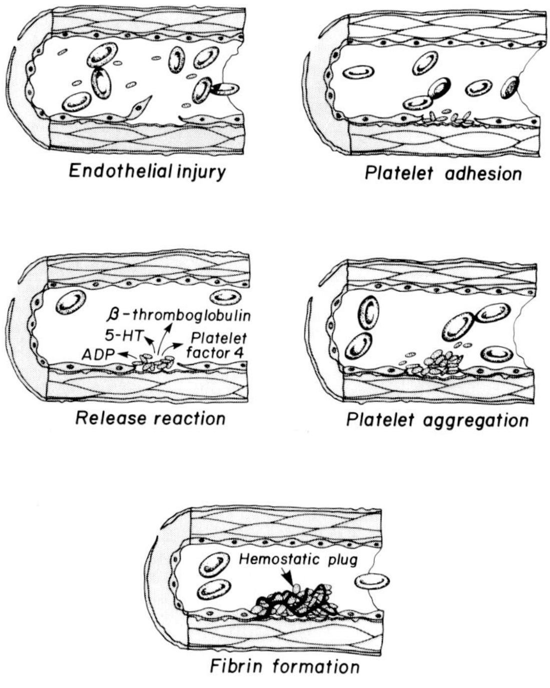

FIG. 17–2. The formation of a hemostatic plug proceeds through a series of rapidly successive events, starting with injury to the endothelium, followed by platelet adhesion, the release of platelet constituents, platelet aggregation, and eventually fibrin formation.

Risk Factors

Common risk factors are associated with alterations in function of the platelets and integrity of the endothelium. Thus cigarette smoking increases platelet aggregation and shortens the platelet survival time in man (768), as does hypercholesterolemia (773). Diabetes is also associated with increased platelet aggregation as well as with a frequent occurrence of coronary artery disease (450). Circulating catecholamines liberated during stress activate platelet aggregation and enhance the formation of thrombin (435,921).

PLATELET CONSTITUENTS AND VASCULAR WALL

Platelets are cytoplasmatic fragments with a complex function (see 1113; Chapters 4, 7, and 15). In humans, the most important platelet constituents that allow estimation of the turnover of platelets and their interactions with the blood vessel wall include the following:

Platelet Factor 4

Platelet factor 4 is a tetramer of basic peptides that is present in the α-granules (926). It readily binds to and neutralizes heparin. As there is no circulating heparin in man, the platelet factor 4 released from platelets binds to other acid mucopolysaccharides, e.g., to heparan sulfate in endothelial cells. The presumed function of heparan sulfate is to neutralize thrombin by catalyzing circulating antithrombin III. Released platelet factor 4 thus may delay inhibition of thrombin. The low concentration of platelet factor 4 in plasma probably reflects its strong binding to endothelial cells.

β-Thromboglobulin

β-Thromboglobulin is another tetramer of basic peptides with lower affinity for heparin than platelet factor 4. It might impede the generation of prostacyclin by endothelial cells.

Growth Factor

A platelet-derived growth-promoting factor is released from the α-granules of the platelets; it binds to specific receptors of cells, such as smooth muscle cells and fibroblasts, which can undergo mitosis (470,919). This cationic protein increases the synthesis of proteins and ribonucleic acid (RNA) and augments the number of receptors for low-density lipoproteins (LDL) on the cell membrane. The interaction of stimulated platelets, monocytes, and macrophages, which also produce their own growth factor(s), may be crucial in the development of atherosclerosis.

Adenine Nucleotides

The dense granules in platelets contain adenosine nucleotides. Adenyl compounds increase blood flow in most vascular beds (83,346), but ATP usually contracts isolated blood vessels. In canine coronary and dog femoral arteries ADP and ATP require the presence of endothelial cells to cause relaxation, while adenosine and AMP do so in the absence of the intimal layer (261,497).

Serotonin

Stored serotonin is released from stimulated platelets. It has weak direct platelet-aggregating properties but remarkably enhances the aggregation induced by other stimuli (233; see Chapters 7 and 15). The infusion of serotonin can reduce capillary fragility in thrombocytopenia. Platelets may release continuously small amounts of serotonin, which allow the endothelium of the small vessels to maintain their structural integrity (1012). The effects of serotonin that result in changes in vascular diameters are discussed elsewhere (see Chapters 9, 10, and 11).

Thrombospondin

Thrombospondin is a high-molecular-weight (450,000 dalton) protein that is released from thrombin-stimulated platelets and binds to heparin (619). Antibodies to thromboglobulin and anti-platelet factor 4 do not recognize thrombospondin.

Prostanoids

Activated platelets also produce cyclic endoperoxides that can be taken up by endothelial cells and transformed into prostacyclin. Platelets also contain enzymes to transform the cyclic endoperoxides into the pro-aggregatory and vasoconstrictor substance thromboxane A_2 (744).

PLATELET TURNOVER IN PATIENTS WITH ATHEROSCLEROSIS

If constituents of platelets and products of coagulation play a role in atherosclerosis and its thromboembolic complications, one would expect to find evidence of activation of the platelets and the coagulation process in patients with obstructive vascular disease. There is substantial evidence that this is the case, and it is tempting to ascribe a causal relationship to this association.

Platelet Factor 4 and β-Thromboglobulin

Of the various platelet constituents released by activated platelets, the most studied are platelet factor 4 and β-thromboglobulin because they can be measured by radioimmunoassay. β-Thromboglobulin probably is the best marker of the platelet-release reaction, because its concentration is 300,000 times higher in the α-granules than in other tissues (665) and because it is cleared from the plasma with a half-life of 100 min (667). The clearance of platelet factor 4 is much faster and averages 5 min (226).

The plasma levels of β-thromboglobulin and platelet factor 4 increase with age in apparently healthy individuals (269,1176). There is an inverse correlation between the levels of β-thromboglobulin and the lifespan of the platelets in normal subjects (665). Thus activation of platelets may occur with increasing age, even in the absence of clinical symptoms of atherosclerosis. The finding that plasma levels of β-thromboglobulin and of platelet factor 4 are higher in old healthy females than males is compatible with a higher incidence of coronary heart disease in the former (418).

Several investigators (51,85,155,556), but not all (1054), found increased levels of β-thromboglobulin in patients with peripheral arterial insufficiency. One group found a concomitant increase of platelet factor 4 in the same patients (51), while another did not (556). In the studies where a significant increase in the plasma levels of β-thromboglobulin was found in patients with intermittent claudication, no separation of individual patients from controls was possible because of a fair degree of overlap; moreover, there was no correlation between the increased levels of the platelet product and the severity of the atherosclerotic lesions. Treadmill tests performed by one investigator caused a further increase in the level of β-thromboglobulin but not in that of platelet factor 4 (51); this difference may be due to the brief half-life of the latter.

When heparin is injected as a bolus in man, there is an immediate release and a subsequent rapid clearance of platelet factor 4. By contrast, β-thromboglobulin and thrombospondin are largely unresponsive to heparin injection. The level of platelet factor 4 increases much more after heparin in patients with extensive atherosclerosis, suggesting that this test is a marker of the interaction of platelets with a seriously damaged blood vessel wall (154). The platelet factor 4 released by heparin originates mainly from the atherosclerotic region(s) where the endothelium, enriched with glycosaminoglycans, stores it.

The plasma levels of β-thromboglobulin are also higher in patients with hyperlipidemia (IIa, IIb, IV, and V), malignant disease, or renal insufficiency (284). By contrast, the clearance of platelet factor 4 and thrombospondin is not influenced by renal function.

Platelet Aggregates

Platelet aggregates are found in the blood of patients with peripheral arterial insufficiency (25,663,962,1169). However, it is still uncertain whether these platelet aggregates circulate *in vivo* or are formed *in vitro* (766,813).

Thromboxane A_2

Platelet-rich plasma of patients with peripheral arterial insufficiency, after stimulation with thrombin, contains more thromboxane B_2 (the stable inactive metabolite of thromboxane A_2) than normal (962). In the intact organism the balance between the pro-aggregating thromboxane and the antithrombotic prostacyclin determines the actual amount of platelet aggregation and thus the potential to evoke arterial thrombosis (744). When the ratio of thromboxane to prostacyclin is constant, arterial thrombosis occurs more readily when the platelets can produce more thromboxane A_2 (493). This suggests that lowering the production of thromboxane A_2 by the platelets would reduce the tendency towards arterial thrombosis, even if the formation of prostacyclin is lowered to some extent.

Serotonin

In healthy volunteers of increasing age, there is a decrease in the platelet content of serotonin, but an increase in plasma levels of 5-hydroxyindoles; these two changes are correlated inversely (199,968). These findings could be of particular relevance in aging patients with atherosclerosis. Indeed, in rabbits (473) and monkeys (469) on a high-cholesterol diet, the limb arteries become markedly supersensitive to the constrictor effects of serotonin, possibly as a result of an altered lipid environment of the serotonergic receptors of the arterial cell membrane.

An increased blood level of serotonin in man may not cause a major direct activation of the platelets, which are notoriously less sensitive to serotonin than those of the cat, pig, and sheep (732). However, the amplifying effect of serotonin may be important also in man (63,232).

Platelet Survival

The measurement of platelet survival time probably is one of the better tests to detect platelet consumption *in vivo*, although the methodology is beset with technical problems (in particular, the mathematical model used for calculation of the data) (228).

Several investigators have reported a shortened platelet survival time in patients with peripheral atherosclerosis (1,454,762,962), although these findings have not been confirmed subsequently (106,609,773,1108). One of the reasons for this discrepancy may be that platelets, after activation in the process of thrombosis or atherosclerosis, can recirculate and survive with a normal lifespan (886). Hence, increased levels of platelet products in the circulation may not be associated necessarily with an increased utilization or destruction of platelets.

CAVEAT

There are certain limitations in relating platelet-function tests to symptoms of atherosclerosis, as a small atherosclerotic lesion (e.g., in the cerebral or coronary arteries) may be very important clinically but may not activate enough platelets to cause measurable changes. Since activation of the platelets has not been consistently demonstrated in patients with extensive atherosclerosis of the arteries of the limb, it might be either a minor epiphenomenon or a transient feature. Alternatively, the markers used to assess platelet activation may not be sensitive enough to detect minute changes reliably. Finally, aggregation and coagulation *in vitro* do not necessarily reflect activation of the platelets *in vivo*.

18

Hemorheological Effects of Serotonin and Serotonergic Antagonists

John A. Dormandy

St. James' and St. George's Hospitals, London SW12, England

HEMORHEOLOGY

The abnormal rheological properties of blood are of interest as a possible pathological mechanism in many forms of circulatory disease. Of equal clinical importance is the possibility of hemorheologically oriented treatment of ischemia, even in cases where the primary cause is narrowing of the major vessels. If it may not be possible to increase the caliber of diseased arteries reliably, perhaps the flow through the same narrowed vessels may be augmented by improving the flow properties of the blood. The initial interest of clinicians was in the rheology of the macrocirculation, primarily in the viscosity of bulk blood. Instrumentation that allows accurate measurement of blood viscosity at specified single shear rates is available and permits the carrying out of hemorheological measurements on a large scale. Analysis of abnormalities in whole blood viscosity inevitably led to an increased interest in the factors determining this parameter. These are the physical properties of the cellular components, the viscosity of the plasma in which they are suspended, and, most importantly, the concentration of the red cells in the plasma (169).

Rheology of the Microcirculation

It was originally believed that in the microcirculation the flow properties of the blood are determined principally by the physical characteristics of the individual red cells. Because the diameter of the isolated red cell is so much greater than that of most capillaries, it was thought that the deformability of the red cells was crucial. However, while there is no doubt that impaired red cell deformability markedly increases the viscosity of whole blood, particularly at high shear rates, the pathological role of abnormal red cell deformability in the microcirculation is still uncertain.

Impaired cellular deformability probably does not decrease total flow in a capillary bed, but rather may lead to a maldistribution of flow, favoring the larger through channels. This would lead to a functional underperfusion. The distribution of red cells in the microcirculation is so variable that the term hematocrit, particularly if measured in venous blood, becomes meaningless. The average proportion of plasma in capillary blood is greater than in the major vessels and its viscosity is therefore more important. Moreover, the layer of plasma proteins adsorbed onto the red cells and the endothelial surface may play a role as a lubricant. Indeed, the flow properties of blood are critically dependent on the interactions between individual red cells and between the red cell and the endothelium. These are probably largely dependent on the characterization of the plasma proteins adsorbed onto these surfaces, rather than the properties of the free plasma (192). It is likely that rheological factors become important in the microcirculation only when the perfusion pressure falls. It is under these pathological circumstances that the microrheology of blood may become a limiting factor (947).

Methodology

The techniques for measuring the viscosity of whole blood at defined shear rates can be standardized easily and the results expressed in absolute units (278). By contrast, the assessment of the flow properties of the cellular components of blood is still difficult. So far the emphasis has been on the red cell, because of its numerical superiority; the technique most commonly used in clinical investigations is some variation of a filtration method. Usually filters containing 5-μm diameter pores are used (Fig. 18–1). The details of the individual techniques vary considerably (Fig. 18–2) (279). The current consensus seems to be that if a filtration technique is used to assess the physical properties of the red cell, it is important to eliminate all other cells from the suspension to be filtered. An alternative approach is to use a technique that gives a reading after only a few cells have traversed each pore. The filtration pressures should produce shear stresses comparable to those in the microcirculation. The suspending medium is equally important, as under some conditions the apparent decrease in red cell filterability is secondary to the presence of a pathological plasma factor, as it is not observed when the same red cells are resuspended in a buffer or in a normal plasma (279). Obviously, more attention should be directed at the white cells, although they are

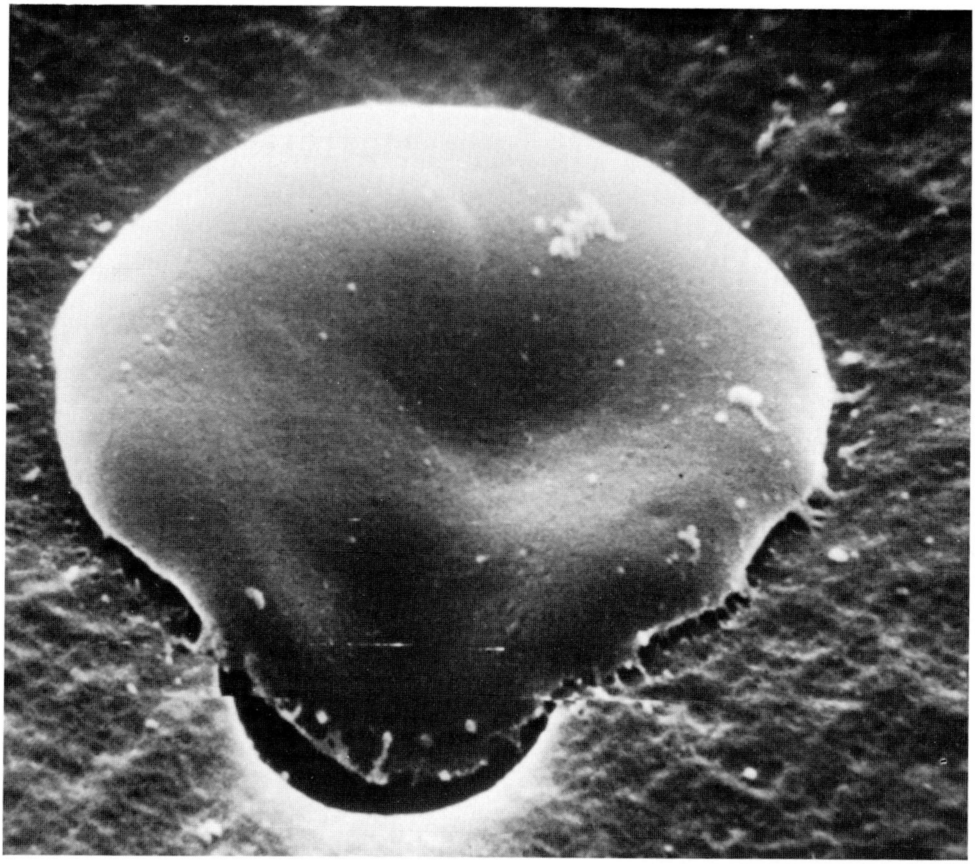

FIG. 18–1. Electron photomicrograph of a red cell entering a 5-μm pore in a typical filter, used in most techniques for assessing red cell deformability.

very much fewer in number than the red cells. Indeed, *in vivo* observations of the microcirculation often suggest that interruption of flow in a capillary is caused by a white cell temporarily stuck at a point of narrowing.

Abnormal Hemorheology

Clinical states characterized by abnormal hemorheology can be classified on a partly etiological basis (Fig. 18–3). The first four groups consist of relatively rare conditions in which specific abnormality in the plasma, the cellular components of blood, or the concentration of the red cells accounts for the hemorheological abnormality. Rheological defects in the red cells can be ascribed to an isolated abnormality in the shape of the cell, its membrane, or the composition of its

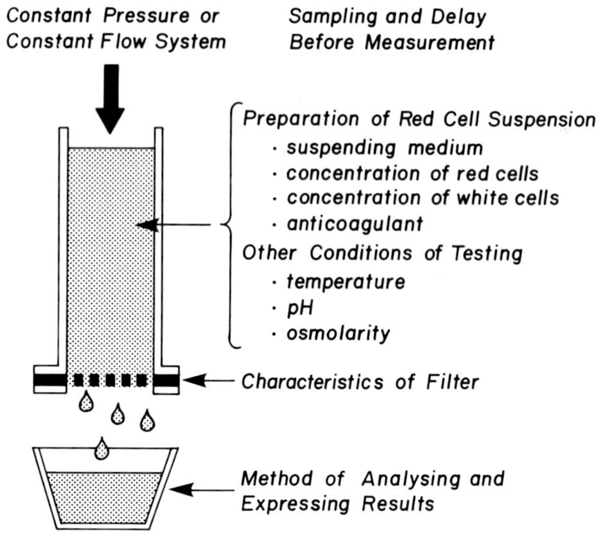

FIG. 18–2. The many variables in filtration techniques for assessing the rheology of the cellular components of blood.

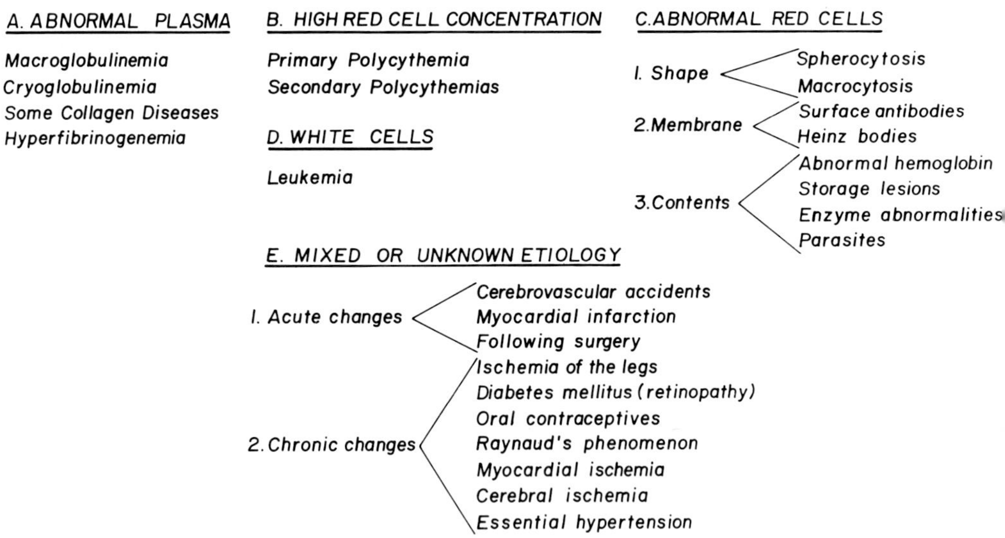

FIG. 18–3. A classification of clinical states associated with hemorheological abnormalities.

content. Many of these conditions are recognized diseases. The most frequent and interesting clinical conditions associated with hemorheological abnormalities fall into the group in which an abnormality can only be found in a proportion of cases and is not due to a single isolated cause (group E, Fig. 18–3;169,282). This group contains most of the common circulatory diseases. Hemorheological abnormalities (usually either raised whole blood viscosity or decreased blood filterability) have been described in almost all ischemic diseases. For instance, there have been at least 10 clinical studies measuring red cell filterability in patients with ischemia of the legs, totalling about 500 patients. In all except one, there was a 15 to 30% decrease in blood filterability compared with controls (279). In the three longitudinal studies where patients were followed for some time after measurement of their blood filterability, the degree of abnormality could be related to the subsequent needs of the patients for surgical intervention (ultimately amputation). While in normal subjects there is no significant difference between the filterability of arterial and venous blood, blood draining the ischemic area in patients is significantly less filterable and the arteriovenous difference is accentuated by a period of exercise. In the case of the heart, acute ischemia is characterized by a marked decrease in blood filterability immediately after the infarction; in most cases blood filterability returned toward normal in the subsequent 2 to 4 days. However, the severity of the initial abnormality, where occasionally blood filterability fell to below 50% of normal, can be closely correlated to the subsequent clinical progress of the patient. Those who developed pulmonary edema or cardiogenic shock had significantly lower blood filterability on admission to the hospital than patients who did not develop these complications (281). Cross-suspension experiments suggest that the decrease in blood filterability is largely due to a plasma factor, which has not been identified (280). When measurements of blood filterability are carried out in patients with chronic myocardial ischemia, the onset of angina is associated with a marked, reversible decrease in red cell filterability (345). In subjects with "noncoronary" angina (patients with all the symptoms and signs of myocardial ischemia including abnormal EKG changes, but in whom coronary angiograms are normal), blood filterability is decreased (537). Abnormalities in the rheological response to low temperature occur in the blood of patients with Raynaud's phenomenon. Plasma exchange in these patients results in a parallel improvement in clinical findings and in blood filterability (273). In young patients with essential hypertension the increased peripheral circulatory resistance is usually assumed to be due to vascular narrowing, although it could equally well be due to an increase in hemorheological resistance. A normal heart presented with hyperviscous blood could only maintain a normal cardiac output by increasing the perfusion pressure (168,277). It is a common error to equate peripheral circulatory resistance with vascular resistance alone, omitting the hemorheological component.

Pathophysiology

The flow properties of blood are determined by a number of factors that interact in a complex manner (Fig. 18–4). For instance, the fibrinogen concentration not only largely determines the viscosity of the plasma, but also affects whole blood viscosity at low shear rates by inducing red cell aggregability. It also plays a role in platelet adhesiveness; in many diseases the fibrinogen concentration correlates with platelet aggregation (663). While augmented red cell aggregability is thought

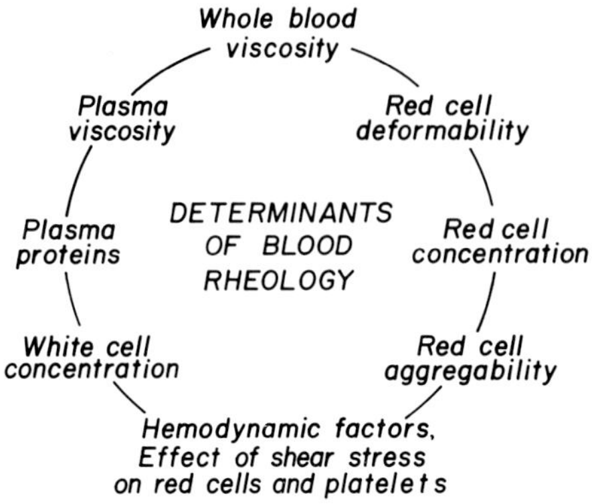

FIG. 18–4. Factors interacting with each other and determining the rheological properties of blood.

to be the principal reason for the increased viscosity of whole blood at low shear or flow rates, at high shear rates the deformability of the red cells is of paramount importance (167). Not only does the shear stress alter the viscosity of whole blood, but it also both influences and is influenced by the concentration of the red cells. The effect of shear forces on individual red cells and platelets may also trigger biochemical reactions (1052). Factors released by platelets or white cells may in turn influence the filterability of whole blood (430,612,827).

A vicious cycle may exist between ischemic changes in the tissues and deterioration in the rheological properties of blood in these areas in need of oxygen. Local changes, such as hypoxia, hyperosmolarity, acidosis, and accumulation of metabolites, impair blood filterability and increase blood viscosity. As a consequence, ischemia decreases the flow properties of the blood locally, which in turn further impairs the circulation, increasing the severity and extent of the ischemia. This hypothesis would explain the prognostic significance of the early hemorheological changes following acute tissue infarction. The hemorheological changes would thus be both a consequence and a cause of the ischemia.

SEROTONIN AND HEMORHEOLOGICAL ABNORMALITIES

There is little information on the direct effect(s) of serotonin on the rheology of blood, except one *in vitro* study showing that serotonin (10^{-9} M) decreases red cell filterability when incubated with blood for 22 hr, but not at 4, 6, or 8 hr (247). The monoamine probably plays no role in determining the normal hemorheology. However, it may become important under pathological situations of ischemia, when the local concentration of plasma serotonin markedly increases as a result of a platelet aggregation (322). Most information on the potential role of serotonin is inferred from studies of the effect of serotonergic antagonists on various hemorheological determinants. A number of possibilities exist for serotonin to affect the flow properties of the blood (Fig. 18–5).

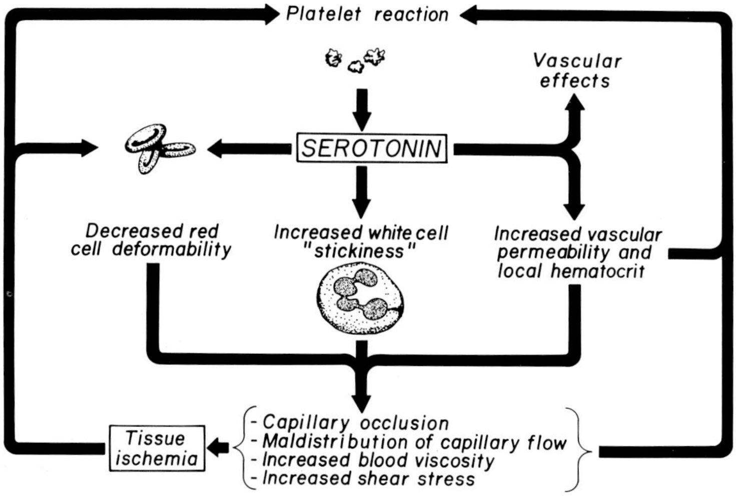

FIG. 18–5. Possible pathways linking serotonin, tissue ischemia, and changes in the rheological properties of the blood.

Does Serotonin Play a Role in the Vicious Cycle of Ischemia?

Serotonin may play a central role in the vicious cycle of local ischemia, resulting in platelet aggregation and release of various substances which favor the lack of oxygen. Tissue ischemia is undoubtedly associated with changes in the flow properties of blood, which in turn may contribute to its perpetuation (345). Hellem (471) suggested that abnormally rigid red cells under high shearing forces may release ADP and initiate platelet reaction. The effect of changes in red cell concentration on platelet adhesion and thrombus formation is well established (1052). It resembles the amplification by serotonin of the contraction of vascular smooth muscle and the aggregation of platelets induced by other agents. The possibility exists that these two amplifying effects are linked. Humoral mediators locally released from either tissues or blood have critical effects on vascular smooth muscle; similarly, the same substances may have important effects on the flow properties of the blood within the vessels. Far less is known about the latter because they are difficult to study, even *in vitro*. For instance, nothing is known of the effect of β-thromboglobulin, thromboxane A_2, or platelet factor 4 on hemorheology.

Is Serotonin the Plasma Factor Affecting Blood Filterability?

The impaired filterability found in many ischemic situations (e.g., recent myocardial infarction) is secondary to the presence of a plasma factor affecting the cellular components of blood. The plasma mediator concerned seems to be released within a few hours (280). Thus it cannot be a factor such as fibrinogen, where the time scale of the changes is measured in days. The available information is compatible with a role for serotonin as the possible mediator of the impaired filterability in ischemic blood. This hypothesis is supported by the *in vitro* studies of De Clerck and Xhonneux (241), who showed that decreased blood filterability could result from a platelet-derived factor, the action of which was blocked by a serotonergic antagonist.

Can Serotonin Amplify Hemorheological Changes in Abnormal Blood?

In normal humans, the direct effect of serotonin on platelets is weak; its influence is indirect, for instance by amplifying the effect of catecholamines (233). Catecholamines also impair the rheological properties of the blood, and serotonin could amplify this effect. A similar amplifying action of serotonin may contribute to the interaction between platelets and the flow behavior of whole blood. For instance, the rate of formation of a platelet thrombus on a subendothelial surface depends on both the shear rate and the hematocrit of the blood (1052). The release of ADP from red cells under increased shear stress may have a role in initiating the platelet reaction. A decreased red cell deformability would predispose to this sequence and the effect may be amplified by serotonin.

Does Serotonin Influence the Calcium Fluxes Across Red Cell Membranes?

The biochemical basis of a possible serotonergic effect on red cell flexibility may be through an increase in calcium influx. Serotonin increases the entry of calcium into endothelial cells (209) as well as into platelets (857). An increase in intracellular Ca^{2+} augments the rigidity of the red blood cells and is probably the pathway through which ATP depletion affects red cell flexibility (167). Certain agents that inhibit calcium influx into red cells may improve microcirculation by reducing blood viscosity (1100).

Does Hyperviscosity Link Serotonin to Hypertension?

Serotonin may play a role in the maintenance of hypertension (185,261; see Chapter 12). Two possible mechanisms could explain the increased peripheral vascular resistance in these patients: an increased contractility of the vascular smooth muscle and an augmented viscosity of the blood. Hyperviscosity (950) and decreased blood filterability (249) occur in essential hypertension. Theoretically, they could be the consequences of an increased responsiveness of the red blood cells to serotonin.

HEMORHEOLOGICAL ACTIONS OF KETANSERIN

Most of the evidence suggesting a possible hemorheological effect of serotonin comes from the use of serotonergic antagonists, in particular ketanserin.

Blood Filtration

De Cree and colleagues (248) first showed an improvement (averaging 14%) in blood filterability in 60 patients following myocardial infarction in a double-blind trial comparing a single oral does of ketanserin (60 mg) to placebo. If the red cell filterability was measured in platelet-rich plasma, ketanserin produced a 30% improvement; there was no change in the placebo group. Using platelet-poor plasma for suspending the red cells, the improvement with ketanserin was only 10% (251). Cross-exchange experiments in the postinfarction period confirmed that the changes in red cell filterability were due mainly to the presence of a plasma factor. Ketanserin (20 mg increasing to 40 mg t.d.s. orally) was given for 3 months and compared to placebo in patients with intermittent claudication. This resulted in a significant increase in blood filterability (246). The mean improvement was approximately 10% irrespective of whether the blood was resuspended in platelet-rich or platelet-poor plasma. The absolute filtration rate, both before and after treatment, was worse in platelet-rich plasma, possibly because these suspensions also contained more white cells. This study, for the first time, reports a hemorheological improvement in a group of patients who have improved clinically. Unfortunately, however, there was no analysis of the correlation between the two on a patient-to-patient basis. A single intravenous dose of ketanserin (10 mg) caused an 18% improvement in blood filterability in patients with essential hypertension, which was accompanied by a normalization of both systolic and diastolic blood pressures (249).

Red Blood Cell Deformability

In the previous studies, the technique used did not distinguish between an effect on the true deformability of red cells (transit time through the pore) and the effect on any particles in the suspension which were progressively blocking an increasing number of pores (clogging effect). A decrease in filtration rate may be due to either or both mechanisms. White cells, for instance, even in very low concentrations, will decrease filtration rate by plugging an increasing proportion of the available pores. Filtration techniques are available that can distinguish between and measure separately the rate of pore clogging and the transit time of cells (mostly red cells) through the pores. This technique is based on measuring the change in the rate of flow at the very start of filtration. The decrease in filtration rate against volume filtered is a measure of the clogging rate, and by extrapolating back to zero time the filtration rate, or transit time of cells in the absence of clogging can be determined. One week of intravenous treatment with ketanserin decreased the pore clogging by 53% (Fig. 18–6). It reduced the viscosity of whole blood at all shear rates, but particularly at low shear rates. The quotient hemoglobin over viscosity takes into account the hemodilution effect and is a better way to present viscosity data, as it assesses the effect of viscosity on oxygen delivery. When the data are expressed in that way, ketanserin causes a significant improvement (17%) at a low shear rate. In these studies the proportion of white cells trapped in the red cell column after centrifugation was significantly decreased by ketanserin, which may be the explanation of the decrease in pore clogging. Serotonin and ketanserin may play a role in altering the "stickiness" of white cells.

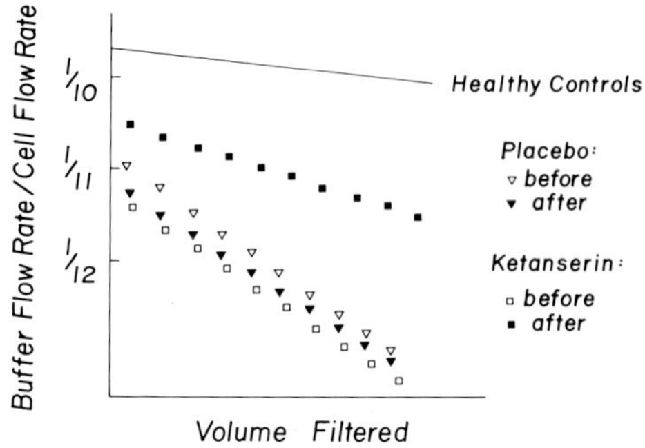

FIG. 18–6. Principle of a new system for distinguishing between cell transit time and filter clogging during blood filtration. The filtration rate is plotted against volume filtered. Pore clogging will be reflected by a linear decrease in filtration rate, while the initial filtration rate (intercept at zero time) will reflect the cell transit time. The blood of patients with leg ischemia has a slower transit time and higher clogging rate than blood of healthy controls. Ketanserin tends to normalize both these changes.

Blood Viscosity

The viscosity of whole blood at high shear rates is also significantly reduced by ketanserin (658). The effect of ketanserin on whole blood viscosity may be particularly relevant in considering its beneficial effect in the treatment of hypertension. A possible hemorheological mechanism underlying the hypotensive action of the drug is suggested by results obtained in old hypertensive dogs (233). The fall in blood pressure caused by ketanserin correlates with the parallel decrease in the viscosity of whole blood. Although in most studies only blood filterability was measured, there is little doubt that this is related to whole blood viscosity. An improvement in blood filterability or red cell deformability would almost certainly cause a reduction in the viscosity of whole blood (315).

19

Serotonin and Raynaud's Phenomenon

James R. Seibold

Division of Rheumatology, University of Medicine and Dentistry of New Jersey—Rutgers Medical School, New Brunswick, New Jersey 08903

Maurice Raynaud (876) first described the clinical syndrome of episodic color changes of the digits in response to cold and, in some patients, to emotional stress. This disorder ranges from a trivial, virtually asymptomatic finding to the pain and disability of digital gangrene (104,876). Raynaud's phenomenon is a prominent clinical feature in diverse conditions (Table 19–1). However, the pathophysiologic mechanism of cold-induced vasospasm in Raynaud's phenomenon is poorly understood. Serotonin released from aggregating platelets may be one of the mediators involved in certain forms of Raynaud's phenomenon (445,944,1159), but foolproof evidence in support of this involvement is lacking. Improved understanding of the function of platelets and the way they affect the blood vessel wall, as well as the recent availability of more selective serotonergic antagonists (637,1093), has allowed a reevaluation of the role of serotonin in peripheral vasospastic events. This chapter focuses on Raynaud's phenomenon seen in the primary form of the disease and that complicating systemic sclerosis. These contrasting forms of the syndrome are paradigms of the role of functional and structural factors, respectively, in the pathogenesis of Raynaud's phenomenon.

PRIMARY RAYNAUD'S PHENOMENON AND SCLERODERMA

Epidemiology

Raynaud's phenomenon is a typical complaint in systemic sclerosis (scleroderma) and a prominent clinical manifestation of systemic lupus erythematosus, inflam-

TABLE 19–1. *Disorders associated with Raynaud's phenomenon*

Vasospastic
Primary (idiopathic) Raynaud's phenomenon
Drug-induced
β-Adrenergic blockers
Ergot
Methysergide
Pheochromocytoma
Variant angina
Migraine
Structural
Vibration syndrome
Arteriosclerosis
Thromboangitis obliterans
Cold injury (frostbite, pernio, immersion foot)
Neurovascular compression (thoracic outlet syndrome, carpal tunnel syndrome, crutch pressure)
Chemotherapy (bleomycin)
Polyvinyl chloride disease
Diffuse connective tissue disease
Systemic sclerosis
Systemic lupus erythematosus
Overlap syndrome
Poly/dermatomyositis
Rheumatoid arthritis
Hemorheologic
Cryoglobulinemia
Cryofibrinogenemia
Cold agglutinin disease
Paraproteinemia (plasma cell dyscrasia)
Polycythemia (essential thrombocythemia, polycythemia vera)

matory myositis, and the overlap syndromes (104,629). Hence, it is an important index characteristic for epidemiologic study of these connective tissue diseases (995). While the exact incidence of Raynaud's phenomenon is not known, population surveys suggest prevalences of 5 to 10% of the nonsmoking population (1007) and up to 22% in premenopausal women (805). These figures contrast with the low incidences of systemic sclerosis (4 to 12 per million) and systemic lupus erythematosus (9 to 27 per hundred thousand) (995). Estimates of the tendency of Raynaud's phenomenon to evolve into connective tissue disease are biased by the selection of cases of relative clinical severity, which are usually secondary forms of the syndrome (629).

Clinical Features

Primary Raynaud's Phenomenon

Primary Raynaud's phenomenon (idiopathic Raynaud's phenomenon, Raynaud's disease) is a functional vasospastic disorder afflicting predominantly women. The criteria for diagnosis include the clinical features proposed by Allen and Brown (16;Table 19–2). The absence of serum antinuclear and anticentromere antibodies

TABLE 19–2. *Criteria for diagnosis of primary Raynaud's phenomenon*

1. At least two of three characteristic episodic color changes in response to *cold*
2. At least 2-year history
3. Bilaterally
4. Normal radial pulsations
5. Absence of clinical evidence of causal disease
6. Absence of digital gangrene
7. Absence of serum antinuclear antibody
8. Absence of serum anticentromere antibody
9. Normal nailfold capillaroscopy
10. Physiologic evidence of normal maximal digital arterial flow
11. No evidence of *in vivo* platelet activation

(554) and normal nailfold capillaroscopy (689) exclude subtle or early cases of systemic lupus erythematosus and systemic sclerosis. Individuals with primary Raynaud's phenomenon tend to have a high frequency of daily attacks (typically greater than 10 per day) precipitated by cold and emotion (354). The symptoms of individual episodes are mild, digital ischemic sequelae are rare, and the impact on occupation and lifestyle is relatively minor.

Systemic Sclerosis

Systemic sclerosis is characterized by thickening and fibrosis of the skin (scleroderma) and by distinctive forms of visceral involvement, including the gastrointestinal tract, heart, lungs, and kidneys (629). The etiology of systemic sclerosis is unknown and effective treatment is lacking. A vascular pathogenesis is possible (143,629). Raynaud's phenomenon is the initial manifestation in 70% of the patients, and eventually is present in nearly 95% of them. The digital arteries of patients with systemic sclerosis exhibit severe intimal hyperplasia consisting predominantly of collagen and to a lesser degree of ground substance (907). Medial changes are inconspicuous but adventitial fibrosis is seen in 40% of cases. Severe narrowing (more than 75%) of the arterial lumen results, which may be of sufficient severity to account for Raynaud's phenomenon (907). Normal vasoconstrictor responses to cold or to emotional stimuli, superimposed on the anatomical obstruction, could cause complete or near complete occlusion of the lumen.

Similar histopathologic changes are seen in the small arteries and arterioles of affected organs (790). Although variable degrees of tissue fibrosis and inflammation are seen, the fibrotic arteriosclerosis is the hallmark of systemic sclerosis (790). Clinical syndromes such as the renal crisis and pulmonary hypertension are due principally to the vascular lesions (629). A Raynaud-like response to cold of the visceral vasculature may explain the increased incidence of renal crisis in cold weather (1049). Radionuclide scanning techniques reveal that perfusion of the myocardium (344), lungs (375), and kidneys (592) decreases during cold pressor tests. Cold challenge induces echocardiographic regional wall motion abnormalities in patients with systemic sclerosis and does not lead to the expected increase in pulmonary diffusing capacity (1163). No vasospastic responses of the intestinal vasculature have been demonstrated.

Distinctive changes at the microcirculatory level are prominent (629). The nail-fold capillaries exhibit varying degrees of capillary loop enlargement interspersed with areas of loss of capillaries. Capillary proliferation is noted clinically as cutaneous telangiectasias. Prospective studies in patients with Raynaud's phenomenon suggest that the capillary abnormalities are an early finding in systemic sclerosis (689) and that the degree of capillary derangement correlates with the extent of internal organ involvement (688).

PATHOPHYSIOLOGY

Adrenergic Mechanisms

Primary Raynaud's Phenomenon

There is substantial evidence that heightened sympathetic tone is not of primary etiologic importance in Raynaud's phenomenon (583,1007). Although sympathetic tone is not increased in patients with primary Raynaud's phenomenon in comparison to normals (583), this does not preclude a contribution of the adrenergic nervous system to the provocation of individual vasospastic episodes. About one-third of attacks in patients with primary Raynaud's phenomenon are associated with tachycardia and increased stress ratings (354). The beneficial therapeutic effects of biofeedback, sympatholytics, Pavlovian conditioning, and placebo all support the concept that increases in sympathetic activity participate in the occurrence of cold-induced digital vasospasm. For example, local cooling of the hand followed by induction of reflex vasoconstriction by general body cooling reliably evokes an augmentation of peripheral vasospasm in patients with primary Raynaud's phenomenon (522). These observations would be explained if the postjunctional α-adrenoceptors of the digital arteries were more sensitive to catecholamines at lower temperatures (1080).

Systemic Sclerosis

A role of the sympathetic nervous system in Raynaud's phenomenon of patients with systemic sclerosis is even more dubious. The responsiveness to catecholamines of arterial smooth muscle obtained from such patients is reduced at lower temperatures (1160). In patients with Raynaud's phenomenon complicating systemic sclerosis, peripheral vasoconstrictor responses to reflex increases in sympathetic activity (e.g., placing an ice pack on the neck, sudden sharp noises) are either normal or reduced (522) and provocation of attacks by emotion is uncommon (354). The electrical resistance of the skin is either normal or slightly decreased in the fingertips of patients with systemic sclerosis, indicating reduced sympathetic activity (361). Detailed histopathologic studies demonstrate the destruction or even the absence of peripheral sympathetic ganglia (829). Controlled studies with reserpine show a lack of benefit in patients with systemic sclerosis (972). Surgical sympathectomy is relatively ineffective in patients with Raynaud's phenomenon complicating systemic sclerosis (1007).

Serotonergic Mechanisms

The infusion of serotonin into the human brachial artery causes a rapid decrease in digital temperature as well as the characteristic sequential color changes of Raynaud's phenomenon (445). The nonspecific serotoninergic antagonist 1-methyl-D-lysergic acid butanolamide modifies the cold-induced cyanosis in patients with Raynaud's phenomenon (445). Subcutaneous injection of serotonin produces prolonged vasoconstriction in the hands of individuals with systemic sclerosis (944). Vascular smooth muscle isolated from the subcutaneous blood vessels of patients with systemic sclerosis exhibits an enhanced responsiveness to serotonin (1159). These observations suggest that antagonizing the vasoconstrictor effect of serotonin might be beneficial to patients with Raynaud's phenomenon. Ketanserin in a 4-week-long open pilot trial of patients with Raynaud's phenomenon caused moderate or marked relief in 83% of patients with systemic sclerosis but in only 33% of patients with Raynaud's phenomenon of other etiology (957). These subjective ratings are supported by the results of digital blood flow measurements during controlled cold challenges (Fig. 19–1;957). Therapy with ketanserin, whether given orally or intravenously, improved digital artery blood flow at 38°C in all forms of

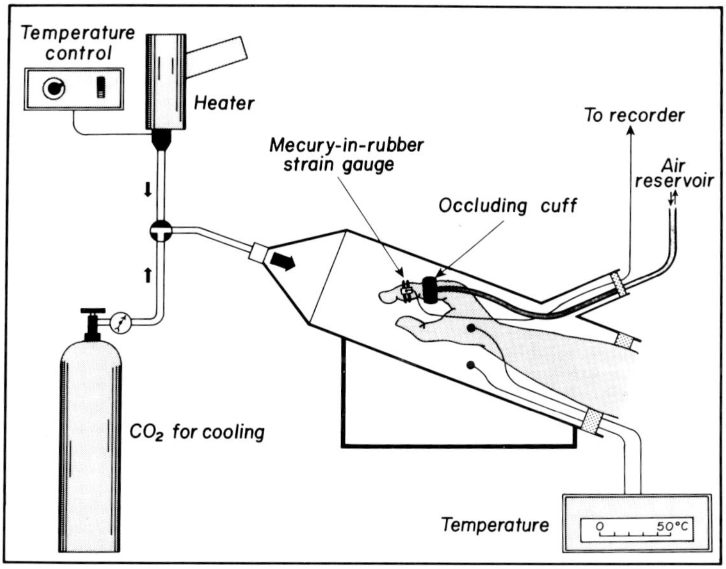

FIG. 19–1. Plethysmographic apparatus (mercury-in silastic strain gauge) employed in measurement of digital blood flow at various temperatures (957).

Raynaud's phenomenon. However, during subsequent controlled cold challenge, ketanserin prevented the abnormal vasoconstriction only in patients with systemic sclerosis (Fig. 19–2;957). To the extent that the effect of ketanserin is due solely to its inhibitory action on S_2-serotonergic receptors, the improved digital flow obtained at all temperatures may imply that the peripheral vasculature of patients with systemic sclerosis is continually exposed to serotonin. Intravenous therapy with ketanserin also improves microvascular flow in patients with systemic sclerosis (531).

The relative lack of efficacy of ketanserin in patients with primary Raynaud's phenomenon is not a uniform finding (1006). A double-blind study, comparing ketanserin with placebo in women with primary Raynaud's phenomenon, shows that if the serotonergic antagonist is given intravenously prior to the cold challenge, it does not improve the ability to maintain digital blood flow (Fig. 19–3; 959).

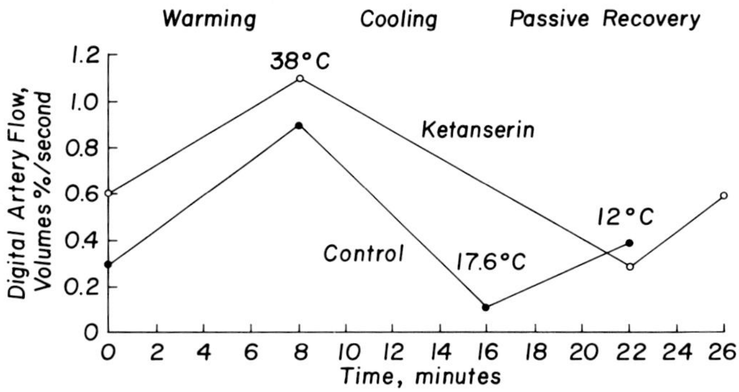

FIG. 19–2. Digital artery flow monitored by strain gauge plethysmography in a patient with Raynaud's phenomenon complicating systemic sclerosis. The intravenous administration of ketanserin (10 mg) improves digital perfusion at all temperatures as well as the ability to withstand cold challenge. (Data from ref. 957, with permission.)

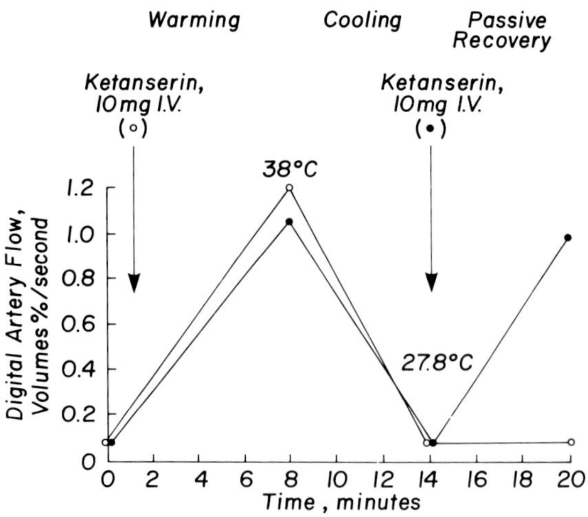

FIG. 19–3. The effect of time of administration of intravenous ketanserin in a patient with primary Raynaud's phenomenon. Ketanserin relieves but does not prevent cold-induced vasoconstriction (959).

Likewise during cold challenge, the drug does not alter the finger temperature associated with loss of digital blood flow (959). However, ketanserin administered at the time of cold-induced vasoconstriction promptly improves digital arterial flow (Fig. 19–3). Again, if the action of ketanserin is due to inhibition of serotonergic receptors, these findings suggest that serotonin is a factor in the maintenance of cold-induced vasoconstriction in primary Raynaud's phenomenon. However, other factors, possibly adrenergic stimulation, are involved in the initiation of the vasoconstrictor response.

Platelet Turnover and Raynaud's Phenomenon

Aggregating platelets release serotonin in amounts sufficient to provoke contraction of isolated blood vessels (see Chapters 9, 10, and 15;240,756,1078). Addition of platelets both 1 and 20 min following induced aggregation causes brisk contraction of isolated human digital arteries (756). Treatment of the preparations with ketanserin attenuates both contractile responses. In arteries taken from animals, moderate cooling augments the contractions evoked by either exogenous serotonin or aggregating platelets (649).

Systemic Sclerosis

In patients with systemic sclerosis, the evidence suggesting that platelets are activated includes elevated plasma levels of β-thromboglobulin (506,550,955) and increased numbers of circulating platelet aggregates (550). Spontaneous platelet aggregation occurs *in vitro* in plasma from these patients (555,898). The aggregatory responses to ADP, epinephrine and collagen are enhanced (730,1175). Platelet aggregation induced by exposure to serotonin is increased in patients with early (less than 3 years duration) systemic sclerosis, although the initial slope of the aggregatory response to the monoamine is comparable to that obtained with platelets of normal subjects (356). In platelets of patients with early systemic sclerosis, the amplifying effect of serotonin on the aggregation induced by ADP, epinephrine, and collagen is also enhanced *(unpublished observations)*. The factor(s) responsible for the abnormal platelet activation in systemic sclerosis include (a) circulating immune complexes (717,958) and (b) interaction with damaged endothelium (549) or exposed subendothelial collagen (717). The increased platelet activation is present at the earliest recognizable stages of disease (956). An abnormally high degree of platelet aggregation would provide the source of serotonin which would serve to enhance constrictor responses of the digital artery to cold.

Primary Raynaud's Phenomenon

In patients with primary Raynaud's phenomenon, levels of plasma β-thromboglobulin are comparable to those of normals (955). *In vitro* aggregatory responses to ADP, epinephrine, collagen, and serotonin are also normal (956). The failure to detect abnormal *in vivo* activation of platelets does not exclude the participation of platelet-derived vasoconstrictor substances, including serotonin, in the pathogenesis of cold-induced vasospasm. The number of α-adrenoceptors is increased in platelets of patients with primary Raynaud's phenomenon (563). The same may be true for the postjunctional α-adrenoceptors of the vascular smooth muscle of these patients. Thus the cold-induced vasospasm of primary Raynaud's phenomenon might be initiated by α-adrenergic stimulation, with subsequent superimposed platelet activation due either to altered platelet reactivity at colder temperatures or to physical interactions with the constricted blood vessels. Local release of thromboxane A_2 and serotonin, perhaps potentiated at lower temperatures (1091), ensues, leading to persistent vasospasm and further local platelet activation. This hypothesis would explain the effect of ketanserin in relieving vasoconstriction and would be compatible with the failure of this drug to prevent cold-induced vasospasm (959). Again, primary Raynaud's phenomenon may be different from systemic sclerosis in this regard, since in patients with the latter disorder, no further increase in local platelet activation is provoked by episodes of Raynaud's phenomenon (506).

SEROTONIN IN THE PATHOGENESIS OF SYSTEMIC SCLEROSIS

The combination in systemic sclerosis of diffuse structural arterial change (790), ongoing endothelial damage (549), and platelet activation (506,550,956) implies that continuous release of serotonin may occur in the peripheral circulation. This interpretation is supported by the improvement in digital blood flow (531,957) and digital microcirculation (531) in response to the selective S_2-serotonergic antagonist ketanserin. The other major pharmacologic action of ketanserin is to inhibit α_1-adrenoceptors (1078). This effect of ketanserin may not be important in patients with systemic sclerosis since this disorder is characterized by relative hyporesponsiveness to α_1-adrenergic stimulation (361,829,957).

Serotonin may be involved not only in the vasospastic phenomena of systemic sclerosis, but also in the pathogenesis of the dermal and vascular fibrosis of this disorder (957). Serotonin, in concentrations likely to occur at sites of platelet aggregation (10^{-7}–10^{-6} M), enhances proliferation of, and collagen synthesis by, cultured fibroblasts (118). Certain patients with carcinoid syndrome develop truncal skin changes (and telangiectasias) of histopathologic similarity to those observed in systemic sclerosis (362). Prolonged infusion of serotonin into the pulmonary and coronary arteries of rabbits and dogs results in the development of fibrotic arteriosclerotic lesions comparable to those of systemic sclerosis (1179). Clinical correlates in man are the fibrotic lesions of the right ventricle and pulmonary circulation in carcinoid syndrome. The subcutaneous administration of serotonin in immature rats induces edema of the skin followed by proliferation of fibroblasts, accumulation of collagen, and increased vascularization, a sequence similar to that observed in systemic sclerosis (670). Development of fibrosis in systemic sclerosis may be related also to tissue hypoxemia, to the production of a variety of lymphocyte- and monocyte-derived factors (629), as well as to the release of platelet-derived growth factors (151). Thus while it is tempting to speculate that serotonin plays a role in the pathogenesis and the expression of systemic sclerosis, many other factors are likely to be involved as well.

ACKNOWLEDGMENT

The author wishes to thank Ms. Camille Vanderzee for her secretarial assistance. Studies described in this chapter were supported in part by grants from the Scleroderma Society, Inc., New York, and the Scleroderma Research Foundation, Columbus, NJ.

20

Serotonin and the Carcinoid Syndrome

Håkan Ahlman

Department of Surgery I and Institute of Neurobiology, University of Göteborg, S-413 45 Göteborg, Sweden

SEROTONIN AND GASTROINTESTINAL PHYSIOLOGY

APUD Cells

Pearse (832) coined the term cells with Amine Precursor Uptake and Decarboxylation (APUD cells) to describe the handling of amine by endocrine cells widely distributed in the body, which are particularly numerous in the gastrointestinal tract. The endocrine cells of the gut originally were thought to emerge from the neural crest to migrate later into the gastrointestinal tract during embryogenesis. However, removal of the neural fold in 9-day-old rat embryos does not prevent the appearance of endocrine cells in the gut and the pancreas (928). Exclusion of the neural crest cells by combining quail entoderm with chick mesoderm also does not abolish endocrine cells in explants or chorioallantoic grafts (31). These observations indicate that APUD cells are derived from specialized ectoderm and epiblast, programmed for neuroendocrine function (833). They may constitute a third division of the nervous system (cf. 366). Strictly speaking, the ability to take up and decarboxylate amine precursors is not an appropriate description, since other protein-synthesizing cells, i.e., exocrine pancreatic cells, can do so. However, the capacity also to synthesize the amine precursor is specific for neurons or APUD cells producing peptide hormones (616).

Enterochromaffin Cells

The concept of APUD cells implies that biologically active peptides and amines are synthesized by the same endocrine cells. This is the case for the enterochromaffin cells, which are the major source of intestinal serotonin (319) and also of substance P in many species (860). Immunocytochemical studies indicate that serotonin and substance P are contained within the same population of enterochromaffin cells, perhaps even within the same secretory granules (22). This close physical relationship suggests that the two compounds share the same release mechanisms. Other peptides, motilin and enkephalin, coexist with serotonin in subpopulations of enterochromaffin cells (23,860). In addition, both serotonin and substance P are localized in certain enteric neurons (393,483;Fig. 20–1). In many species serotonin may also be stored together with histamine in mast cells of the gut mucosa (1158).

Because they contain serotonin, enterochromaffin cells and mast cells of the intestinal mucosa can be visualized by the formaldehyde-induced fluorescence method (Fig. 20–1). Specific antisera for the immunocytochemical detection of serotonin also are available (618,1001). The enterochromaffin cells contain secretory granules of varying electron density, size, and shape (Fig. 20–1). They form close contacts (neuroendocrine complexes) with nerve elements of both the afferent and efferent types adjacent to the basal lamina of the mucosa (811). True synapses also occur on both enterochromaffin cells and mucosal mast cells (cf. 8; Fig. 20–2).

Release of Serotonin

Serotonin is released from enterochromaffin cells by various stimuli, including luminal acidification, hypertonic sugar solution, cholera toxin, or extrinsic nerve stimulation (8,290,520,781). It is likely that the release of serotonin into the portal circulation is controlled by adrenoceptors, which possibly are activated by norepinephrine released from adrenergic nerves (8).

Most of the release of the amine takes place across the basolateral membrane of the enterochromaffin cell, by the emeiocytosis of electron-dense granules (578), which results in elevated levels of serotonin in portal blood. A substantial release of nongranular serotonin also occurs through the apical cytoplasmic processes into the intestinal lumen (Fig. 20–1). *In vivo*, substance P and serotonin are released together into the lumen upon vagal stimulation, suggesting again that they share a common release mechanism (8). The control of the release of serotonin into the lumen of the intestine seems to differ from that into the portal circulation, since the former is blocked by cholinergic antagonists but the latter by adrenergic inhibitors (428). The functional role of serotonin released endoluminally is poorly understood. Many of the effects (postprandial vasodilatation, secretion, and hypermotility) of intraluminally applied serotonin seem to be mediated by intrinsic reflexes activated from the mucosa (Fig. 20–3). Removal of the mucosa, local treatment with tetrodotoxin, local anesthesia, or ganglionic blockade abolish these

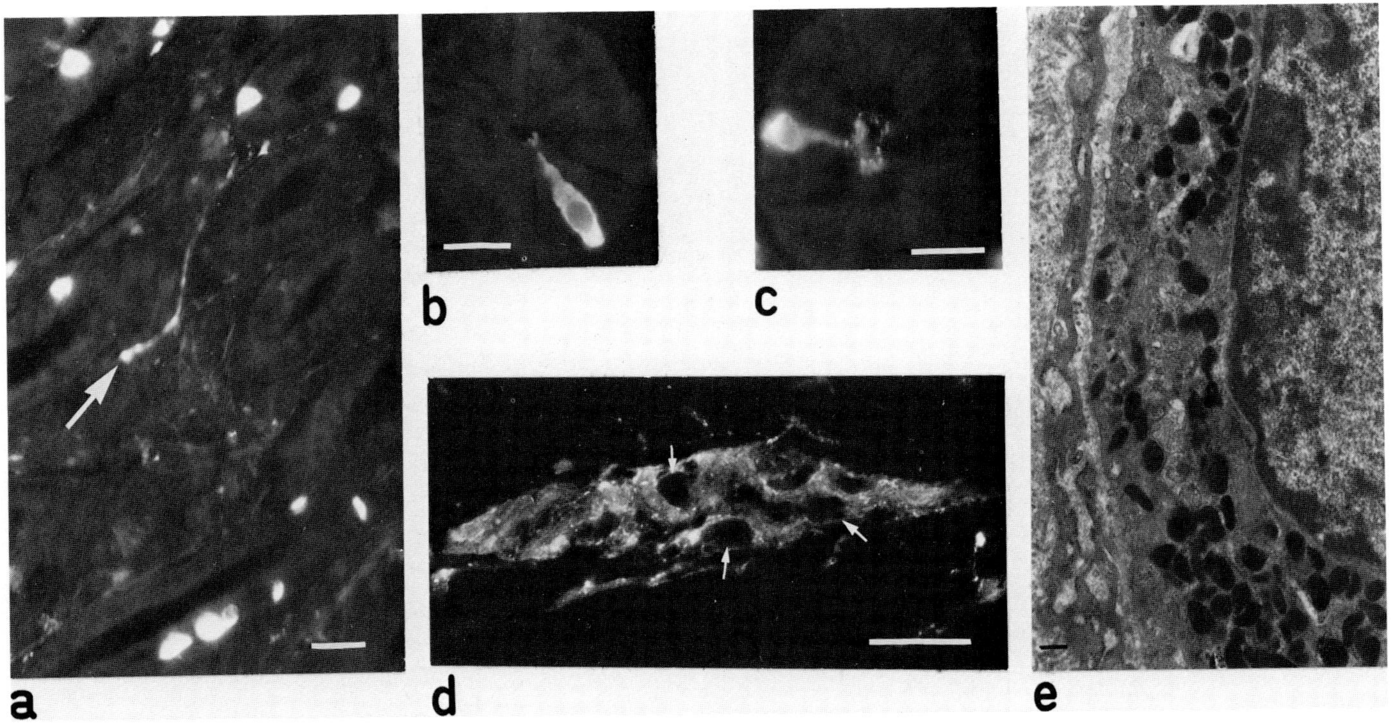

FIG. 20–1. **a:** Enterochromaffin cells visualized by formaldehyde-induced fluorescence. The *arrow* indicates a slender green fluorescent adrenergic nerve terminal running to intestinal crypts. In epithelium of the crypt numerous yellow fluorescent serotonin-containing enterochromaffin cells are seen. With this reaction yellow fluorescence is located to granules located basally. **b** and **c:** Enterochromaffin cells visualized by immunocytochemistry. Serotonin-positive immunoreactivity located to basal granules as well as to the apical portion of the cell. In **(c)** apical release of serotonin-like immunoreactivity into the lumen of the crypt is shown. **d:** Serotonin-immunoreactive nerve terminals of the myenteric plexus surrounding nonfluorescent ganglion cells *(arrows)*. **e:** Ultrastructural detail of an enterochromaffin cell with numerous electron-dense secretory granules of different sizes and shapes. *White bars*, 25 μm; *black bar*, 200 nm.

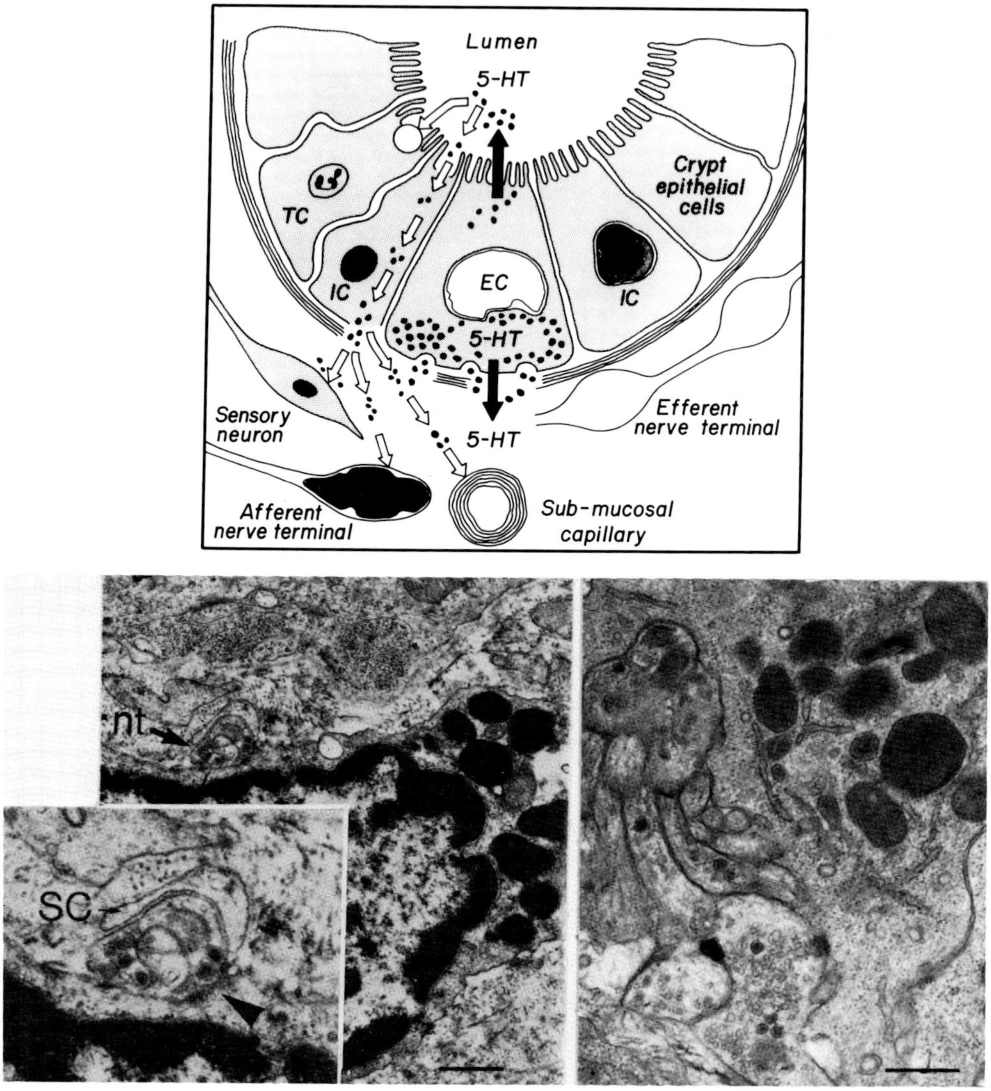

FIG. 20–2. Top: Close relationship (neuroendocrine complex) between nerve terminals, sensory neurons, and enterochromaffin cells (EC). The epithelium also contains intestinocytes (IC) and hypothetical taste cells (TC) (8). *Large arrows* indicate the two types of release of serotonin from the enterochromaffin cells: emeiocytosis via the basolateral membrane and apical endoluminal release (see Fig. 20–1c). Once released into the lumen, serotonin may activate luminal receptors or be absorbed *(small arrows)* across the epithelium to reach submucosal capillaries. **Bottom left:** Nerve terminal (nt) with surrounding Schwann cell cover (SC) located adjacent to the cell membrane of a gut mucosal mast cell. **Bottom right:** Efferent nerve terminal in direct synaptic contact with enterochromaffin cell. *Bars*, 100 nm.

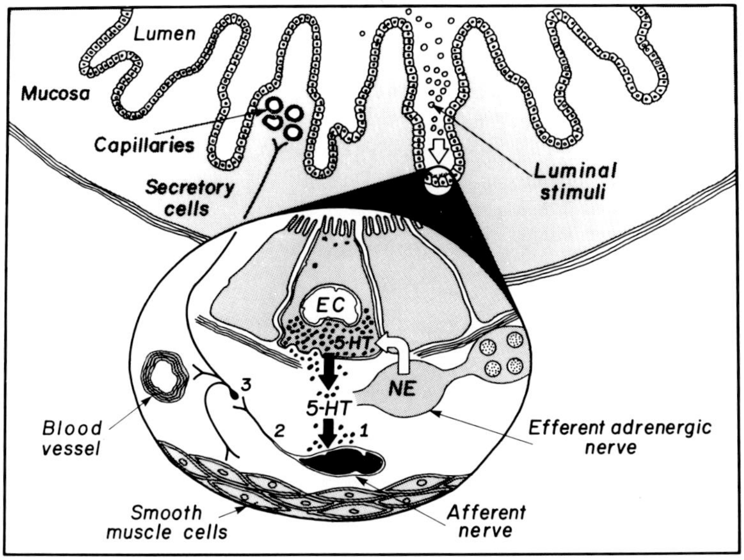

FIG. 20–3. Relation between nerves and enterochromaffin cells (EC) and the proposed contribution of serotonin, released from enterochromaffin cells, in intrinsic reflexes leading to secretion, hyperemia, and motility of the intestine. Luminal stimulation of enterochromaffin cells can be evoked by hypertonic glucose, acid pH, and cholera toxin. The effector responses can be antagonized by (a) tachyphylaxis against serotonin; (b) lidocaine or tetrodotoxin; or (c) hexamethonium. Release of serotonin from enterochromaffin cell can also be caused by activation of adrenergic nerves in close contact with these cells (8).

effects (150). A submucosal barrier may protect intramural serotonergic neurons against overactivation by serotonin reaching them from the intestinal lumen (cf. 393; see Chapter 2).

Functional Role

Serotonin inhibits the secretion of gastric acid under basal conditions and during stimulation with insulin, histamine, and pentagastrin (96,460); in contrast, it stimulates pepsin secretion and mucus production (727). Serotonin increases the motility of the small intestine (132,227) and the secretion of water and electrolytes (275,569). The exocrine pancreatic secretion is stimulated (291) but the bile flow reduced (671). Substance P exerts similar actions (584,646,647,1041,1122).

Meals with a high protein and fat content release serotonin into the peripheral circulation and cause subsequent inhibition of gastric acid secretion (566); they also increase the levels of substance P (14). Similar degrees of gastric acid inhibition are obtained with endogenously released or exogenously administered serotonin (566). When exogenous serotonin or substance P is applied intraluminally, in doses comparable to those released by physiological stimuli, a localized vasodilatation occurs in the mucusa and submucosa (429). Thus, food stimulates the release of serotonin from enterochromaffin cells by several mechanisms (chemical, osmotic, indirectly neural). Once released, serotonin prepares the intestine for optimal handling of the food by causing local hyperemia, sustaining secretion for absorptive and digestive purposes, and inducing motility for the propulsion of the bolus.

THE CARCINOID TUMOR

Historical Notes

The classical description of an ileal carcinoid tumor can be credited to Lubarsch (664). Oberndorfer (800) introduced the term "Karzinoide," implying that these tumors resemble carcinomas but have a benign clinical course. Ciaccio (171) first suggested the endocrine nature of the tumors, but proposed epinephrine as the substance being released. Huebschmann (500) postulated that the tumor is derived from the enterochromaffin cells.

Lembeck actually assayed serotonin in a carcinoid tumor (627). Page and colleagues (821) demonstrated elevated levels of 5-hydroxyindoleacetic acid in a patient with carcinoid syndrome. Sandler and Snow (937) reported a case of gastric carcinoid syndrome with atypical facial flush and elevated levels of histamine and 5-hydroxytryptophan. Oates and colleagues (796) proposed that carcinoid tumors also liberate kallikrein, which by activating bradykinin causes facial flush. Pearse (832) referred to intestinal carcinoids as "gut APUD-omas," which could be derived from multiple stem cells. In humans at least five different types have the capacity to produce serotonin (577).

Histopathology, Classification

The typical carcinoid of the small bowel is a rather small, firm submucosal nodule with yellow surface. The tumor is often surrounded by an intense serosal desmoplasia, muscular hypertrophy (754), and retraction of mesenteric blood vessels (108), which explains the mechanical obstruction of the intestine often seen in these patients. Microscopically the tumor forms nests and ribbons of palisaded polygonal granular cells. These tumors contain several neuropeptides as well (24).

All carcinoid cells have affinity for chromate salts, hence the term "enterochromaffin." However, the classical silver staining reactions of carcinoids depend on their embryologic origin. The *argentaffin* (699) reaction is a direct reduction of silver salts to metallic silver, in the absence of other reducing agents, by a constituent of the secretory granules, most likely serotonin. Carcinoids of the midgut, occurring in the small intestine and proximal colon, are typically argentaffin and besides serotonin produce substance P (691). The *argyrophil* reaction (427) is a two-step reduction of silver salts, where the tissue is treated with additional reducing agents. Carcinoids of the foregut are generally argyrophil but nonargentaffin, indicating that substances other than serotonin are responsible for the atypical foregut carcinoid syndrome (1152). Carcinoids of the hindgut may be negative for both the argyrophil and argentaffin reaction and do not generally evoke a carcinoid syndrome. Carcinoids of both the foregut and hindgut variety contain several peptides, including gastrin, somatostatin, pancreatic polypeptide, and glucagon (21,24). Carcinoids of the midgut contain mainly substance P as peptide hormone (691). In the diagnosis of doubtful cases of carcinoid tumors, the identification of typical secretory granules at the ultrastructural level is helpful. One isomer of the glycolytic enzyme enolase also can be used as a specific marker for neurons and neuroendocrine cells (1034).

Malignancy and Metastasis

Although carcinoids were long considered to be harmless or semimalignant, this is no longer the case (741,835). The 5-year survival time varies between 50 and 90%, but is reduced to 20% in cases with distant metastases (417,749). An exception with good prognosis is the carcinoid of the appendix (740), although cases with metastases and development of a carcinoid syndrome can occur (690).

The histologic growth pattern of carcinoids may be related to survival time (988). There are five major growth patterns: insular, trabecular, glandular, mixed, and undifferentiated. Prognostic differences exist between insular and undifferentiated growth patterns (536).

Metastases in bone and skin are rare but may occur late in the disease. Carcinoids of the foregut variety seem to be more aggressive and metastasize more often to bone (1152). In occasional female patients' breast metastases with identical cytochemical properties to the primary tumor have been reported (11).

Incidence and Distribution

The true incidence of endocrine tumors of the gut is unknown. In a Swedish study, 199 cases were found among 16,194 autopsies performed over a 12-year period (78). With the inclusion of surgically diagnosed tumors, the calculated annual incidence was 8.4 per 100,000 inhabitants. This figure is probably of little clinical relevance, since it includes a large percentage of benign carcinoids of the appendix. Despite their relative infrequency, carcinoids are the most common tumors of the small intestine (10). The majority of carcinoids are of the midgut variety. They increase in frequency in the distal small bowel and are usually multiple (1155). Carcinoids are often small (less than 1 cm in diameter); an increasing size is related to higher frequency of metastases. At discovery 30 to 70% are metastatic (691,1008). Only 6 to 9% are associated with a carcinoid syndrome. The syndrome occurs most commonly with a primary tumor of the small bowel (67%) and only rarely appears with lesions located elsewhere in the gastrointestinal tract (1155). Ovarian forms are associated with an early carcinoid syndrome in 12% of the cases, because the ovarian lesion releases serotonin and other secretory products directly into the systemic venous circulation bypassing the hepatic filter. Patients with neoplasia of serotonin-containing cells in the thyroid gland associated with medullary thyroid carcinoma may also develop the syndrome (739). Carcinoids may coexist with other endocrine disorders (multiple endocrine adenomatosis of type I and II) in certain predisposed individuals (1141). Patients with midgut carcinoids often have elevated levels of growth hormone, but manifest acromegaly is rare (Öberg, *personal communication*).

Diagnosis

Clinical Diagnosis

The clinical diagnosis is easy only after a carcinoid syndrome is evident. The patients in whom the diagnosis is made during surgery frequently undergo it for appendicitis or obstruction of the small bowel. Obstruction may be due to luminal constriction or intussusception of the tumor, or to serosal fibrosis and intestinal ischemia caused by humoral agents secreted by the carcinoid cells. Gastrointestinal bleeding is rare. Carcinoids often give nonspecific symptoms because of the adequate hepatic inactivation of serotonin at the early stages of disease (1008). The mean duration of symptoms related to the tumor is 4 years at the time of diagnosis (741).

Laboratory Tests

Urinary 5-hydroxyindoleacetic acid

The most reliable laboratory test to detect a carcinoid tumor is the repeated determination of the urinary content of 5-hydroxyindoleacetic acid. A rough correlation exists between urinary levels of 5-hydroxyindoleacetic acid and the tumor mass (1008). In all patients with metastatic carcinoids the urinary excretion of 5-hydroxyindoleacetic acid is elevated, while the level of serotonin in the serum is normal in a few cases; there is only rarely urinary excretion of serotonin, tryptamine, and tyramine, but 5-hydroxytryptophan is excreted in 80% of the cases (333). This is probably due to a relative deficiency of decarboxylase, which even may be completely lacking in cases with carcinoids of the foregut variety (797). The 5-hydroxytryptophan released from such tumors may (a) be decarboxylated extrarenally and oxidized to 5-hydroxyindoleacetic acid in the urine, (b) be decarboxylated in the kidney but escape oxidation and be excreted as serotonin, or (c) escape decarboxylation and be excreted as 5-hydroxytryptophan. A 24-hr excretion in excess of 9 mg of 5-hydroxyindoleacetic acid (or more than 6 mmol/mol creatinine) in patients without malabsorption suggests the presence of a carcinoid. In carcinoid cases with malabsorption, higher levels of 5-hydroxyindoleacetic acid may be found (1043). In patients with celiac disease (and malabsorption), an elevated excretion of 5-hydroxyindoleacetic acid in the urine may occur (423). The ingestion of food rich in serotonin (bananas, tomatoes, pineapple, walnuts) and certain medicines (phenotiazines, glycerol, guaiacolate, or reserpine) may increase the urinary excretion of 5-hydroxyindoleacetic acid.

Levels of serotonin in the blood

The measurement of serotonin levels in peripheral blood yields varying results in carcinoids of the midgut variety (306,333). Strodel and colleagues (1008) suggested that there may be a mechanism early in the disease with hypereffective serotonin metabolism, resulting in stable blood levels of serotonin but elevated levels of 5-hydroxyindoleacetic acid, rendering the patient symptom-free during this period. It may be better to analyze serotonin in platelet-poor plasma (306). The serotonin produced by the tumor first might occupy all available binding sites on platelets before it circulates as free serotonin. However, it has been argued that the platelet-poor fraction may contain mainly serotonin from platelets disintegrated during centrifugation or from platelets aggregating during the withdrawal of the sample (221,501,565). In peripheral whole blood of patients with a metastatic carcinoid tumor of the midgut variety the level of serotonin (252 ± 56 ng/ml) is increased compared to that in normal subjects (46 ± 10 ng/ml) (9;Fig. 20–4).

Provocation tests

In doubtful cases provocation tests can be performed by injecting epinephrine (or norepinephrine) (631), calcium (557), or pentagastrin (363). These injections may induce carcinoid symptoms and elevate the blood levels of the tumoral mediators. Thus, pentagastrin induces flushing or gastrointestinal symptoms in all patients tested and increases the blood level of serotonin to 422 ± 71 ng/ml within the first 5 min. Calcium infusion is only positive in one-third of the patients (Figs. 20–4 and 20–5).

Radiography

The radiographic appearance of carcinoids of the small bowel may be typical with luminal filling defects due to fibrosis and shortening of the mesentery, mesenteric calcification and rigidity, and kinking of bowel loops (57). Angiography of the superior mesenteric artery may give additional information, unless the tumor is too poorly vascularized. The angiographic signs include stellate arrangement of mesenteric arcade arteries, narrowing of the peripheral branches together with poor accumulation of contrast medium, and poor venous drainage of the tumor area

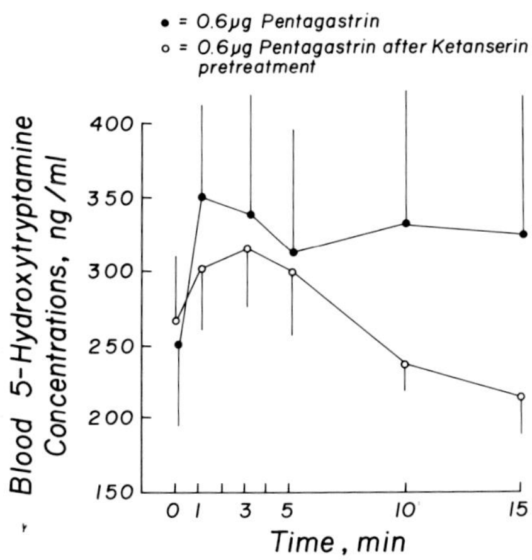

FIG. 20–4. Serotonin levels in peripheral blood following intravenous pentagastrin provocation in patients with a metastatic midgut carcinoid ($N = 16$) before treatment (●) and after treatment with ketanserin (○) on 2 successive days. The basal levels of serotonin are significantly elevated and increase significantly during provocation on both days.

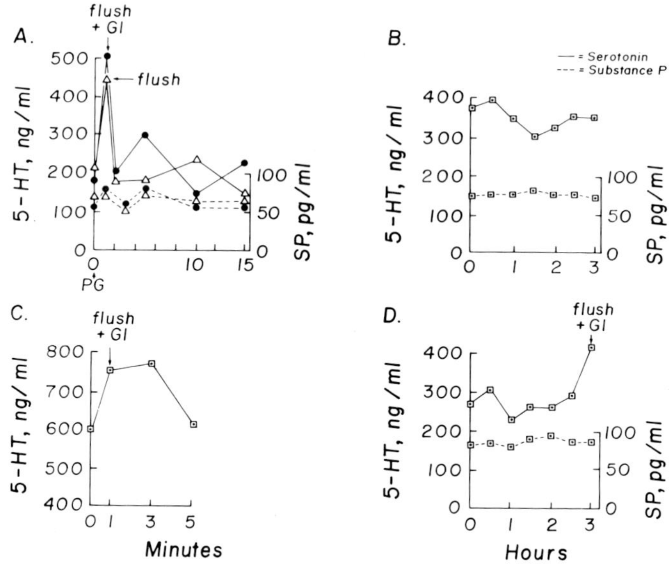

FIG. 20–5. Individual pentagastrin tests with concomitant determination of the peripheral levels of serotonin and substance P in comparison with a calcium test. **A:** Pentagastrin provoked a pronounced serotonin peak (●) associated with facial flush and gastrointestinal symptoms (GI). After treatment with ketanserin (△) a similar peak was obtained, now associated only with flushing. Peripheral levels of substance P were unchanged. **B:** In the same patient (⊡) the infusion of calcium provoked no symptoms despite constantly elevated peripheral levels of serotonin and substance P. **C:** A third patient had a positive pentagastrin-provocation test with very high basal levels of serotonin and symptoms related to the serotonin peak. **D:** However, in another patient calcium induced both flushing and gastrointestinal symptoms associated with an elevated level of serotonin, while that of substance P was constant. This patient also had a positive pentagastrin provocation for serotonin.

(108). Selective catheterizations of mesenteric and hepatic veins may be performed with concomitant blood sampling to detect high levels of serotonin or substance P in veins draining the suspect lesions (784,881). Elevated levels of serotonin in hepatic veins together with the absence of a gradient for serotonin across the liver strongly indicate a hepatic involvement (Fig. 20–6). Angiography is superior to computed tomography in detecting hepatic disease, especially in cases with a fine nodular spread. In a comparative study the computed tomography missed nearly 40% of angiopositive cases (932). A complete angiographic examination also includes confirmation of the patency of the portal vein, which is a prerequisite for treatment of hepatic disease by dearterialization.

PATHOPHYSIOLOGY

A carcinoid syndrome usually occurs either when the hepatic inactivation of the amine is exceeded by the massive secretion from the primary tumor, or when the tumoral mass involves the liver, which not only reduces the metabolizing capacity but also permits a direct release of secretory products into the hepatic veins.

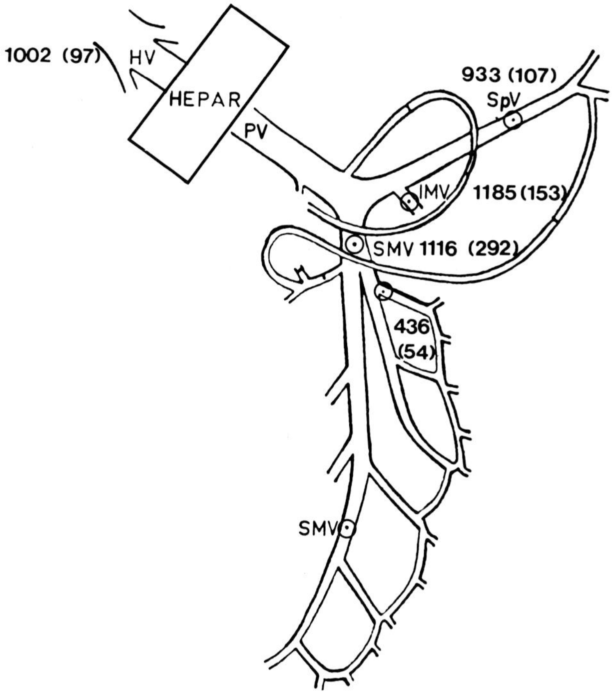

FIG. 20–6. The patient C in Fig. 20–6 had undergone a previous resection of a midgut carcinoid tumor. He had carcinoid symptoms, slightly elevated urinary excretion of 5-hydroxyindoleacetic acid, and a positive pentagastrin test. The diagram illustrates the levels of serotonin in whole blood and platelet-poor plasma (within parentheses) from different regions during selective catheterizations of his hepatic and mesenteric veins. All levels were markedly elevated without evident regional differences, thus not allowing the localization of a distinct tumor. No gradient of serotonin levels across the liver was found, indicating hepatic involvement in spite of negative angiographic findings. A laparotomy demonstrated generalized peritoneal carcinoidosis with miliary spread to the liver. HV = Hepatic vein, PV = portal vein, SpV = splenic vein, IMV = inferior mesenteric vein, SMV = superior mesenteric vein.

However, the syndrome can occur also in the absence of hepatic disease, when the tumors can secrete their products directly into the systemic circulation (e.g., ovarian carcinoids or retroperitoneal tumors). The systemic manifestations of the syndrome are quite variable (977). The most frequent symptoms include diarrhea, hepatomegaly, and flushing (in about 80% of the cases), followed by valvular disease of the right heart (50%) and asthma (25%) (912). More unusual are malabsorption and pellagra due to overconsumption of tryptophan by the tumor (423).

Gastrointestinal Symptoms

Diarrhea

The episodes of diarrhea probably are due to the overproduction of serotonin (423). Measurements of fluid and electrolyte secretion and of gastrointestinal motility suggest that it is due to the combination of an increased intestinal secretion, reduced colonic absorption, and accelerated gastrointestinal transit (220,274,870,877). Physiological studies in rats and rabbits indicate activation of serotonergic receptors for secretion. These may be located directly on the enterocytes (569,1181). This interpretation is supported by studies with the S_2-serotonergic antagonist ketanserin. In patients subjected to the pentagastrin test, ketanserin alleviated the gastrointestinal symptoms, while facial flush remained unchanged (9). The drug did not notably influence the release of serotonin into the peripheral blood (Fig. 20–4). Chronic treatment with ketanserin provided objective good results in one-fifth of the patients, subjective relief in another fifth, and no obvious results in the others. Blockade of S_1-serotonergic receptors was not carried out, since serotonin seems to exert its secretory effect via activation of S_2-serotonergic receptors (1181). The clinical findings suggest that other humoral agents may cause the diarrhea. Peptides that would do so include calcitonin, motilin, vasoactive intestinal polypeptide, and neurotensin (738). The finding that carcinoid patients with proven hypersecretion of serotonin do not always suffer from diarrhea also implies that serotonin does not play an obligatory role in the diarrhea of the carcinoid syndrome. In carcinoid patients somatostatin reverses the chloride secretion to absorption and enhances water and sodium absorption (220). The cellular site of action of somatostatin is not known, but it may either inhibit the effects of serotonin on the gastrointestinal wall or reduce production and/or release of the monoamine from the tumor.

The level of circulating serotonin is also elevated in patients with celiac disease (854). Serotonin may activate phospholipase in cell membranes, which in turn hydrolyzes phosphatidyl inositol (325) and initiates the formation of prostaglandins. The fasting jejunal fluid of patients with a carcinoid tumor or with celiac disease contains high levels of prostaglandin E_2. Thus serotonin could possibly induce hypersecretion by accelerating the metabolism of arachidonic acid.

Substance P may be present in carcinoid tumors. The levels are elevated in the plasma of patients with carcinoids of the midgut variety (691). Substance P stimulates intestinal smooth muscle both directly and indirectly (cf. 646). The peptide decreases sodium absorption in the small bowel and induces net secretion (1122). Theoretically, it thus could cause diarrhea. However, during pentagastrin provocation, most patients develop gastrointestinal symptoms without a concomitant increase in the plasma level of substance P (Fig. 20–5), which argues against the hypothesis that the peptide plays a major role in the diarrhea of the carcinoid syndrome.

Abdominal Crisis

Patients with carcinoid tumors may suffer severe abdominal pain without intestinal obstruction (carcinoid abdominal crisis). This is probably caused by intestinal ischemia induced by agents released from the tumor (1013). Serotonin in high concentration causes constriction of mesenteric blood vessels (976). The pain may be due to obstruction of the venous drainage (954), but obstruction of mesenteric arteries also occurs (272). In these patients the arteries are narrowed by severe fibroelastosis (765). Further narrowing of the lumen of such diseased arteries during a vasospastic episode (e.g., induced by serotonin) may cause necrosis of the small intestine.

Flushing

There is no temporal relationship between diarrhea and flushing in patients with carcinoid syndrome, and one symptom may occur in the absence of the other. Intravenous infusion of serotonin does not induce flushing, and nonselective serotonergic antagonists are of little therapeutic value in preventing it (796,824,901). Ketanserin does not prevent flushing. Thus it is unlikely that serotonin contributes to the phenomenon. Substance P is a potent vasodilator, and the intravenous infusion of it in man induces tachycardia, hypotension, and vasodilatation in skin and muscle (302). Tachykinins other than substance P, i.e., substance K and eledoisin, have similar peripheral actions and have been identified in midgut carcinoids (Norheim, *personal communication*). However, patients with carcinoid syndrome develop facial flush without changes in the peripheral levels of the peptide (Fig. 20–4). Thus substance P cannot be the sole mediator of the flushing.

The typical midgut carcinoid flush is brief, localized to the face and neck, and tricolor (pink, red, and blue). It is associated with a slight decrease in blood pressure and hyperventilation. The flush is due to venous constriction and pooling of blood in passively dilated capillaries. When hemoglobin is reduced in stagnant blood the initial red flush turns to blue. It is usually provoked by alcohol, stress, physical exercise, or certain compounds (catecholamines, pentagastrin, calcium), which release the mediator from the tumor. Evidence for flush as a catecholamine-induced event was obtained from the relative therapeutic success with α-adrenoceptor blocking agents (cf. 423,1157). Another tentative mediator is bradykinin, since catecholamines may release the enzyme kallikrein from the tumor, which in turn transfers kininogen to bradykinin (796).

Valvular Lesions

The valvular lesions of the heart are characterized by a severe fibrosis, which is located underneath an intact endocardium and does not penetrate into the internal elastic lamina. It is frequently accompanied by an increased number of mast cells (423). The fibrotic changes are similar to those seen in patients with retroperitoneal fibrosis and fibrosis around mesenteric blood vessels. The chronic infusion of serotonin in animals did not reproduce the valvular lesions (1120) unless large amounts of serotonin were given in combination with a hepatotoxic agent to tryptophan-deficient guinea pigs (993). Cardiac lesions generally occur in patients with hepatic disease, in which the right side of the heart is exposed to high circulatory levels of serotonin while the left side is protected by pulmonary deactivation (408; see Chapter 3). Left heart lesions and subsequent pulmonary edema are relatively more common in patients with bronchial carcinoids (1043). Endomyocardial fibrosis is not uncommon in populations with a serotonin-rich diet (976). Methysergide and its derivatives, which possess both agonistic and antagonistic actions at serotonergic receptors, can cause retroperitoneal fibrosis (147). These different observations are consistent with the hypothesis that serotonin might be responsible for the valvular fibrosis observed in certain patients with a carcinoid tumor.

Asthma

Asthma can occur in patients with a carcinoid syndrome. Serotonin, bradykinin, and substance P all cause bronchoconstriction that can lead to asthma (423,668,796). Hyperventilation occurs commonly during flushing. This may be due to the elevated levels of serotonin, since the monoamine can cause hyperventilation (901).

Central Nervous System

Patients with extensive carcinoid disease may develop psychiatric symptoms leading eventually to stupor. This has been attributed to an overconsumption of tryptophan by the tumor cells, resulting in low levels of serotonin in the central nervous system. The administration of exogenous tryptophan may alleviate these symptoms (626).

TREATMENT

Surgery is the method of choice in the primary treatment of small bowel carcinoids and limited hepatic lesions (1180). Inextirpable carcinoids are generally insensitive to irradiation and chemotherapy (cf. 456). Streptozotocin, a drug with rather selective effect on pancreatic endocrine tumors, has proven to be of very limited value, especially in cases with midgut carcinoids (799). More encouraging results in this group of patients were reported for interferon. The release of serotonin can be reduced by inhibiting its synthesis with α-methyldopa or *p*-chlorophenylalanine (310,978). Serotonergic antagonists (cyproheptadine, methysergide, or ketanserin) may have antidiarrheogenic effects, but are of little value in controlling flush (9,423). A better control of this symptom is obtained with the α-adrenergic antagonist phenoxybenzamine (1157). The infusion of somatostatin can treat both the diarrhea and the flushing (220,363,572,657). Histaminergic antagonists have also been used against flush (517,867).

In order to control hepatic disease, surgical dearterialization of the liver has been tried in cases with verified patency of the portal vein (72). An alternative nonsurgical dearterialization can be achieved by angiographic hepatic artery embolization (17). To protect these patients from the early reaction on release of tumor products and from the late septic complications, treatment with inhibitors of serotonin synthesis, serotonergic antagonists, inhibitors of kallikrein, steroids, and broad-spectrum antibiotics is recommended.

ACKNOWLEDGMENT

This work was supported by the Swedish MRC 5220.

— 21 —

Serotonin and Acute Respiratory Failure

Herbert B. Hechtman

Department of Surgery, Brigham and Women's Hospital, and Harvard Medical School, Boston, Massachusetts 02115

One of the functions of the lungs is to clear serotonin from blood. Between 60 and 90% of the monoamine is removed by the pulmonary vasculature during a single passage in experimental dogs or humans (408,1149). In addition to clearance, a second mechanism whereby the lungs counter the effects of serotonin could be by producing a physiological antagonist such as a vasodilator prostaglandin. If these functions fail, the levels of serotonin may increase to the point that the smooth muscles of the pulmonary blood vessels and the airways contract. This chapter reviews the pulmonary effect of serotonin and its potential role in certain respiratory diseases.

PHYSIOLOGY

Pulmonary Metabolic Activity

The ability of the lung to clear serotonin and other amines is quantitatively similar in dogs and humans (905; see Chapter 3). It largely reflects the properties of pulmonary endothelial cells, especially those of the microvasculature. The pulmonary extraction of serotonin is rapid, while the uptake by platelets is relatively slow. Thus the half-life of injected serotonin in plasma is 1 to 2 min, and the uptake by platelets accounts for less than 11% of the removal (896). A reduced serotonin extraction occurs in lung damage secondary to (a) hyperoxia (101); (b)

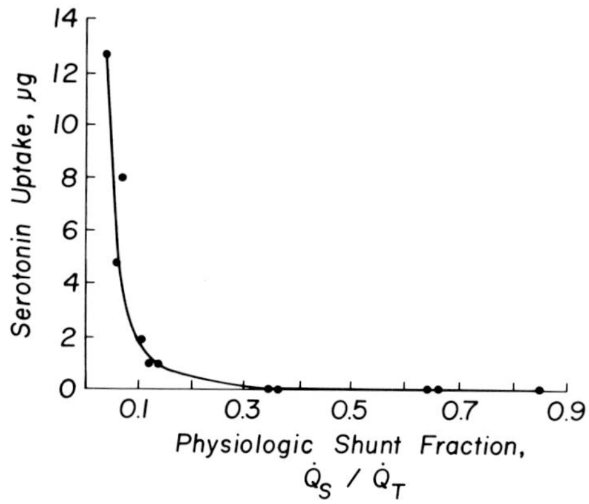

FIG. 21–1. Lung damage occurs during pump perfusion resulting in an increase in the physiologic shunt, as well as an inhibition of serotonin clearance by the lung.

pulmonary hypertension; and (c) after cardiopulmonary bypass (407; Fig. 21–1). Abnormalities in the clearance of serotonin may lead to altered levels of this vasoactive amine in the systemic circulation with resulting variations in blood flow and its distribution (467,608). The experience of Starling and Verney (996) who used the lung as a metabolic filter supports this hypothesis. In their experiments, renal vascular resistance declined and urine flow was maintained when a lung was included in the perfusion circuit. However, the experimental studies designed to test the functional importance of the pulmonary clearance of serotonin yield suggestive but not definitive information. Thus, in the perfused canine lung, the extraction of infused serotonin by the microvasculature is not a major factor controlling the degree of constriction of the pulmonary veins (896). Although the venoconstrictor response to serotonin is relatively small in this preparation, it may still be a determinant of elevations in capillary pressure and therefore of edema formation. High concentrations of serotonin lead to pulmonary venoconstriction in intact sheep and dogs (513).

Prostacyclin Synthesis

The endothelium is a major site of synthesis of prostacyclin (138). This prostaglandin is produced in large amounts by the lungs (509). It antagonizes a number of the vaso- and bronchoconstrictor effects of serotonin (992). Prostacyclin is also synthesized by smooth muscle, an event stimulated by low concentrations of serotonin and platelet-derived growth factor (195). The stimulation by serotonin is mediated by specific serotonergic receptors on the smooth muscle cells, although it is not certain what subtype(s) of serotonergic receptors are involved. Serotonin and platelet-derived growth factor, released from the dense bodies and α-granules of the platelets, may provide a signal to increase the synthesis of prostacyclin by the neointima or the smooth muscle of the media underlying an area of endothelial damage. This may then protect against an uncontrolled accumulation of platelets and an occlusive thrombus. Alternatively, serotonin may possibly affect vascular tone by stimulating the production of the vasodilator prostacyclin.

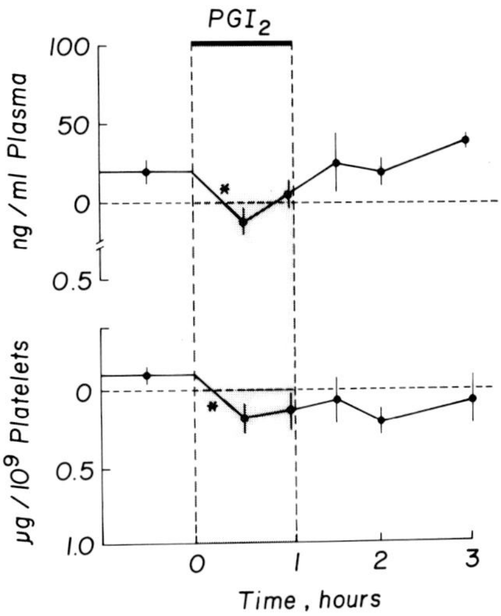

FIG. 21–2. Prostacyclin infusion in a dose of 100 ng/kg·min will block lung extraction of serotonin. The ordinate gives the transpulmonary difference, that is, arterial minus pulmonary arterial concentrations. The effect on platelets is long-lasting. *Asterisks* indicate a significant change from baseline values.

Prostacyclin helps to regulate the transport of serotonin. In normal pulmonary arterial blood the concentration of serotonin in plasma and platelets exceeds that of systemic arterial blood (Fig. 21–2). The infusion of prostacyclin reduces the level of serotonin in plasma but augments the level in whole blood and platelets. Prostacyclin also inhibits the pulmonary endothelial clearance of serotonin. The effect on platelets lasts more than 2 hr after the prostacyclin infusion is stopped. Thus prostacyclin probably inhibits the action of serotonin by at least two mechanisms: (a) relaxation of smooth muscle of pulmonary blood vessels and bronchi; and (b) modulation of the transport of serotonin.

Pulmonary Vasoconstriction

Physiological concentrations of serotonin probably do not affect the resting tone of the smooth muscle of the pulmonary vasculature. However, higher concentrations of the monoamine activate pulmonary vascular smooth muscle. Comroe et al. (187) first described the pulmonary vasoconstrictor effects of serotonin. The systemic cardiovascular effects of this monoamine are complicated by the occurrence of a variety of pressor and depressor reflexes. Serotonin causes pulmonary vasoconstriction in all species studied; the effect is more prominent in the dog than in humans (109). All segments of the pulmonary vasculature constrict during the response to serotonin, even the venules (91,513), although they contract very slowly. Thus the infusion of large doses of serotonin (75 μg/kg·min) in anesthetized dogs increases the mean pulmonary arterial pressure from 20 to 30 mm Hg. Furthermore, the pressure gradient between the small pulmonary veins and the left atrium increases, which confirms that pulmonary venoconstriction occurs (608). The prolonged infusion of low doses of serotonin in awake sheep increases the lymph flow and decreases the ratio between the protein concentrations of lymph and plasma, which

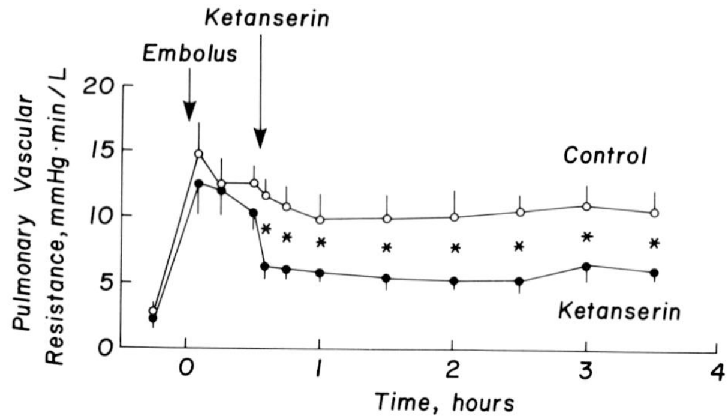

FIG. 21–3. Embolization leads to a rise in pulmonary vascular resistance that is largely reversed by ketanserin. *Asterisks* indicate that the difference from untreated embolized animals is statistically significant.

suggests that the pulmonary transvascular pressure, but not the permeability, increases (127).

Bronchoconstriction

Several reflexes arise from sensory receptors in the lung (256,413). Serotonin may stimulate these receptors. Such an action would explain why the bradycardia, apnea, and hypotension that follow (within half a second) the injection of serotonin into the right ventricle can be prevented by vagal blockade (187). Atropine inhibits the bronchoconstriction but not the vasoconstriction induced by serotonin in the dog (515). Thus the bronchoconstriction evoked by the monoamine involves vagal afferent nerves. Since atropine prevents serotonin-mediated bronchoconstriction also in isolated perfused lungs, a nonreflex cholinergic mechanism also contributes (90). Finally, serotonin can potentiate the vagal actions on airway smooth muscle by activating efferent pathways (442).

The peripheral airways, which are perfused by the pulmonary circulation, are composed of respiratory bronchioles and alveolar ducts. Together with the surrounding alveoli they comprise the distensible part of the lung. Although the vagal nerves have a major influence on the dimensions of larger bronchi, their action on peripheral airways is weak or absent (804). Constriction of the central conducting airways increases pulmonary resistance and decreases the anatomical dead space, but does not restrict the distension of the lungs. In contrast, peripheral airway constriction expels gas from the lungs and decreases the functional residual capacity as well as the pulmonary compliance. Serotonin (3–35 μg/kg·min) given to vagotomized cats causes constriction of both central and peripheral airways (182). However, the net effect of serotonin infusion depends on the dose infused. Small amounts can induce air expulsion, while larger doses induce air retention because of the constriction of the large airways. In sheep the infusion of a moderate dose of serotonin (4 μg/kg·min) usually causes a fall in partial pressure of oxygen in arterial blood from 80 to 70 mm Hg. The decrease in oxygen tension is due to an

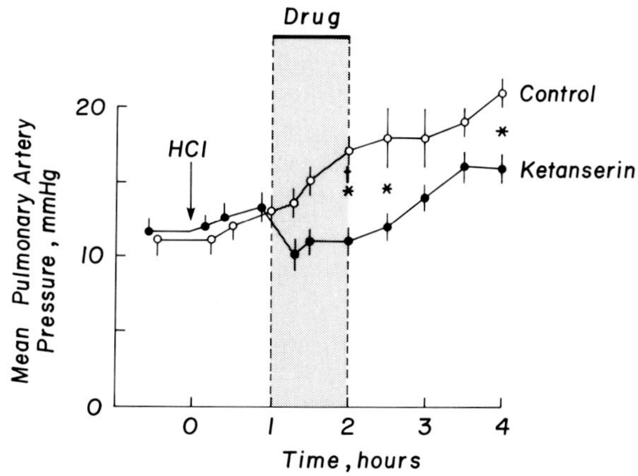

FIG. 21–4. After acid aspiration the serotonergic inhibitor ketanserin lowers pulmonary arterial pressure. The effect is seen primarily during the period of drug infusion, although pressure remains low throughout the 4-hr period of observation. *Asterisks* indicate a statistically significant difference from compared untreated control animals.

imbalance in ventilation/perfusion ratio since there is no evidence of pulmonary edema.

PATHOPHYSIOLOGY

Platelet Serotonin

When they are activated, platelets aggregate and may be entrapped in the lungs. They release their content of serotonin, which can achieve high local concentrations. For example, in pulmonary embolism thrombin activates the platelets and causes their sequestration in the lungs (997). The periphery of the embolus is encrusted with multitudes of aggregating platelets. The functional consequences depend on the action and the quantity of the released vasoconstrictor products. Experimental emboli lead to transient but significant decreases in thoracic gas volume, falls in compliance, and increases in airway resistance. These events can be prevented by pharmacological agents that induce thrombocytopenia, inactivate thrombin, prevent platelet aggregation, or antagonize serotonin. The beneficial effects of the S_2-serotonergic antagonist ketanserin (525,637,642) confirm the importance of the vaso- and bronchospastic effects of serotonin released from platelets. Thus 3 hr following experimental pulmonary embolization in dogs, the physiological shunt increases from 12 to 31%. Ketanserin given as an intravenous bolus 1 half-hour after embolization decreased the shunt to 9%. The compound also reduces pulmonary vascular resistance by more than 50% (Fig. 21–3). The involvement of platelets and of the serotonin that they release in response to pulmonary embolism is suggested further by the reduction in the number of circulating platelets (from

186 to $148 \times 10^3/mm^3$) and the decrease in serotonin content (from 1.71 to 1.51 $\mu g/10^9$ platelets) observed after the embolization (511).

Acid Aspiration

Within 5 min after experimental instillation of hydrochloric acid (0.1 N), severe thrombocytopenia develops along with the appearance of platelet microaggregates in blood smears. The platelet counts in systemic arterial blood are lower than those in pulmonary arterial blood, which indicates that platelet sequestration occurs in the lungs (1063). In addition, the circulating platelets are activated since they synthesize thromboxane A_2 and release serotonin. The level of serotonin in the platelets falls progressively (from 2.09 to 0.82 $\mu g/10^9$ platelets) over the course of the 4 hr following the aspiration.

The pulmonary arterial pressure progressively rises after aspiration, presumably because serotonin released from the platelets induces vasospasm. This is suggested by the ability of ketanserin to reverse the phenomenon. After the administration of ketanserin is stopped, the pulmonary arterial pressure rises but remains below the value observed in untreated animals (Fig. 21–4). The pulmonary arterial wedge pressure remains stable at 3 mm Hg. Pulmonary edema is observed 2 hr after aspiration. Although the granulocytes probably mediate the changes in microvascular permeability, which cause the edema (507), a reduction in pulmonary pressure also decreases the extravasation of fluid. For example, ketanserin reduces mean pulmonary arterial pressure by 24% and the edema formation by 60%. Sodium nitroprusside causes vasodilatation with a 20% fall in mean pulmonary arterial pressure and a 44% decrease in edema formation. The nonselective nature of the vasodilator effect of sodium nitroprusside is illustrated by a larger fall (30%) in mean systemic arterial pressure than that observed with ketanserin (11%).

The direct relationship between the pulmonary vascular pressure and the occurrence of lung edema is usually evident only at very high pressures. However, if the microvascular permeability increases, such a relationship is obtained even at low pressures. The favorable effect of serotonergic inhibitors on gas exchanges and on physiological shunting following acid aspiration may be due both to the lessening of pulmonary edema and the reversal of bronchospasm.

Acid aspiration localized to one lung stimulates a generalized response that leads to the entrapment of similar numbers of platelets in both lungs. The entrapment occurs within 5 min. Five minutes after aspiration the shunt rises to 47% in the aspirated side and to 43% in the nonaspirated lung, from a baseline value of 12%. It is possible that platelet-activating factor is synthesized by alveolar macrophages in response to the acid stimulus (38). This potent chemotoxin and aggregator of platelets and white blood cells may underlie the generalized pulmonary response to localized injury.

Experimental Adult Respiratory Distress Syndrome

Platelets are entrapped in the lungs of experimental animals infused with oleic acid or live *Escherichia coli* (466). However, the relationship of cell entrapment to abnormalities in lung function is not clear, and the release of serotonin may not be causally related to the pulmonary dysfunction. At least one study suggests that the monoamine is responsible for the acute alterations in lung mechanics observed during septicemia (998). However, the release of serotonin is not likely to explain the rise in airway resistance or the fall in lung compliance evoked by endotoxin (764). Indeed, endotoxin does not cause the release of serotonin *in vitro*, despite the formation of large platelet aggregates. Furthermore, the infusion of endotoxin in dogs decreases the platelet count by 95% but increases the plasma level of serotonin by less than 10%. To reproduce the alterations in lung function caused by endotoxin, the injection of a dose of serotonin is required that corresponds to 25% of the amount contained in circulating platelets. Since the acute pulmonary hypertensive response to endotoxin is mediated by thromboxane A_2 (455), this prostanoid may also be responsible for the abnormalities in lung mechanics (991). The late effects of endotoxin on permeability are not mediated by thromboxane A_2. Although it is not likely that serotonin increases the permeability of the capillary-alveolar membrane, its concentration increases in the lymph draining the lung (509). Thus it may be the mediator of the delayed secondary abnormalities in lung mechanics and the later increase in pulmonary arterial pressure evoked by endotoxin (991).

Serotonin has toxic effects on the pulmonary vasculature (608). In contrast to histamine, it does not directly increase the permeability of the pulmonary microvessels (852). However, serotonin can act together with mediators of endothelial cell damage. Thus the monoamine amplifies the destructive interaction of granulocytes and endothelial cells in culture (111). Such an interaction could explain the observations that, in the dog, an infusion of a very large dose of serotonin (75 μg/kg·min) leads to respiratory failure after 2 hr, with a histological picture similar to that observed following 2 hr of severe hemorrhagic hypotension. The effects of the hemorrhagic hypotension can be prevented by methysergide (608).

ADULT RESPIRATORY DISTRESS SYNDROME

Pulmonary hypertension frequently accompanies the adult respiratory distress syndrome. It probably reflects the pulmonary entrapment of platelets and platelet aggregates that lead to the intrapulmonary release of serotonin and pulmonary vasoconstriction. However, in patients with the adult respiratory distress syndrome, entrapment of platelets is frequent but not obligatory. The patients with an increase in physiological shunting entrap more platelets than the others (466). This interaction between platelet entrapment and physiological shunting does not correlate with other measures of pulmonary dysfunction. In another study of respiratory failure, platelet sequestration in the lungs was not observed, although a relationship exists between the pulmonary hypertension and the concentration of serotonin in the serum (969). Unfortunately, no information seems available as to the relationship between platelet sequestration in the lungs, levels of serotonin, and the time of onset of acute respiratory failure.

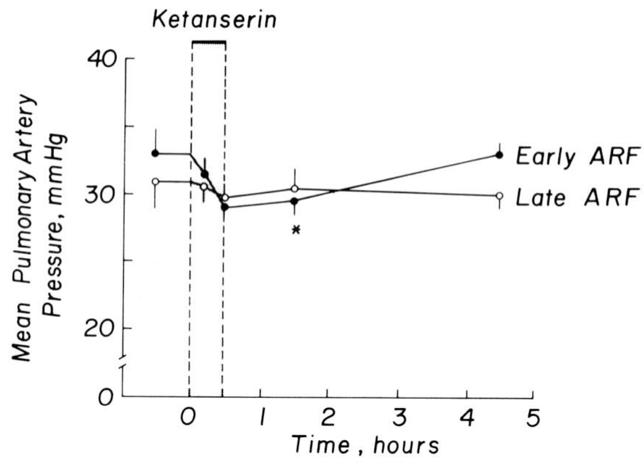

FIG. 21–5. Among patients given ketanserin early in the course of their respiratory failure, treatment resulted in a significant reduction in mean pulmonary arterial pressure. The serotonergic antagonist had no effect on pulmonary pressure in patients with longer-standing acute respiratory failure (ARF).

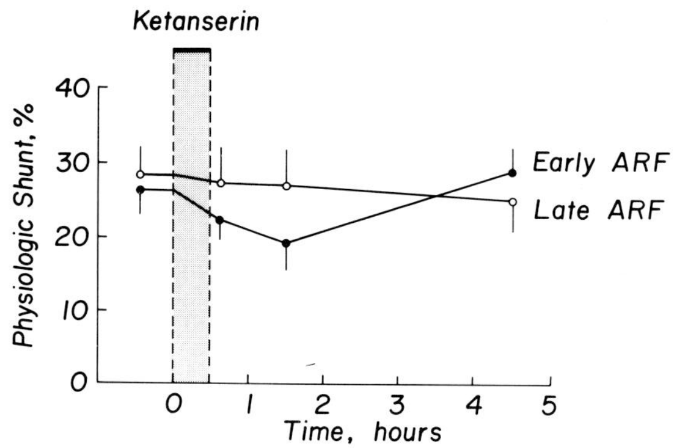

FIG. 21–6. Patients in the early acute respiratory failure (ARF) group had a decline in physiologic shunt following ketanserin administration. Improvements were greatest 1½ hr after the start of treatment.

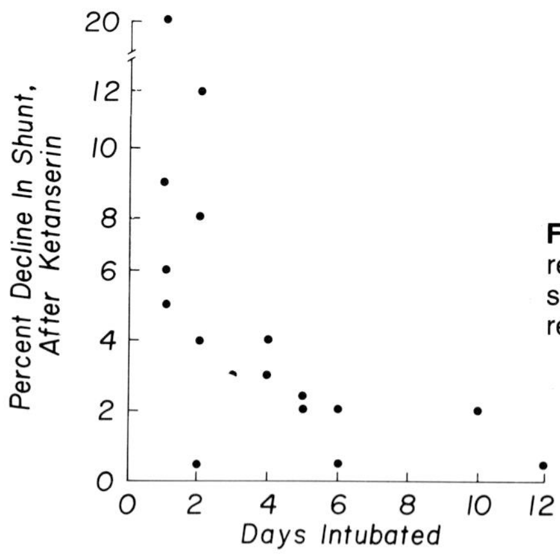

FIG. 21–7. Among patients from both early and late respiratory failure groups, the decline in physiologic shunting following ketanserin infusion was inversely related to the number of days of intubation.

In a small group of patients with the adult respiratory distress syndrome, the intravenous administration of ketanserin (10 mg as a bolus followed by an infusion at 33 μg/min) reduced mean arterial blood pressure by 10 mm Hg (12%) and mean pulmonary arterial pressure by 3 mm Hg (11%). The drug did not affect physiological shunting (28%), heart rate, or cardiac index (1115). A larger, more detailed study of the effect of the serotonergic inhibitor ketanserin was reported by Huval et al. (510). The onset of respiratory failure was defined as the time that endotracheal intubation and mechanical ventilatory support were required. Early in the course of respiratory failure (within 4 days from the time of intubation), the platelet count in pulmonary arterial blood exceeded that in systemic arterial blood by 26,300/mm^3, which indicates entrapment of platelets in the lungs. In contrast, later in the course of respiratory failure the patients showed no pulmonary entrapment of platelets. The level of serotonin in platelets was 55 ng/10^9 platelets early in respiratory failure, a value significantly lower than that noted later in the illness (95 ng/10^9) or that observed in normal humans (290 ng/10^9). Ketanserin was given as an intravenous bolus of 0.1 mg/kg followed by a 30-min infusion of 0.08 mg/kg. The mean arterial pressure fell within minutes by 13 mm Hg (15%), without significant changes in heart rate, cardiac index, or pulmonary arterial wedge pressure. After 1½ hr of ketanserin infusion, arterial pressure had returned to baseline. At this point, patients with early respiratory failure showed a 3 mm Hg (10%) decline in mean pulmonary arterial pressure (Fig. 21–5). The temporal disparity in the effect of ketanserin in the systemic and pulmonary circuits suggests different mechanisms of action. The early systemic hypotensive effect is likely due to α-adrenergic blockade, whereas the delayed pulmonary effect probably reflects inhibition of serotonergic receptors.

Other beneficial actions of ketanserin in patients with early respiratory failure were a modest but significant lowering of the peak inspiratory pressure (from 35 to 32 cm H_2O), an event consistent with a reversal of serotonin-induced bronchoconstriction. In addition, the physiological shunt decreased significantly (from 26 to 19%; Fig. 21–6). This also may reflect in part a reversal of serotonin-induced bronchospasm with a resulting improvement in ventilation. Alternatively, the lowering of the pulmonary arterial pressure may have led to a decrease in the formation of pulmonary edema, thereby improving ventilation and reducing the physiological shunt. Four hours after the end of the infusion, the changes induced by ketanserin had faded away completely. Ketanserin was not beneficial in patients with late respiratory failure. Indeed, the greatest improvement in the physiological shunting occurred within the first 2 days of respiratory distress (Fig. 21–7).

ACKNOWLEDGMENT

This work was supported in part by National Institutes of Health grant GM24891-06, the Brigham Surgical Group, Inc., and the Trauma Research Foundation.

— 22 —

Serotonin and Shock

Ulf Haglund

Department of Surgery, University of Lund, Malmö General Hospital, Malmö, Sweden

The term "shock" has been used for more than two centuries to indicate a cardiovascular collapse following disease or trauma. Today the term covers various conditions characterized by a disturbed microcirculation and often accompanied by hypovolemia. These conditions lead to changes in cellular metabolism which, if not effectively counteracted within a short period of time, are followed by collapse and death. Depending on the initiating event, shock is subdivided into cardiac, hypovolemic, neurogenic, septic, traumatic, and anaphylactic shock. Although the initial event causing the circulatory collapse differs in the various forms of shock, the development is caused by similar mechanisms. The most frequently studied are the hypovolemic (in particular hemorrhagic) and septic shocks. Experimentally, shock can be induced by causing major ischemia in the splanchnic circulation. Shock is rapidly lethal if volume substitution is not applied (1040). Indeed, rapid substitution with large volumes of fluid is very important following major trauma or surgery and is the initial therapy in sepsis and septic shock. Together with faster transportation to hospitals for effective monitoring and surgical intervention, this substitution therapy minimizes the frequency of renal complications following shock, while pulmonary complications become more prominent (747,1042). Thus the adult respiratory distress syndrome ("shock-lung") contributes to the overall failure that can complicate shock and trauma (116,364). The risk to develop failure of several organs is especially high under septic conditions and regardless of whether the septic shock is primary, secondary, or superimposed on trauma or complicating surgery. The organs involved include the lungs, heart, kidneys, liver, and gastrointestinal tract. A number of humoral events and factors are also altered, in particular blood coagulation, fibrinolysis, and the kallikrein-kinin and complement systems.

The pathophysiology of this syndrome involving so many organs and systems is poorly understood. It probably is multifactorial and may be due to the release of various substances from ischemic or traumatized tissue. Among those substances, this chapter reviews the possible role of serotonin in the development of severe shock.

RELEASE OF SEROTONIN

Shock causes platelet aggregation (652) and initiates the platelet-release reaction (484). The concentrations of serotonin in portal venous plasma increase following regional intestinal hypotension (331,579). Following hemorrhage, the level of serotonin in portal venous plasma increases (331) while that of systemic venous plasma can either increase (1011) or decrease (913). Serotonin is released in plasma during septic states (464,593). Incubation of platelet-rich plasma or whole blood with endotoxin initiates the rapid release of serotonin (222). The release of serotonin from platelets induced by endotoxin requires the presence of calcium and a plasma protein factor, probably prothrombin (786). Ischemia may cause the release of serotonin not only from platelets but also from the intestinal wall (331). Release of large amounts of serotonin could contribute to the cardiovascular decompensation that is characteristic of the shocked patient. Furthermore, the aggregation of platelets during shock, in particular septic shock, results in sequestration of platelet aggregates in the pulmonary vasculature (100,770,900). Hence, pulmonary dysfunction may be the consequence of the release of serotonin (see Chapter 21).

CARDIOVASCULAR DECOMPENSATION

Blood Vessels

Hemorrhage and hypovolemia reduce the systemic arterial blood pressure, which in turn increases sympathoadrenal activity. As a consequence the arterioles and the venules constrict, which increases peripheral resistance and venous return, respectively, in an attempt to restore arterial blood pressure (343,964). However, the peripheral vasoconstriction, even with moderate blood losses, endangers an adequate perfusion of tissues, in particular the skin and skeletal muscles. In response to the increased sympathetic activity the pre- to postcapillary resistance ratio changes and the mean hydrostatic capillary pressure decreases. Thus the movement of fluid from the tissue to blood increases in an attempt to restore the circulating blood volume. This occurs mainly in skeletal muscles (725) but not in the small intestine (329,441).

In septic states the initial cardiovascular response is hyperkinetic with an increased cardiac output and peripheral vasodilatation. The mechanism underlying this response is unknown, although tissue hypoxia may contribute. Later on in the development of the shock, cardiac function is depressed, reflex peripheral vasoconstriction ensues, and tissue acidosis develops (1137,1139). At this later stage septic and hemorrhagic shock become very similar hemodynamically. This biphasic response is difficult to reproduce in experimental models of septic shock, although cardiac depression is present late in sepsis (330,679).

The responses of the blood vessels during the further development of shock differ from tissue to tissue, as they are the result of opposing forces: the reflex increase in sympathetic tone and other homeostatic regulators, which tend to maintain vasoconstriction, and the local accumulation of vasodilator metabolites. Among the homeostatic neurohumoral mediators, norepinephrine and epinephrine play a primary role. However, other vasoactive factors may play a role in the remote control of vascular function, including angiotensin II, vasopressin, and serotonin (579,714). The local control is exerted mainly by the metabolites that accumulate in the tissues when the microcirculation is disturbed. The exact nature of these metabolites remains uncertain (343,440). Vasoactive amines such as histamine, kinins, or vasoactive polypeptides might contribute. These local factors are powerful vasodilators and in many tissues they dominate the microcirculation as shock progresses (437,441,725). Again, the small intestine may differ from skeletal muscle as its vascular resistance to flow remains in the normal range throughout a period of hemorrhagic or septic shock. In particular, there is no tendency to vasoconstriction when the systemic arterial blood pressure falls (329,437,441).

The intraarterial or intravenous injection of serotonin causes intestinal vasodilatation (92). When ketanserin is given intravenously together with serotonin, the intestinal vasodilatation still occurs (40). Thus the intestinal vasodilator response to serotonin is not mediated by S_2-serotonergic receptors (see Chapters 9 and 10). Serotonin may release other vasoactive substances such as vasoactive intestinal polypeptide (303). Ketanserin does not influence the tendency toward a reduced intestinal vascular resistance during a 2-hr period of profound septic shock (40), showing that, if serotonin is involved in that reduction, it does not act on S_2-serotonergic receptors.

In experimental intestinal ischemia, hemorrhagic shock, and septic shock, microscopic damage of the mucosa of the small intestine is observed within 1 or 2 hr (Fig. 22–1; 120,440). Similar changes are seen in man following shock (439). The mucosal damage is caused by tissue hypoxia that results from reduced blood flow velocity in the villous vascular loops (440). The occurrence of these mucosal lesions in the intestine probably is of major importance for the transition of the shock to irreversibility because of local fluid loss and the release of cardiotoxic material (440). The development of the mucosal lesions during intestinal ischemia or septic shock is not influenced by ketanserin or dihydroergotamine (41,436). Despite the fact that sequestration of aggregated platelets occurs in the small intestine during ischemia, the mucosal damage is not affected by depletion of the serotonin content of the platelets *(unpublished observations)*. Thus it appears unlikely that serotonin plays a major role in ischemic changes of the intestinal wall.

Cardiac Effects

Early in hypovolemic shock the increased sympathetic tone has a positive chronotropic and inotropic effect. However, cardiac function is impaired if the shock is

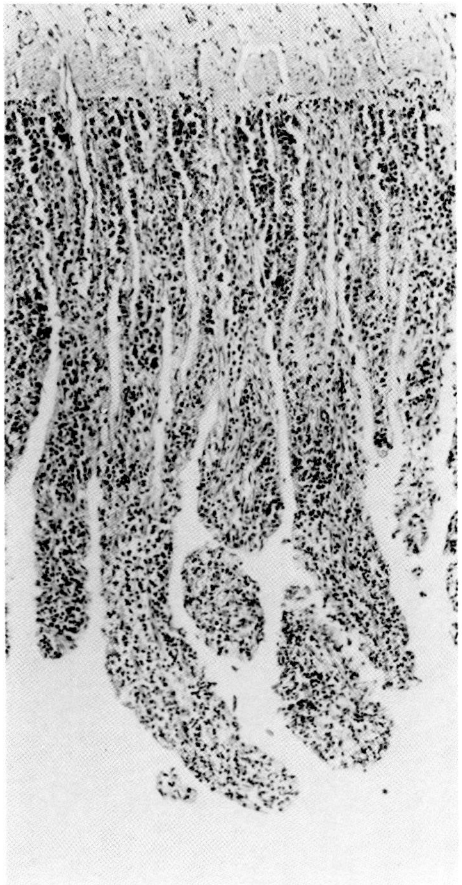

A

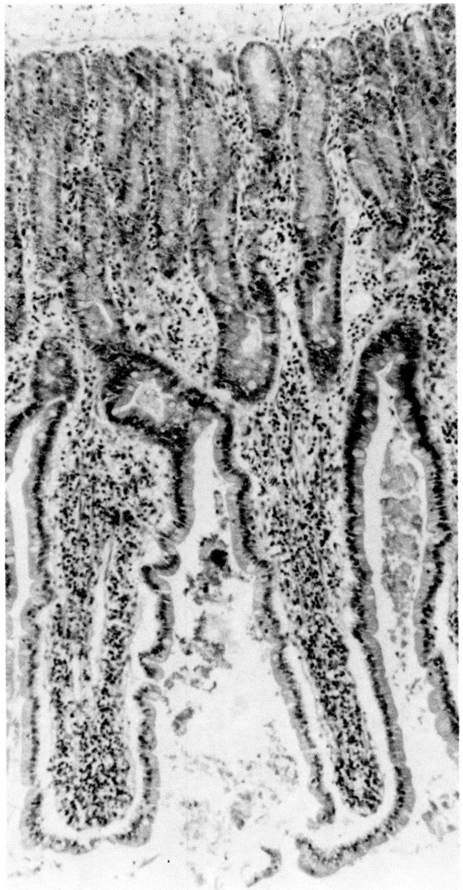

B

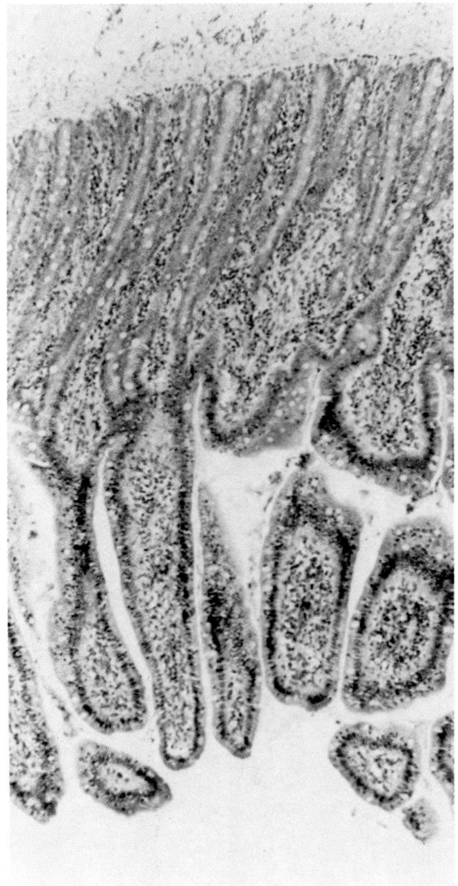

C

prolonged. The impaired cardiac function may be due to: (a) reduced coronary perfusion; (b) metabolic abnormalities of the myocardium; (c) direct myocardial depression caused by bacteria or bacterial toxins; and (d) release of cardiotoxic factors mainly from the small intestine and the pancreas (286,438,623). Several of these mechanisms probably can be activated at the same time. One of the myocardial depressant factors is generated by the ischemic pancreas (623). The ischemia is caused primarily by the profound reflex vasoconstriction, but to some extent it may be exacerbated by the depressant factor itself. The other mediators causing pancreatic ischemia in shock are not well defined, and serotonin cannot be ruled out as a contributing factor.

None of the proposed cardiotoxic shock factors have been chemically identified. Most of these factors have a molecular weight larger than 500 daltons, which excludes serotonin. The production of intestinally released cardiotoxic factors by the small bowel is associated with the mucosal damage which is not influenced by S_2-serotonergic blockade (440). Thus it seems unlikely that the S_2-serotonergic component of the action of serotonin plays a significant role in the cardiac derangement in shock.

PULMONARY DYSFUNCTION

Pulmonary dysfunction can lead to adult respiratory distress syndrome. This is found following traumatic and septic shock but also in various other conditions such as overhydration, thoracic trauma, oxygen toxicity, aspiration, and inhalation

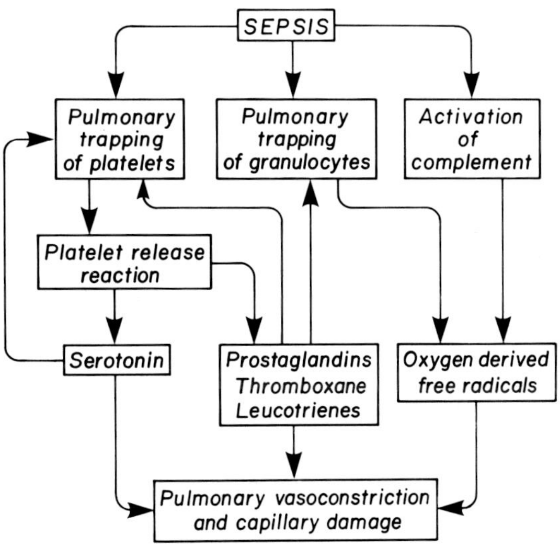

FIG. 22–2. Possible role of serotonin in the pathophysiology of pulmonary damage during shock.

FIG. 22–1. The small intestinal mucosal damage characteristic for shock states and ischemia. Panel **A** illustrates normal mucosa, and **B** moderate damage with the lesions confined to the tips of the villi. Panel **C** illustrates profound damage with destruction of villi. Even in severe mucosal damage as in panel **C** the deeper layers of the small intestine have a normal appearance.

of irritating gases (188; Chapter 21). The development of adult respiratory distress syndrome following septic conditions occurs in several phases (747). In a few cases arterial hypoxia and interstitial pulmonary edema ultimately develop, which precede the cardiorespiratory collapse. In animals the lung injury is characterized by the trapping of platelets and leukocytes in the pulmonary vascular bed and an increase in pulmonary vascular resistance. A high volume of protein-rich lymph flowing from the lung indicates endothelial damage (126,982).

Several mechanisms may contribute to the occurrence of an acute respiratory distress syndrome. These include: (a) activation of complement; (b) generation of oxygen-free radicals; (c) metabolism of arachidonic acid; (d) increase in capillary permeability by fibrin degradation products; and (e) release of serotonin (Fig. 22–2; 116,608,677,969).

The rapid pulmonary trapping of platelets and white blood cells triggers the chain of reactions that ultimately leads to adult respiratory distress syndrome (100,464,652,770,900). This is due not to the mechanical effects of the sequestrated aggregates but rather to the release of the vasoactive substances that they contain (868). Serotonin induces pulmonary vasoconstriction, as do endotoxin and live bacteria (40,329,607,608), and the former is proposed as the likely mediator of the pulmonary response to the latter two.

The intravenous injection of live *Escherichia coli* or endotoxin evokes not only an increase in pulmonary arterial pressure and pulmonary vascular resistance, but also a rapid decrease in systemic arterial blood pressure and a rapid reduction of the number of circulating platelets and white blood cells (329,607). In the cat and the pig this response is not affected by ketanserin (Fig. 22–3), dihydroergotamine, or methysergide (40,41,55). By contrast, UK 38485, an inhibitor of thromboxane synthesis, reduces the increase in pulmonary arterial pressure caused by live *Escherichia coli (unpublished observation)*. Ketanserin also does not prevent or reduce the activation of coagulation and fibrinolysis induced by the infusion of bacteria. However, in the dog ketanserin attenuates the increase in pulmonary

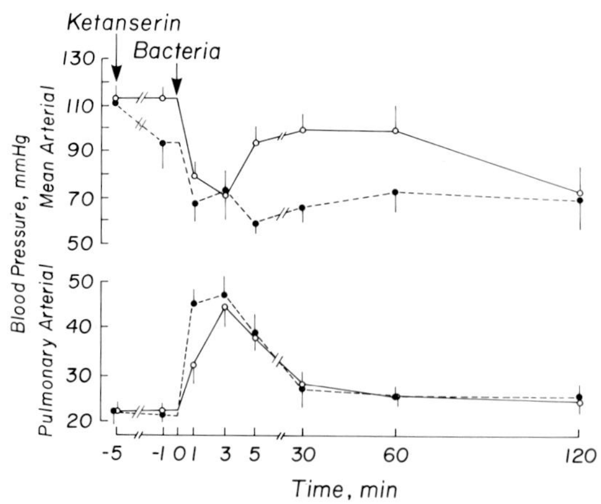

FIG. 22–3. Mean systemic arterial blood pressure and pulmonary arterial blood pressure before and during intravenous infusion of live *Escherichia coli* bacteria in control *(open symbols)* and in ketanserin-treated cats *(filled symbols)*. Note that there was no difference in the acute pulmonary arterial blood pressure response between the two groups. (From ref. 40, with permission.)

vascular resistance caused by endotoxin injection (677,729); cyproheptadine, a combined S_2-serotonergic and H_1-histaminergic inhibitor, prevents the pulmonary trapping of platelets (19).

The reason for these species differences is not clear. The release of serotonin may play a more important role than the metabolism of arachidonic acid in the dog, while the reverse is true in cats and pigs. In acute cases of the respiratory distress syndrome in humans, serotonin may play a similar role as in the dog (see Chapter 21).

References

1. Abrahamsen, A. F. (1968): Platelet survival studies in man. *Scand. J. Haematol.*, 3:7–53.
2. Acharya, S. B., and McIntyre, D. E. (1983): Platelet products and vascular PGI_2 production. *Prostaglandins Leukotrienes Med.*, 10:73–81.
3. Adair, J. R., Hamilton, B. L., Scappaticci, K. A., Helke, C. J., and Gillis, R. A. (1977): Cardiovascular responses to electrical stimulation of the medullary raphe area of the cat. *Brain Res.*, 128:141–145.
4. Adler, S. (1977): Serotonin and the kidney. In: *Serotonin in Health and Disease*, edited by W. B. Essman, pp. 99–137. Spectrum Publishers, Inc., New York.
5. Affolter, H., Erne, P., Buhler, F. R., and Pletscher, A. (1984): Increased sensitivity of blood platelets to 5-hydroxytryptamine in essential hypertension *(submitted for publication)*.
6. Agranoff, B. W., Murthy, P., and Seguin, E. B. (1983): Thrombin-induced phosphodiesteratic cleavage of phosphatidylinositol bisphosphate in human platelets. *J. Biol. Chem.*, 258:2076–2078.
7. Ahlenius, S., Larsson, K., Svensson, L., Hjorth, S., Carlsson, A., Lindberg, P., Wikström, H., Sanchez, D., Arvidsson, L.-E., Hacksell, U., and Nilsson, J. L. G. (1981): Effects of a new type of 5-HT receptor agonist on male rat sexual behavior. *Pharmacol. Biochem. Behav.*, 15:785–792.
8. Ahlman, H., and Dahlström, A. (1983): Vagal mechanisms controlling serotonin release from the gastrointestinal tract and pyloric motor function. *J. Auton. Nerv. Syst.*, 9:119–140.
9. Ahlman, H., Dahlström, A., Grönstad, K., Tisell, L.-E., Öberg, K., Zinner, M. J., and Jaffe, B. M. (1984): The pentagastrin test in the diagnosis of the carcinoid syndrome: Blockade of gastrointestinal symptoms by ketanserin *(submitted for publication)*.
10. Ahlman, H., and Kjellström, T. (1981): Clinically diagnosed small intestinal tumors in an urban Swedish area. *Acta Chir. Scand.*, 147:371–376.
11. Ahlman, H., Larsson, J., Grönstad, K., Tisell, L.-E., and Dahlström, A. (1984): A case of midgut carcinoid with breast metastasis *(submitted for publication)*.
12. Ahlman, H., Lundberg, J., Dahlström, A., and Kewenter, J. (1976): A possible vagal adrenergic release of serotonin from the enterochromaffin cells in the cat. *Acta Physiol. Scand.*, 98:366–375.
13. Ahlund, L., Lundgren, Y., Sjoberg, B., and Weiss, L. (1977): Vascular reactivity to 5-hydroxytryptamine (5-HT) in hindquarter vascular beds, aortic strips and portal veins from spontaneously hypertensive and normotensive rats. *Acta Physiol. Scand.*, 101:489–492.
14. Akande, B., Reilly, P., Modlin, I. M., and Jaffe, B. M. (1981): Radioimmunoassay measurement of SP release following a meat meal. *Surgery*, 89:378–383.
15. Alabaster, V.A. (1977): Inactivation of endogenous amines in the lungs. In: *Metabolic Functions of the Lung, Vol. 4: Lung Biology in Health and Disease*, edited by C. Lenfant, pp. 3–31. Marcel Dekker, Inc., New York.
16. Allen, E., and Brown, G. E. (1932): Raynaud's disease affecting men. *Ann. Intern. Med.*, 5:1384–1386.
17. Allison, D. J., Modlin, I. M., and Jenkins, W. J. (1977): Treatment of carcinoid liver metastases by hepatic artery embolisation. *Lancet*, 2:1323–1325.
18. Al-Mefty, O., Marano, G., Rajaraman, S., Nugent, G. R., and Rodman, N. (1979): Transient ischemic attacks due to increased platelet aggregation and adhesiveness. Ultrastructural and functional correlation. *J. Neurosurg.*, 50:449–453.
19. Almqvist, P., Kuenzig, M., and Schwartz, S. I. (1983): The effect of naloxone and cyproheptadine

on pulmonary platelet trapping, hypotension, and platelet aggregability in traumatized dogs. *J. Trauma*, 23:405–407.

20. Altura, B. M., and Zweifach, B. W. (1965): Pharmacologic properties of antihistamines in relation to vascular reactivity. *Am. J. Physiol.*, 209:550–556.
21. Alumets, J., Alm, P., Falkmer, S., Håkansson, R., Ljungberg, O., Mårtensson, H., Sundler, F., and Tibblin, S. (1981): Immunohistochemical evidence of peptide hormones in endocrine tumors of the rectum. *Cancer*, 48:2409–2412.
22. Alumets, J., Håkansson, R., Ingemansson, D., and Sundler, F. (1977): Substance P and 5-HT in granules isolated from an intestinal argentaffin carcinoid. *Histochemistry*, 52:217–222.
23. Alumets, J., Håkansson, R., Sundler, F., and Chang, K.-J. (1978): Leu-enkephalin-like material in nerves and enterochromaffin cells in the gut. An immunocytochemical study. *Histochemistry*, 56:187–196.
24. Alumets, J., Sundler, F., Falkmer, S., Ljungberg, O., Håkansson, R., Mårtensson, H., Nobin, A., and Lasson, Å. (1983): Neurohormonal peptides in endocrine tumors of the pancreas, stomach and upper small intestine: I. An immunohistochemical study of 27 cases. *Ultrastruct. Pathol.*, 5:469–486.
25. Amadeo, P. F., Angelico, F., Arca, M., and Ricci, G. (1980): Circulating platelet aggregates in an adult population sample. *Atherosclerosis*, 35:375–381.
26. Anderson, A., Morgan, T., and Dumpys, R. (1983): The effect of ketanserin in essential hypertension. *Clin. Exp. Pharmacol. Physiol.*, 10:331–333.
27. Anderson, G. H., Hellums, J. D., Moake, J., and Alfrey, C. P. (1978): Platelet response to shear stress: Changes in serotonin uptake, serotonin releases and ADP induced aggregation. *Thromb. Res.*, 13:1039–1047.
28. Anderson, G. M., Young, J. G., and Batter, D. K. (1981): Determination of indoles and catechols in rat brain and pineal gland using liquid chromatography with fluorometric and amperometric detection. *J. Chromatogr.*, 223:315–320.
29. Anderson, J. L. (1983): Serotonin receptor changes after chronic antidepressant treatments: Ligand binding, electrophysiological and behavioural studies. *Life Sci.*, 32:1791–1801.
30. Andren, L., Svensson, A., Dahlöf, B., Eggertsen, R., and Hansson, L. (1983): Ketanserin in hypertension. Early clinical evaluation and dose finding study of a new 5-HT_2 receptor antagonist. *Acta Med. Scand.*, 214:125–130.
31. Andrew, A., Kramer, B., and Rowdon, B. B. (1980): The embryonic origin of gut endocrine cells. *Regul. Pept. (Suppl.)*, 1:S5.
32. Antonsen, S., Hansen, M. G. J., Bukhave, K., and Rask-Madsen, J. The influence of a new selective 5HT_2 receptor antagonist on jejunal prostaglandin E2 release and ion secretion due to malignant carcinoid syndrome *(unpublished study)*.
33. Apperley, E., Feniuk, W., Humphrey, P. P. A., and Levy, G. P. (1980): Evidence for two types of excitatory receptors for 5-hydroxytryptamine in dog isolated vasculature. *Br. J. Pharmacol.*, 68:215–224.
34. Apperley, E., Humphrey, P. P. A., and Levy, G. P. (1976): Receptors for 5-hydroxytryptamine and noradrenaline in rabbit isolated ear artery and aorta. *Br. J. Pharmacol.*, 58:212–221.
35. Apperley, E., Humphrey, P. P. A., and Levy, G. P. (1977): Two types of excitatory receptor for 5-hydroxytryptamine in dog vasculature? *Proc. Br. Pharmacol. Soc.*, 7–9 September, 465P.
36. Arias, F., Andrinopoulos, G., and Zamora, J. (1979): Whole blood fibrinolytic activity in normal and hypertensive pregnancies and its relation to placental concentration of urokinase inhibitor. *Am. J. Obstet. Gynecol.*, 133:624.
37. Arias, F., and Mancilla-Jimenez, R. (1976): Hepatic fibrinogen deposits in preeclampsia—Immunofluorescent evidence. *N. Engl. J. Med.*, 295:578.
38. Arnoux, B., Durand, J., Rigaud, M., Vargaftig, B. B., and Benveniste, J. (1981): Release of platelet activating factors (PAF-acether) and arachidonic acid metabolites from alveolar macrophages. *Agents Actions*, 11:555–556.
39. Arriaga, F., Leitao, J., Mills, F. J., Padua, J., Ruiz, I., Tropa, G., and Soura, M. P. (1984): R 55 667, an effective non-benzodiazepine anxiolytic. *Collegium Internationale Neuro-Psychopharmacologicum, 14th CINP Congress, Book of Abstracts*, p. 726.
40. Arvidsson, S., Falk, A., Haglind, E., and Haglund, U. (1983): The role of 5-hydroxytryptamine in the feline response to intravenous infusion of live *E. coli*. *Br. J. Pharmacol.*, 79:711–718.
41. Arvidsson, S., Lindblad, B., Esquivel, C., Fält, K., Lindström, C., Bergqvist, D., and Haglund,

U. (1984): The effects of dihydroergotamine on the feline cardiovascular response to i.v. infusion of live *E. coli* bacteria. *Eur. Surg. Res. (in press).*
42. Ashford, T. (1968): Platelet aggregation at sites of minimal endothelial injury. *Am. J. Pathol.*, 53:599–607.
43. Askenase, P. W., Bursztajn, S., Gershon, M. D., and Gershon, R. K. (1980): T cell-dependent mast cell degranulation and release of serotonin in murine delayed-type hypersensitivity. *J. Exp. Med.*, 152:1358–1374.
44. Astrom, A., and Samelius, M. (1957): The action of 5-hydroxytryptamine and some of its antagonists on the umbilical vessels of the human placenta. *Br. J. Pharmacol.*, 12:410.
45. Atack, C. (1977): Measurement of biogenic amines using cation exchange chromatography with fluorometric assay. *Acta Physiol. Scand. (Suppl.* 451).
46. Atack, C., and Magnusson, T. (1978): A procedure for the isolation of noradrenaline (together with adrenaline), dopamine, 5-hydroxytryptamine, and histamine from the same tissue sample using a single column of strongly acidic cation exchange resin. *Acta Pharmacol. Toxicol.*, 42:35–57.
47. Awouters, F., Leysen, J. E., De Clerck, F., and Van Nueten, J. M. (1982): General pharmacological profile of ketanserin (R 41 468), a selective 5-HT_2 receptor antagonist. In: *5-Hydroxytryptamine in Peripheral Reactions*, edited by F. De Clerck and P. M. Vanhoutte, pp. 193–197. Raven Press, New York.
48. Axelrod, J. (1962): The enzymatic *N*-methylation of serotonin and other amines. *J. Pharmacol. Exp. Ther.*, 138:28–33.
49. Axelrod, J., and Inscoe, J. K. (1963): The uptake and binding of circulating serotonin and the effect of drugs. *J. Pharmacol. Exp. Ther.*, 141:161–165.
50. Azmitia, E. C., and Segal, M. (1978): An autoradiographic analysis of the differential ascending projections of the dorsal and median raphe nuclei in the rat. *J. Comp. Neurol.*, 179:641–668.
51. Baele, G., Bogaerts, H., Clement, D. L., Pannier, R., and Barbier, F. (1981): Platelet activation during treadmill exercise in patients with chronic peripheral arterial disease. *Thromb. Res.*, 23:215–223.
52. Baer, P. G., Kauker, M. L., and McGiff, J. C. (1979): Prostacyclin effects on renal hemodynamic and excretory functions in the rat. *J. Pharmacol. Exp. Ther.*, 208:294–299.
53. Bakhle, Y. S. (1976): The inactivation of endogenous amines. In: *Agents Actions*, 6:505–509.
54. Bakhle, Y. S., and Vane, J. R. (1974): Pharmacokinetic function of the pulmonary circulation. *Physiol. Rev.*, 54:1007–1045.
55. Ball, H. A., Parratt, J. R., and Rodger, I. W. (1983): The effect of a selective 5-HT_2 antagonist, ketanserin, on the pulmonary responses to *Escherichia coli* endotoxin. *Br. J. Pharmacol.*, 80:295–301.
56. Ball, S. G., Zabludowski, J. R., and Robertson, J. I. S. (1983): Mechanism of antihypertensive action of ketanserin in man. *Br. Med. J.*, 287:1065.
57. Bancks, N. H., Goldstein, H. M., and Dodd, G. D. (1975): The roentgenologic spectrum of small intestinal carcinoid tumors. *Am. J. Roentgenol.*, 123:274–280.
58. Barbaccia, M. L., Gandolfi, O., Chuang, D.-M., and Costa, E. (1983): Differences in the regulatory adaptation of the $5HT_2$ recognition sites labelled by ^{3}H-mianserin and ^{3}H-ketanserin. *Neuropharmacology*, 22:123–126.
59. Barger, A. C., Beeuwkes, R., Lainey, L. L., and Silverman, K. J. (1984): Vasa vasorum and neovascularization of human coronary arteries: A possible role in the pathophysiology of atherosclerosis. *N. Engl. J. Med.*, 310:175–178.
60. Basran, G. S., Dewar, A., Morley, J., Page, C. P., Paul, W., and Wood, L. (1982): Properties of platelet activating factor (PAF-acether) which suggest involvement in chronic inflammation and persistent asthma. *Br. J. Pharmacol.*, 77:437P.
61. Baum, T., and Shropshire, A. T. (1975): Inhibition of efferent sympathetic nerve activity by 5-hydroxytryptophan and centrally administered 5-hydroxytryptamine. *Neuropharmacology*, 14:227–233.
62. Baumgarten, H. G., and Björklund, A. (1976): Neurotoxic indoleamines and monoamine neurons. *Annu. Rev. Pharmacol. Toxicol.*, 16:101–111.
63. Baumgartner, H. R., and Born, G. V. R. (1968): Effects of 5-hydroxytryptamine on platelet aggregation. *Nature*, 218:137–141.
64. Beauvallet, M., Godefroy, F., Weil-Fugazza, J. (1968): Modification de la teneur en 5-hydroxy-

tryptamine du coeur au cours d'une surcharge du regime en chlorure de sodium. *C. R. Soc. Biol. (Paris)*, 162:2085–2088.
65. Beetens, J. R., Van Hove, C., Rampart, M., and Herman, A. G. (1983): Acetylcholine stimulates the release of prostacyclin by rabbit aorta endothelium. *J. Pharm. Pharmacol.*, 35:251–252.
66. Beets, J. L., and Paul, W. (1980): Actions of locally administered adrenoreceptor agonists on increased plasma protein extravasation and blood flow in guinea-pig skin. *Br. J. Pharmacol.*, 70:461–467.
67. Beitz, A. J. (1982): The sites of origin of brain stem neurotensin and serotonin projections to the rodent nucleus raphe magnus. *J. Neurosci.*, 2(7):829–842.
68. Belcher, G., Ryall, R. W., and Schaffner, R. (1978): The differential effects of 5-hydroxytryptamine, noradrenaline and raphe stimulation on nociceptive and non-nociceptive dorsal horn interneurones in the cat. *Brain Res.*, 151:307–321.
69. Benarroch, E. E., Balda, M. S., Finkielman, S., and Nahmod, V. E. (1983): Neurogenic hypertension after depletion of norepinephrine in anterior hypothalamus induced by 6-hydroxydopamine administration into the ventral pons: Role of serotonin. *Neuropharmacology*, 22:29–34.
70. Benditt, E. P., Wong, R. L., Arase, M., and Roeper, E. (1955): 5-Hydroxytryptamine in mast cells. *Proc. Soc. Exp. Biol. Med.*, 90:303–304.
71. Benfey, B. G., Cohen, J., Kunos, G., and Vermes-Kunos, I. (1974): Dissociation of 5-hydroxytryptamine: Effects on myocardial contractility and cyclic AMP accumulation. *Br. J. Pharmacol.*, 50:581–585.
72. Bengmark, S., Ericsson, M., Lunderquist, A., Mårtensson, H., Nobin, A., and Sako, M. (1982): Temporary liver dearterialization in patients with metastatic carcinoid disease. *World J. Surg.*, 6:46–53.
73. Benirschke, K. (1962): A review of the pathologic anatomy of the human placenta. *Am. J. Obstet. Gynecol.*, 84:1595.
74. Bennett, J. P., Jr., and Snyder, S. H. (1975): Stereospecific binding of *d*-lysergic acid diethylamide (LSD) to brain membranes: Relationship to serotonin receptors. *Brain Res.*, 94:523–544.
75. Bennett, J. P., Jr., and Snyder, S. H. (1976): Serotonin and lysergic acid diethylamide binding in rat brain membranes: Relationship to post-synaptic serotonin receptors. *Mol. Pharmacol.*, 12:373–389.
76. Bentley, K. I., and Henry, P. D. (1981): Serotonin, a potent constrictor of human coronary artery. *Circulation*, 64:446A (abstract).
77. Berecek, K. H., Schwertschlag, U., and Gross, F. (1980): Alterations in renal vascular resistance and reactivity in spontaneous hypertension of rats. *Am. J. Physiol.*, 238:H287–H293.
78. Berge, T., and Linell, F. (1976): Carcinoid tumors. *Acta Pathol. Microbiol. Scand. A*, 84:322–330.
79. Berk, M. L., and Finkelstein, J. A. (1981): Afferent projections to the preoptic area and hypothalamic regions in the rat brain. *Neuroscience*, 6(8):1601–1624.
80. Berkowitz, B. A., Lee, C. H., and Spector, S. (1974): Disposition of serotonin in the rat blood vessels and heart. *Clin. Exp. Pharmacol. Physiol.*, 1:397–400.
81. Bernd, P., Gershon, M. D., Nunez, E. A., and Tamir, H. (1981): Separation of dissociated thyroid follicular and parafollicular cells: Association of serotonin binding protein with parafollicular cells. *J. Cell Biol.*, 88:499–508.
82. Bernd, P., Gershon, M. D., Nunez, E. A., and Tamir, H. (1983): Separation of parafollicular cells from thyroid follicular cells by affinity chromatography using thyroglobulin-coupled sepharose. In: *Cell Separation: Methods and Selected Application, Vol. 2*, pp. 1–14.
83. Berne, R. M., and Rubio, R. (1979): Coronary circulation. In: *Handbook of Physiology, Section 2: The Cardiovascular System, Vol. 1*, edited by R. M. Berne, N. Sperelakis, and S. R. Geiger, pp. 873–952. The American Physiological Society, Bethesda, MD.
84. Bertler, A., Falck, B., and Owman, C. (1964): Studies on 5-hydroxytryptamine stores in pineal gland of rat. *Acta Physiol. Scand.*, 63(*Suppl.* 239):1–18.
85. Bevan, D., Yates, C. J. P., Dormandy, J. A., and Flute, P. T. (1979): Beta-thromboglobulin in peripheral artery disease. *Thromb. Haemost.*, 42:147 (abstract).
86. Bevan, J. A. (1983): Diltiazem selectively inhibits cerebrovascular extrinsic but not intrinsic myogenic tone. A review. *Circ. Res.*, 52:104–109.
87. Bevan, J. A., and Heptinstall, S. (1983): Effects of ketanserin and mepyramine on platelet aggregation and on the uptake of 5-hydroxytryptamine into platelets. *Thromb. Res.*, 30:415–423.

88. Bhargava, K. P., Rhina, N., Misra, N., Shanker, K., and Vrat, S. (1979): Uptake of serotonin by human platelets and its relevance to CNS involvement in hypertension. *Life Sci.*, 25:195–199.
89. Bhargava, K. P., and Tangri, K. K. (1959): The central vasomotor effects of 5-hydroxytryptamine. *Br. J. Pharmacol. Chem.*, 14:411–414.
90. Bhattacharya, B. K. (1955): A pharmacological study on the effect of 5-hydroxytryptamine and its antagonist on the bronchial musculature. *Arch. Intern Pharmacodyn.*, 103:357–369.
91. Bhattacharya, J., Nanjo, S., and Staub, N. C. (1982): Micropuncture measurement of lung microvascular pressure during 5HT infusion. *J. Appl. Physiol.*, 52:634–637.
92. Biber, B., Fara, J., and Lundgren, O. (1973): Intestinal vascular responses to 5-hydroxytryptamine. *Acta Physiol. Scand.*, 87:526–534.
93. Billah, M. M., and Lapetina, E. G. (1982): Rapid decrease of phosphatidylinositol 4,5 bisphosphate in thrombin-stimulated platelets. *J. Biol. Chem.*, 257:12705–12708.
94. Billah, M. M., Lapetina, E. G., and Cuatrecasas, P. (1981): Phospholipase A_2 activity specific for phosphatidic acid. *J. Biol. Chem.*, 256:5399–5403.
95. Binet, L., and Burstein, M. (1955): Sur les effets vasculaires peripherigues du 5-hydroxytryptamine creatinine sulfate (serotonine). *Societe de Biologie* 12 November, pp. 1860–1866.
96. Black, J., Fisher, E., and Smith, A. (1959): The effect of precursors of 5-HT on gastric secretion in anesthetized dogs. *J. Physiol. (Lond.)*, 140:10–17.
97. Black, J. L., French, R. J., and Mylecharane, E. J. (1981): Receptor mechanisms for 5-hydroxytryptamine in rabbit arteries. *Br. J. Pharmacol.*, 74:619–626.
98. Blackshear, J. L., Orlandi, C., Garnic, J. D., and Hollenberg, N. K. (1983): Serotonin induces large artery spasm *in vivo* via 5HT-2 receptors. *Circulation* (II), 68:III–197.
99. Blackshear, J. L., Orlandi, C., and Hollenberg, N. K. (1983): Modification of the renal vascular response to serotonin by indomethacin. Presented at the American Society of Nephrology, 16th Annual Meeting, December 4–7, Washington,DC *Kidney Int. (in press).*
100. Blaisdell, F. W., Lim, R. C., Jr., and Stallone, R. J. (1970): The mechanism of pulmonary damage following traumatic shock. *Surg. Gynecol. Obstet.*, 130:15–22.
101. Block, E. R., and Fisher, A. B. (1977): Depression of serotonin clearance by rat lungs during oxygen exposure. *J. Appl. Physiol.*, 42:33–38.
102. Block, E. R., and Stalcup, S. A. (1982): Metabolic functions of the lung: Of what clinical relevance? *Chest*, 81:215–223.
103. Bloom, A. L., and Thomas, D. P. (1981): *Haemostasis and Thrombosis*, pp. 593–795. Churchill Livingstone, Edinburgh.
104. Blunt, R. J., and Porter, J. M. (1981): Raynaud syndrome. *Semin. Arthritis Rheum.*, 10:282–308.
105. Bobillier, P., Seguin, S., Petitjean, F., Salvert, D., Touret, M., and Jouvet, M. (1976): The raphe nuclei of the cat brainstem: A topographical atlas of their efferent projections as revealed by autoradiography. *Brain Res.*, 113:449–486.
106. Bogdanski, D., Pletscher, A., Brodie, B., and Udenfriend, S. (1956): Identification and assay of serotonin in brain. *J. Pharmacol. Exp. Ther.*, 117:82–88.
107. Bohr, D. F., and Elliot, J. (1962): 5-Hydroxytryptamine and the contraction of smooth muscle from resistance vessels. In: *Metabolismus Parietis Vasorum*, edited by B. Prusik, Z. Reinis, and O. Riedl, pp. 337–345. Statni Zdravotnicke Nakladatelstvi, Prague.
108. Boijsen, E., Kaude, J., and Tylén, U. (1974): Radiologic diagnosis of ileal carcinoid tumours. *Acta Radiol.*, 15:65–82.
109. Bojs, G. (1961): An experimental study on serotonin (5HT) on man (with special reference to renal function). *Scand. J. Clin. Lab. Invest.*, 13:188.
110. Boneu, B., Fournié, A., Sci, P., Grandjean, H., Bierme, R., and Pontonnier, G. (1980): Platelet production time, uricemia and some hemostasis tests in pre-eclampsia. *Eur. J. Obstet. Gynecol. Reprod. Biol.*, 11:85–94.
111. Boogaerts, M. A., Yamada, O., Jacobs, H. S., and Moldow, C. F. (1982): Enhancement of granulocyte-endothelial cell adherence and granulocyte induced cytotoxicity by platelet release products. *Proc. Natl. Acad. Sci. U.S.A.*, 79:7019–7023.
112. Born, G. V. R., and Bricknell, J. (1959): The uptake of 5-hydroxytryptamine by blood platelets in the cold. *J. Physiol. (Lond.)*, 147:153–161.
113. Born, G. V. R., and Gilson, R. R. (1959): Studies on the uptake of 5-hydroxytryptamine by blood platelets. *J. Physiol. (Lond.)*, 146:472–491.
114. Born, G. V. R., Juengjaroen, K., and Michal, F. (1972): Relative activities of and uptake by

human blood platelets of 5-hydroxytryptamine and several analogues. *Br. J. Pharmacol.*, 44:117–139.

115. Born, G. V. R., and Michal, F. (1975): 5-Hydroxytryptamine receptors of platelets. In: *Biochemistry and Pharmacology of Platelets: Ciba Foundation Symposium 35*, pp. 287–307. Elsevier-Excerpta Medica-North-Holland, Amsterdam, Oxford, New York.
116. Borzotta, A. P., and Polk, H. C., Jr. (1983): Multiple system organ failure. *Surg. Clin. North Am.*, 63:315–336.
117. Bosin, T. R., and Lahr, P. D. (1981): Mechanisms influencing the disposition of serotonin in mouse lung. *Biochem. Pharmacol.*, 30:3187–3193.
118. Boucek, R. J., Speropoulus, A. J., and Noble, N. C. (1972): Serotonin and ribonucleic acid and collagen metabolism of fibroblasts *in vitro*. *Proc. Soc. Exp. Biol. Med.*, 140:599–603.
119. Bouchard, S., and Roberge, A. G. (1979): Biochemical properties and kinetic parameters of dihydroxyphenylalanine-5-hydroxytryptophan decarboxylase in brain, liver, and adrenals of cat. *Can. J. Biochem.*, 57:1014–1018.
120. Bounous, G. (1969): "Tryptic enteritis": its role in the pathogenesis of stress ulcer and shock. *Can. J. Surg.*, 12:397–409.
121. Bove, A. A., and Dewey, J. D. (1983): Effects of serotonin and histamine on proximal and distal coronary vasculature in dogs: Comparison with alpha-adrenergic stimulation. *Am. J. Cardiol.*, 52:1333–1339.
122. Bowker, R. M., Westlund, K. N., and Coulter, J. D. (1981): Origins of serotonergic projections to the spinal cord in rat: An immunocytochemical-retrograde transport study. *Brain Res.*, 226:187–199.
123. Bowker, R. M., Westlund, K. N., Sullivan, M. C., Wilber, J. F., and Coulter, J. P. (1983): Descending serotonergic, peptidergic and cholinergic pathways from the raphe nuclei: A multiple transmitter complex. *Brain Res.*, 288:33–48.
124. Branchek, T., Kates, M., and Gershon, M. D. (1984): Enteric receptors for 5-hydroxytryptamine. *Brain Res. (in press)*.
125. Braunstein, P. W., Cuénoud, H. F., Joris, I., and Majno, G. (1980): Platelets, fibroblasts, and inflammation. *Am. J. Pathol.*, 99:53–66.
126. Brigham, K. L., Bowers, R. E., and Haynes, J. (1979): Increased lung vascular permeability caused by *Escherichia coli* endotoxin. *Circ. Res.*, 45:292–297.
127. Brigham, K. L., and Owen, P. J. (1975): Increased sheep lung vascular permeability caused by histamine. *Circ. Res.*, 37:647–657.
128. Brigham, K. L., and Owen, P. J. (1975): Mechanism of serotonin effect on lung transvascular fluid and protein movement in awake sheep. *Circ. Res.*, 36:761–770.
129. British Coronary Prevention Group (1982): Prevention of coronary heart disease in the United Kingdom. *Lancet*, 2:846–847.
130. Broekman, M. J., Ward, J. W., and Marcus, A. J. (1980): Phospholipid metabolism of human platelets. Changes in phosphatidylinositol, phosphatidic acid and lysophospholipids. *J. Clin. Invest.*, 66:275–283.
131. Brown, H. (1977): Serotonin-producing tumors. In: *Serotonin in Health and Disease, Vol. 4*, edited by W. B. Essman, pp. 393–423. Spectrum Publications, New York.
132. Brownlee, G., and Johnson, E. S. (1963): The site of 5-hydroxytryptamine receptor on the intramural nervous plexus of the guinea-pig isolated ileum. *Br. J. Pharmacol.*, 21:306–322.
133. Bruinink, A., Lichtensteiger, W., and Schlumpf, M. (1983): Pre- and postnatal ontogeny and characterization of dopaminergic D_2, serotonergic S_2, and spirodecanone binding sites in rat forebrain. *J. Neurochem.*, 40:1227–1236.
134. Buccino, R. A., Covell, J. W., Sonnenblick, E. H., and Braunwald, E. (1967): Effects of serotonin on the contractile state of the myocardium. *Am. J. Physiol.*, 213:483–486.
135. Bulbring, E., and Crema, A. (1958): Observation concerning the action of 5-hydroxytryptamine on the peristaltic reflex. *Br. J. Pharmacol.*, 13:444–457.
136. Bulbring, E., and Crema, A. (1959): The release of 5-hydroxytryptamine in relation to pressure exerted on the intestinal mucosa. *J. Physiol. (Lond.)*, 146:18–28.
137. Bulbring, E. R., Lin, C. Y., and Schofield, G. C. (1958): An investigation of the peristaltic reflex in relation to anatomical observations. *Q. J. Exp. Physiol.*, 43:26–37.
138. Bunting, S., Gryglewski, R. J., Moncada, S., and Vane, J. R. (1976): Arterial walls generate from prostaglandin endoperoxides a substance (prostaglandin X) which relaxes strips of mesenteric and coeliac arteries and inhibits platelet aggregation. *Prostaglandins*, 12:896–913.

139. Bunting, S., Moncada, S., and Vane, J. R. (1977): Antithrombotic properties of vascular endothelium. *Lancet*, 2:1075–1076.
140. Burgen, A. S. V., and Iversen, L. L. (1965): The inhibition of noradrenaline uptake by sympathomimetic amines in the rat isolated heart. *Br. J. Pharmacol.*, 25:34–49.
141. Burrows, A., Chavin, S., and Hockaday, T. (1978): Plasma thromboglobulin concentrations in diabetes mellitus. *Lancet*, 1:235–237.
142. Cameron, H. A., and Ramsey, L. E. (1984): A double-blind placebo controlled study of ketanserin in hypertension. *Br. J. Clin. Pharmacol.*, 18:275P.
143. Campbell, P. M., and LeRoy, E. C. (1975): Pathogenesis of systemic sclerosis: A vascular hypothesis. *Semin. Arthritis Rheum.*, 4:351–368.
144. Cannon, J. G., Costall, B., Laduron, P. M., Leysen, J. E., and Naylor, R. J. (1978): Effects of some derivatives of 2-aminotetralin on dopamine-sensitive adenylate cyclase and on the binding of [^{3}H]haloperidol to neuroleptic receptors in the rat striatum. *Biochem. Pharmacol.*, 27:1417–1420.
145. Caramona, M. M. (1983): Localization of monoamine oxidase of type A and B in blood vessels with different innervation patterns. *Naunyn Schmiedebergs Arch. Pharmacol.*, 32:185–189.
146. Carlsson, A., Shore, P. A., and Brodie, B. B. (1957): Release of serotonin from blood platelets by reserpine *in vitro*. *J. Pharmacol. Exp. Ther.*, 120:334–339.
147. Carr, R. J., and Biswas, B. K. (1966): Methysergide and retroperitoneal fibrosis. *Br. Med. J.*, 2:1116–1117.
148. Carter, F. B., Cherny, W. B., and Crenshaw, C. (1962): Serotonin studies in abnormal pregnancies—A preliminary report. *Am. J. Obstet. Gynecol.*, 84:913.
149. Cassuto, J., Jodal, M., Sjövall, H., and Lundgren, O. (1981): Nervous control of intestinal secretion. In: *Clinical Research Review, Vol. 1 (Suppl. 1): Diarrhoea: New Insights*, edited by N. W. Read, pp. 12–21. Janssen Research Foundation.
150. Cassuto, J., Jodal, M., Tuttle, R., and Lundgren, O. (1981): On the role of intramural nerves in the pathogenesis of choleratoxin-induced intestinal secretion. *Scand. J. Gastroenterol.*, 16:377–384.
151. Castor, C. W., Ritchie, J. C., Williams, C. H., Scott, M. E., Whitney, S. L., Myers, S. L., Sloan, T. B., and Anderson, B. E. (1979): Connective tissue activation. XIV. Composition and actions of a human platelet autacoid mediator. *Arthritis Rheum.*, 22:260–272.
152. Catravas, J. D., and Gillis, C. N. (1980): Pulmonary clearance of ^{14}C-5-hydroxytryptamine and ^{3}H-norepinephrine *in vivo*: Effects of pretreatment with imipramine or cocaine. *J. Pharmacol. Exp. Ther.*, 213:120–127.
153. Catravas, J. D., and Gillis, C. N. (1983): Single-pass removal of [^{14}C]-5-hydroxytryptamine and [^{3}H]-norepinephrine by rabbit lung *in vivo*: Kinetics and sites of removal. *J. Pharmacol. Exp. Ther.*, 224:28–33.
154. Cella, G., Colby, S. I., Taylor, A. D., McCracken, L., Parisi, A. F., and Sasahara, A. A. (1983): Platelet factor 4 (PF4) and heparin-released platelet factor 4 (HR-PF4) in patients with cardiovascular disorders. *Thromb. Res.*, 29:499–510.
155. Cella, G., Zahavi, J., de Haas, H. A., and Kakkar, V. V. (1979): Beta-thromboglobulin, platelet production time and platelet function in vascular disease. *Br. J. Haematol.*, 43:127–136.
156. Ceulemans, D., Hoppenbrouwers, M.-L., Gelders, Y., and Reyntjens, A. (1984): Serotonin blockade or benzodiazepine: What kind of analysis. *Collegium Internationale Neuro-Psychopharmacologicum, 14th CINP Congress, Book of Abstracts*, p. 727.
157. Chahl, L. A. (1976): Interactions of bradykinin, prostaglandin E_1, 5-hydroxytryptamine, histamine and adenosine-5′-triphosphate on dye leakage response in rat skin. *J. Pharm. Pharmacol.*, 28:753–757.
158. Chalmers, J. P. (1975): Brain amines and models of experimental hypertension. *Circ. Res.*, 36:469–480.
159. Chand, N., and Altura, B. M. (1981): Acetylcholine and bradykinin relax intrapulmonary arteries by acting on endothelial cells: Role in lung vascular diseases. *Science*, 213:1376–1379.
160. Chand, N., and Altura, B. M. (1981): Endothelial cells and relaxation of vascular smooth muscle cells: Possible relevance to lipoxygenases and their significance in vascular diseases. *Microcirculation*, 1:297–317.
161. Chand, N., and Altura, B. M. (1981): Inhibition of endothelial cell-dependent relaxations to acetylcholine and bradykinin by lipoxygenase inhibition in canine isolated renal arteries. *Microcirculation*, 1:211–223.

162. Chan-Palay, V. (1975): Fine structure of labelled axons in the cerebellar cortex and nuclei of rodents and primates after intraventricular infusions with tritiated serotonin. *J. Anat. Embryol.*, 148:235–265.
163. Cheng, Y. C., and Prusoff, W. H. (1973): Relationship between the inhibition constant (K_i) and the concentration of inhibitor which causes 50 per cent inhibition (I_{50}) of an enzymatic reaction. *Biochem. Pharmacol.*, 22:3099–3108.
164. Cherry, P. D., Furchgott, R. F., and Zawadski, J. V. (1982): Role of endothelial cells in relaxation of isolated arteries by bradykinin. *Proc. Natl. Acad. Sci. U.S.A.*, 79:2106–2110.
165. Chesley, L. C. (1978): *Hypertensive Disorders in Pregnancy*. Appleton-Century Crofts, New York.
166. Chesley, L. C. (1979): False steps in the study of preeclampsia. *J. Reprod. Med.*, 23:151–158.
167. Chien, S. (1975): Biophysical behavior of red cells in suspensions. In: *The Red Blood Cell, Vol. 2*, edited by Douglas Mac N. Surgenor, pp. 1031–1133. Academic Press, New York.
168. Chien, S. (1977): Blood rheology in hypertension and cardiovascular diseases. *Cardiovasc. Med.*, 356–360.
169. Chien, S. (1981): Haemorheology in disease: Pathophysiological significance and therapeutic implications. *Clin. Haemorheol.*, 1:419–442.
170. Choy, V. J., and Chalmers, J. P. (1984): Importance of central serotonin neurons in the hypotensive action of methyldopa in the rat. *Clin. Exp. Pharmacol. Physiol.*, 11:37–44.
171. Ciaccio, C. (1906): Sur une nouvelle espèce cellulaire dans les glands de Lieberkühn. *C. R. Soc. Biol. Paris*, 1:76–79.
172. Clement, D., Vanhoutte, P., and Leusen, I. (1969): Capacitance reactions of isolated veins to monoamines and acetylcholine. *Arch. Int. Physiol. Biochem.*, 77:73–87.
173. Coccheri, S., and Fiorentini, P. (1971): Platelet adhesiveness and aggregation in hypertensive patients. *Acta Med. Scand.*, 525:273–275.
174. Cocks, T. M., and Angus, J. A. (1983): Endothelium-dependent relaxation of coronary arteries by noradrenaline and serotonin. *Nature (Lond.)*, 305:627–630.
175. Cohen, M. L., Fuller, R. W., and Kurz, K. D. (1983): Evidence that blood pressure reduction by serotonin antagonists is related to alpha receptor blockade in spontaneously hypertensive rats. *Hypertension*, 5:676–681.
176. Cohen, M. L., Fuller, R. W., and Wiley, K. S. (1981): Evidence for 5-HT_2 receptors mediating contraction in vascular smooth muscle. *J. Pharmacol. Exp. Ther.*, 218:421–425.
177. Cohen, M. L., Mason, N., Wiley, K. S., and Fuller, R. W. (1983): Further evidence that vascular serotonin receptors are of the 5-HT_2 type. *Biochem. Pharmacol.*, 32:567–570.
178. Cohen, R. A., and Coffman, J. D. (1984): Digital vasospasm: The pathophysiology of Raynaud's phenomenon. *Int. Angiol.*, 3:47–55.
179. Cohen, R. A., Shepherd, J. T., and Vanhoutte, P. M. (1983): 5-Hydroxytryptamine can mediate endothelium-dependent relaxation of coronary arteries. *Am. J. Physiol.*, 245:H1077–H1080.
180. Cohen, R. A., Shepherd, J. T., and Vanhoutte, P. M. (1983): Inhibitory role of the endothelium in the response of isolated coronary arteries to platelets. *Science*, 221:273–274.
181. Cohen, R. A., Shepherd, J. T., and Vanhoutte, P. M. (1984): Endothelium and asymmetrical responses of the coronary arterial wall. *Am. J. Physiol.*, 247:H403–H408.
182. Colebatch, H. J. H., Olsen, C. R., and Nadel, J. A. (1966): Effect of histamine, serotonin and acetylcholine on the peripheral airways. *J. Appl. Physiol.*, 21:217–226.
183. Collis, M. G., DeMey, C., and Vanhoutte, P. M. (1979): Renal vascular reactivity in the young spontaneously hypertensive rat. *Hypertension*, 2:45–52.
184. Collis, M. G., and Vanhoutte, P. M. (1977): Vascular reactivity of isolated perfused kidneys from male and female spontaneously hypertensive rats. *Circ. Res*, 41:759–767.
185. Collis, M. G., and Vanhoutte, P. M. (1981): Tachyphylaxis to 5-hydroxytryptamine in perfused kidneys from spontaneously hypertensive and normotensive rats. *J. Cardiovasc. Pharmacol.*, 3:229–235.
186. Colpaert, F. C., Niemegeers, C. J. E., and Janssen, P. A. J. (1982): A drug discrimination analysis of lysergic acid diethylamide (LSD): *In vivo* agonist and antagonist effects of purported 5-hydroxytryptamine antagonist and of pirenperone, a LSD-antagonist. *J. Pharmacol. Exp. Ther.*, 221:206–214.
187. Comroe, J. H., VanLingen, B., Stroud, R. C., and Roncoroni, A. (1953): Reflex and direct cardiopulmonary effects of 5-OH-tryptamine (serotonin). *Am. J. Physiol.*, 173:379–386.

188. Connors, A. F., McCafree, D. R., and Rogers, R. M. (1981): The adult respiratory distress syndrome. *DM*, 27:(4).
189. Conrad, L. C. A., Leonard, C. M., and Pfaff, D. W. (1974): Connections of the median and dorsal raphe nuclei in the rat: An autoradiographic and degeneration study. *J. Comp. Neurol.*, 156:179–206.
190. Coote, J. H., and Macleod, V. H. (1974): The influence of bulbospinal monoaminergic pathways on sympathetic nerve activity. *J. Physiol. (Lond.)*, 241:453–475.
191. Coote, J. H., Macleod, V. H., Fleetwood-Walker, S., and Gilbey, M. P. (1981): The response of individual sympathetic preganglionic neurones to microelectrophoretically applied endogenous monoamines. *Brain Res.*, 215:135–145.
192. Copley, A. L., and Seaman, G. V. F., editors (1982): *Surface Phenomena in Haemorheology: Their Theoretical, Experimental and Clinical Aspects*. Symposium of New York Academy of Sciences, October *(to be published* by New York Academy of Sciences).
193. Cortellazzo, S., Viero, P., Barbui, T., and deGaetano, G. (1982): Clinical syndromes related to platelet activation *in vivo*. In: *5-Hydroxytryptamine in Peripheral Reactions*, edited by F. De Clerck and P. M. Vanhoutte, pp. 121–127. Raven Press, New York.
194. Costa, M., Furness, J. B., Cuello, A. C., Verhofstad, A. A. J., Steinbusch, H. W. J., and Elde, R. P. (1982): Neurons with 5-hydroxytryptamine-like immunoreactivity in the enteric nervous system: Their visualization and reactions to drug treatment. *Neuroscience*, 7:351–364.
195. Coughlin, S. R., Moskowitz, M. A., Antoniades, H. N., and Levine, L. (1981): Serotonin receptor-mediated stimulation of bovine smooth muscle cell prostacyclin synthesis and its modulation by platelet derived growth factor. *Proc. Natl. Acad. Sci. U.S.A.*, 78:7134–7138.
196. Coughlin, S. R., Moskowitz, M. A., and Levine, L. (1984): Identification of a serotonin type 2 receptor linked to prostacyclin synthesis in vascular smooth muscle cells. *Biochem. Pharmacol.*, 33:692–695.
197. Crawford, N. (1963): Plasma free serotonin (5-hydroxytryptamine). *Clin. Chim. Acta*, 8:39–45.
198. Crawford, N. (1965): Systemic venous platelet bound and plasma free serotonin levels in non carcinoid malignancy. *Clin. Chim. Acta*, 12:274–281.
199. Crawford, N. (1967): Serotonin absorption by normal and "carcinoid" platelets. *Clin. Chim. Acta*, 18:297.
200. Criscione, L., and Hedwall, P. R. (1984): The antihypertensive effect of ketanserin is attributable to α-receptor blockade. *Naunyn Schmiedebergs Arch. Pharmacol.*, 3–5*(Suppl.)*:R55.
201. Crooks, P. A., Dreyer, R. N., Sulens, C. H., Gillis, C. N., and Coward, J. K. (1979): Deamination of 5-hydroxytryptamine metabolites in isolated perfused rabbit lung by high-performance liquid chromatography. *Anal. Biochem.*, 93:143–152.
202. Cross, S. A. M., Alabaster, V. A., Bakhle, Y. S., and Vane, J. R. (1974): Sites of uptake of ^{3}H-5-hydroxytryptamine in rat isolated lung. *Histochemistry*, 39:83–91.
203. Culley, W. J., Saunders, R. N., Mertz, E. T., and Jolly, D. H. (1963): Effect of a tryptophan deficient diet on brain serotonin and plasma tryptophan level. *Proc. Soc. Exp. Biol. Med.*, 113:645–648.
204. Culvenor, A. J., and Lovenberg, W. (1983): Analysis of aromatic L-amino acid decarboxylase. In: *Methods in Biogenic Amine Research*, edited by S. Parvez, T. Nagatsu, I., Nagatsu, and H. Parvez, pp. 375–384. Elsevier Science Publishers B.V., The Netherlands.
205. Curro, F. A., Greenberg, S., Verbeuren, T. S., and Vanhoutte, P. M. (1978): Interaction between alpha-adrenergic and serotonergic activation of canine saphenous veins. *J. Pharmacol. Exp. Ther.*, 207:936–949.
206. Dahlen, S. E., Björk, J., Hedqvist, P., Arfors, K. E., Hammerström, S., Lindgren, J. A., and Samuelsson, B. (1981): Leukotrienes promote plasma leakage and leukocyte adhesion in postcapillary venules: *In vivo* effects with relevance to the acute inflammatory response. *Proc. Natl. Acad. Sci. U.S.A.*, 78:3887–3891.
207. Dahlstrom, A., and Fuxe, K. (1964): Evidence for the existence of monoamine-containing neurons in the central nervous system. I. Demonstration of monoamines in the cell bodies of brain stem neurons. *Acta Physiol. Scand.*, 62(*Suppl.* 232):1–55.
208. D'Amore, P., and Shepro, D. (1977): Stimulation of growth and calcium influx in cultured bovine, aortic endothelial cells by platelets and vasoactive substances. *J. Cell. Physiol.*, 92:177–184.
209. D'Amore, P., and Shepro, D. (1982): Captation of 5-hydroxytryptamine and effects on endothelial

cell metabolism. In: *5-Hydroxytryptamine in Peripheral Reactions*, edited by F. De Clerck and P. M. Vanhoutte, pp. 37–47. Raven Press, New York.

210. Dampney, R. A. L. (1981): Functional organization of central cardiovascular pathways. *Clin. Exp. Pharmacol. Physiol.*, 8:241–259.
211. Daniel, J. L., Molish, I., Rigmaiden, M., and Stewart, G. (1984): Role of myosin phosphorylation in the initiation of the platelet shape change response. *J. Biol. Chem.*, 259 *(in press).*
212. Da Prada, M., Bartholini, G., and Pletscher, A. (1965): Formation of 5-hydroxytryptophol by blood platelets after thrombin and reserpine. *Experientia*, 21:135–137.
213. Da Prada, M., and Picotti, G. B. (1979): Content and subcellular localization of catecholamines and 5-hydroxytryptamine in human and animal blood platelets: Monoamine distribution between platelets and plasma. *Br. J. Pharmacol.*, 65:653–662.
214. Da Prada, M., Picotti, G. B., Kettler, R., and Launay, J. M. (1980): Serotonin, histamine, catecholamine, normetanephrine and octopamine in blood platelets. In: *Platelets: Cellular Response Mechanisms and Their Biological Significance*, edited by A. Rotman, F. A. Meyer, G. Gitler, and A. Silberberg, pp. 277–288. John Wiley and Sons Ltd, New York.
215. Da Prada, M., and Pletscher, A. (1969): Different localization of reserpine and tyramine within the 5-hydroxytryptamine organelles of blood platelets. *Experientia*, 25:923–924.
216. Da Prada, M., and Pletscher, A. (1969): Storage of exogenous monoamines and reserpine in 5-hydroxytryptamine organelles of blood platelets. *Eur. J. Pharmacol.*, 7:45–49.
217. Da Prada, M., Richards, J. G., and Kettler, R. (1981): Amine storage organelles in platelets. In: *Platelets in Biology and Pathology—2*, edited by J. L. Gordon, pp. 107–146. Elsevier-North Holland Biomedical Press, Amsterdam.
218. Da Prada, M., Tranzer, J. P., and Pletscher, A. (1972): Storage of 5-hydroxytryptamine in human platelets. *Experientia*, 28:1328–1329.
219. David, J. L., and De Clerck, F. (1982): Detection of platelet activation *in vivo*: Significance and limitations of the available tests. In: *5-Hydroxytryptamine in Peripheral Reactions*, edited by F. De Clerck and P. M. Vanhoutte, pp. 61–70. Raven Press, New York.
220. Davis, G. R., Camp, R. C., Raskin, P., and Krejs, G. J. (1980): Effect of somatostatin infusion on jejunal water and electrolyte transport in a patient with secretory diarrhea due to a malignant carcinoid syndrome. *Gastroenterology*, 78:346–349.
221. Davis, R. B. (1959): The concentration of serotonin in normal human serum as determined by an improved method. *J. Lab. Clin. Med.*, 54:344–350.
222. Davis, R. B., Meeker, W. R., and Bailey, W. L. (1961): Serotonin release by bacterial endotoxin. *Proc. Soc. Exp. Biol. Med.*, 108:774–776.
223. Davis, T. P., Gehrke, C. W., Gehrke, C. W., Jr., Cunningham, T. D., Kuo, K. C., Gerhardt, K. O., Johnson, H. D., and Williams, C. H. (1979): High-performance liquid chromatographic assay of biogenic amines in biological materials as *O*-phthalaldehyde derivatives. *J. Chromatogr.*, 162:293–310.
224. Davis, V. E., Brown, H., Huff, J. A., and Cashaw, J. L. (1967): The alteration of serotonin metabolism to 5-hydroxytryptophol by ethanol ingestion in man. *J. Lab. Clin. Med.*, 69:132–140.
225. Davis, V. E., Huff, J. A., and Brown, H. (1966): Isolation of serotonin-*O*-sulfate from human urine. *Clin. Chim. Acta*, 13:380–382.
226. Dawes, J., Smith, R. C., and Pepper, D. S. (1978): The release, distribution and clearance of human β-thromboglobulin and platelet factor 4. *Thromb. Res.*, 12:851–881.
227. Day, M., and Vane, J. R. (1963): An analysis of the direct and indirect actions of drugs on the isolated guinea-pig ileum. *Br. J. Pharmacol.*, 20:150–170.
228. De Boer, A. C., Han, P., Turpie, A. G. G., Butt, R., Gent, M., and Genton, E. (1983): Platelet test and antiplatelet drugs in coronary artery disease. *Circulation*, 67:500–511.
229. De Chaffoy de Courcelles, D., De Clerck, F., Leysen, J. E., Van Belle, H., and Janssen, P. A. J. (1984): Evidence that phospholipid turnover is the signal transducing system coupled to serotonin-S_2 receptor sites. *J. Biol. Chem. (in press).*
230. De Chaffoy de Courcelles, D., Roevens, P., and Van Belle, H. (1984): Stimulation by serotonin of 40 K and 20 K protein phosphorylation in human platelets. *FEBS Lett.*, 171:289–292.
231. De Clerck, F., and David, J. L. (1981): Pharmacological control of platelet and red blood cell function in the microcirculation. *J. Cardiovasc. Pharmacol.*, 3:1388–1412.
232. De Clerck, F., David, J. L., and Janssen, P. A. J. (1982): Inhibition of 5-hydroxytryptamine-

induced and -amplified human platelet aggregation by ketanserin (R 41 468), a selective 5-HT_2 receptor antagonist. *Agents Actions*, 12:388–397.

233. De Clerck, F., David, J. L., and Janssen, P. A. J. (1982): Serotonic amplification mechanisms in blood platelets. In: *5-Hydroxytryptamine in Peripheral Reactions*, edited by F. De Clerck and P. M. Vanhoutte, pp. 83–94. Raven Press, New York.
234. De Clerck, F., De Brabander, M., Neels, H., and Vandevelde, V. (1981): Direct evidence for the contractile capacity of endothelial cells. *Thromb. Res.*, 23:505–520.
235. De Clerck, F., and Herman, A. G. (1983): 5-Hydroxytryptamine and platelet aggregation. *Fed. Proc.*, 42:228–232.
236. De Clerck, F., and Reneman, R. S. (1984): Serotonin and microvascular permeability. *Int. J. Microcirc. Clin. Exp.* (abstract) *(in press)*.
237. De Clerck, F., Somers, Y., and Van Gorp, L. (1984): Platelet-vessel wall interactions during haemostasis: Implication of 5-hydroxytryptamine. *Thromb. Haemost. (submitted)*.
238. De Clerck, F., and Van Nueten, J. M. (1982): Platelet-mediated vascular contractions: Inhibition of the serotonergic component by ketanserin. *Thromb. Res.*, 27:713–727.
239. De Clerck, F., and Van Nueten, J. M. (1983): Platelet-mediated vascular contractions. Inhibition by flunarizine, a calcium-entry blocker. *Biochem. Pharmacol.*, 32:765–771.
240. De Clerck, F., and Van Nueten, J. M. (1984): Platelet-mediated vascular contractions: Inhibition of the serotonergic component by ketanserin. *Thromb. Res. (in press)*.
241. De Clerck, F., and Xhonneux, B. (1983): The effect of ketanserin on the red blood cell; platelet interaction in shear flow *in vitro*. Janssen Preclinical Research Report No. R.41 468/40.
242. De Clerck, F., Xhonneux, B., Leysen, J. E., and Janssen, P. A. J. (1984): Evidence for the functional 5-HT_2 receptor sites on human blood platelets. *Biochem. Pharmacol. (in press)*.
243. De Clerck, F., Xhonneux, B., Leysen, J., and Janssen, P. A. J. (1984): The involvement of 5-HT_2-receptor sites in the activation of cat platelets. *Thromb. Res.*, 33:305–321.
244. De Cree, J. (1981): Red blood cell deformability in patients with essential hypertension and acute myocardial ischaemia. Improvement by treatment with selective 5-hydroxytryptamine-2 antagonist (R 41468). Abstracts of the Second European Conference on Clinical Haemorheology, London, p. 31.
245. De Cree, J., Leempoels, J., De Cock, W., Geukens, H., and Verhaegen, H. (1981): The antihypertensive effects of a pure and selective serotonin-receptors blocking agent (R41 468) in elderly patients. *Angiology*, 32:137–144.
246. De Cree, J., Leempoels, J., Geukens, H., and Verhaegen, H. (1983): Placebo controlled double-blind trial of ketanserin in the treatment of intermittent claudication. *Lancet*, ii 775–779.
247. De Cree, J., Leempoels, J., Demoen, B., and Verhaegen, H. (1984): Possible interaction of 5HT with Na/K ATPase activity: Influence of ketanserin, a 5HT receptor antagonist. Abstracts of the Satellite Symposium to the 10th Scientific Meeting of the International Society of Hypertension, "Serotonergic Mechanisms in the Cardiovascular System," Zurich, 1984, p. 70.
248. De Cree, J., Leempoels, J., Geukens, H., De Cock, W., and Verhaegen, H. (1981): The antihypertensive effects of ketanserin (R41 468), a novel 5-hydroxytryptamine-blocking agent, in patients with essential hypertension. *Clin. Sci.*, 61:473–476S.
249. De Cree, J., Leempoels, J., Geukens, H., De Cock, W., and Verhaegen, H. (1982): Are serotonergic mechanisms involved in high blood pressure? In: *5-Hydroxytryptamine in Peripheral Reactions*, edited by F. De Clerck and P. M. Vanhoutte, pp. 183–192. Raven Press, New York.
250. De Cree, J., Leempoels, J., Geukens, H., De Cock, W., and Verhaegen, H. (1983): Oral treatment with ketanserin in intermittent claudication. Abstracts of the Satellite Symposium to the First European Meeting on Hypertension, "Serotonin and Cardiovascular Diseases," Milan, 1983.
251. De Cree, J., Leempoels, J., Geukens, H., and Verhaegen, H. (1984): Comparison of blood pressure ratio by Doppler velocimetry and by plethysmography during treatment with ketanserin, a 5-HT_2 receptor antagonist: A measure of improved collateral flow. Abstracts of the Satellite Symposium to the 10th Scientific Meeting of the International Society of Hypertension, "Serotonergic Mechanisms in the Cardiovascular System," Zurich, 1984, p. 51.
252. De Cree, J., Verhaegen, H., and Symoens, J. (1981): Acute blood pressure lowering effect of ketanserin. *Lancet*, 1:1161–1162.
253. De Felipe, M. C., Fuentes, J. A., and Drummond, A. H. (1982): Specific 5-hydroxytryptamine binding to rat platelets as a system to evaluate tricyclic antidepressants in plasma. *Biochem. Pharmacol.*, 31:1661–1663.

254. DeGroat, W. C., and Ryall, R. W. (1967): An excitatory action of 5-hydroxytryptamine on sympathetic preganglionic neurones. *Exp. Brain Res.*, 3:299–305.
255. Deguchi, T. (1977): Tryptophan hydroxylase in pineal gland of rat: Postsynaptic localization and absence of circadian change. *J. Neurochem.*, 28:667–668.
256. DeKoch, M. A., Nadel, J. A., Zwi, S., Colebatch, H. J. H., and Olsen, C. R. (1966): New methods for perfusing bronchial arteries: Histamine bronchoconstriction and apnea. *J. Appl. Physiol.*, 21:185–194.
257. De La Lande, I. S., Cannel, V. A., and Waterson, J. G. (1966): The interaction of serotonin and noradrenaline on the perfused artery. *Br. J. Pharmacol. Chemother.*, 28:255–272.
258. De La Lande, I. S., Frewin, S., Waterson, J. G., and Cannel, V. A. (1967): Factors influencing supersensitivity to noradrenaline in the isolated perfused artery, comparative effects of cocaine, denervation and serotonin. *Circ. Res.*, 20–21 (*Suppl.* III):177–181.
259. De Lean, A., Munson, P. J., and Rodbard, D. (1978): Simultaneous analysis of families of sigmoidal curves: Application to bioassay, radioligand assay, and physiological dose-response curves. *Am. J. Physiol.*, 235:E97–E102.
260. De Lean, A., Munson, P. J., and Rodbard, D. (1979): Multi-subsite receptors for multivalent ligands. Application to drugs, hormones and neurotransmitters. *Mol. Pharmacol.*, 15:60–70.
261. DeMey, C., and Vanhoutte, P. M. (1981): Effect of age and spontaneous hypertension on the tachyphylaxis to 5-hydroxytryptamine and angiotensin II in the isolated rat kidney. *Hypertension*, 3:718–724.
262. DeMey, J. G., Claeys, M., and Vanhoutte, P. M. (1982): Endothelium-dependent inhibitory effects of acetylcholine, adenosine triphosphate, thrombin and arachidonic acid in canine femoral artery. *J. Pharmacol. Exp. Ther.*, 222:166–173.
263. DeMey, J. G., and Vanhoutte, P. M. (1981): Role of the intima in cholinergic and purinergic relaxation of isolated canine femoral arteries. *J. Physiol. (Lond.)*, 316:347–355.
264. DeMey, J. G., and Vanhoutte, P. M. (1983): Anoxia and endothelium dependent reactivity of the canine femoral artery. *J. Physiol. (Lond.)*, 335:65–74.
265. Demoulin, J. C., Bertholet, M., Soumagne, D., David, J. L., and Kulbertus, H. E. (1981): 5-HT_2 receptor blockade in the treatment of heart failure—A preliminary study. *Lancet*, 1:1186–1188.
266. Denham, M., Fischer, M., James, G., and Hassan, M. (1977): Beta thromboglobulin and heparin neutralizing activity test in clinical conditions. *Lancet*, 1:1154.
267. Denney, R. M., Fritz, R. R., Patel, N. T., and Abell, C. W. (1982): Human liver MAO-A and MAO-B separated by immunoaffinity chromatography with MAO-B-specific monoclonal antibody. *Science*, 215:1400–1403.
268. De Vries, J. I., Vellenga, E., and Aarnoudse, J. G. (1983): Plasma beta-thromboglobulin in normal pregnancy and pregnancy-induced hypertension. *Eur. J. Obstet. Gynecol. Reprod. Biol.*, 19:209–216.
269. Dewar, H. A., Marshall, T., Weightman, D., Prakash, V., and Boon, P. J. (1979): β-Thromboglobulin in antecubital vein blood. The influence of age, sex and blood group. *Thromb. Haemost.*, 42:1159–1163.
270. Diggorry, G. L., Ceasar, P. M., and Morgan, R. M. (1979): The regional metabolism of 5-hydroxytryptamine in mouse brain *in vitro*. *Life Sci.*, 24:1939–1946.
271. DiMaria, F. J. (1964): Placental monoamine oxidase in normal and toxemic pregnancies. *Am. J. Obstet. Gynecol.*, 88:490.
272. Dockerty, M. B. (1963): Carcinoid tumors. *Calif. Med.*, 99:157–160.
273. Dodds, A. J., O'Reilly, M. J. G., Yates, C. J. P., Cotton, L. T., Flute, P. T., and Dormandy, J. A. (1979): Haemorheological response to plasma exchange in Raynaud's syndrome. *Br. Med. J.*, 2:1186–1187.
274. Donowitz, M., and Binder, H. J. (1975): Jejunal fluid and electrolyte secretion in carcinoid syndrome. *Am. J. Dig. Dis.*, 20:1115–1121.
275. Donowitz, M., Charney, H. J., and Heffernan, J. M. (1977): Effect of serotonin treatment on intestinal transport in the colon. *Am. J. Physiol.*, 232:85–94.
276. Doolittle, R. F., Hunkapiller, M. W., Hodd, L. E., Devare, S. G., Robbins, K. C., Aaronson, S. A., and Antoniades, H. N. (1983): Simian sarcoma virus *onc* gene, v-sis, is derived from the gene(s) encoding a platelet-derived growth factor. *Science*, 221:275–278.
277. Dormandy, J. (1970): Clinical significance of blood viscosity. *Ann. R. Coll. Surg. Engl.*, 47:211–228.

278. Dormandy, J. (1980): Blood viscosity and red cell deformability. In: *Methods in Angiology*, edited by M. Verstraete, pp. 214–266. Martinus Nijhoff, The Hague.
279. Dormandy, J., editor (1983): *Red Cell Deformability and Filterability*. Martinus Nijhoff, The Hague.
280. Dormandy, J., Boyd, M., Ernst, E. (1981): Red cell filterability after myocardial infarction. *Scand. J. Clin. Lab. Invest.*, 41(*Suppl.* 156):195–198.
281. Dormandy, J., Ernst, E., Matrai, A., and Flute, P. T. (1982): Haemorheologic changes following acute myocardial infarction. *Am. Heart J.*, 104:1364–1367.
282. Dormandy, J. A., Yates, C. J. P., and Berent, G. A. (1981): Clinical relevance of blood viscosity and red cell deformability including newer therapeutic aspects. *Angiology*, 32:236–242.
283. Doubilet, P., and Abrams, H. L. (1984): The cost of underutilization: Percutaneous transluminal angioplasty for peripheral vascular disease. *N. Engl. J. Med.*, 310:95–100.
284. Douglas, J. T., Lowe, G. D. O., Forbes, C. D., and Prentice, C. R. M. (1982): β-thromboglobulin and platelet counts. Effect of malignancy, infection, age and obesity. *Thromb. Res.*, 25:459–464.
285. Douglas, J. T., Shaw, M., Lowe, G. D., Belch, J. J., Forbes, C. D., and Prentice, C. R. (1981): Fibrinopeptide A and beta-thromboglobulin levels in preeclampsia and hypertensive pregnancy. *Thromb. Hem.*, 46:8.
286. Downing, E. S., (1983): The heart in shock. In: *Handbook of Shock and Trauma, Vol. 1: Basic Science*, edited by B. M. Altura, A. M. Lefer, and W. Schumer, pp. 5–28. Raven Press, New York.
287. Downing, I., Sheppard, G. L., and Lewis, P. J. (1980): Reduced prostacyclin production in preeclampsia. *Lancet*, 2:1374.
288. Drakontides, A. B., and Gershon, M. D. (1972): Studies of the interaction of 5-hydroxytryptamine (5-HT) and the perivascular innervation of the guinea pig caecum. *Br. J. Pharmacol.*, 45:417–434.
289. Drapanas, T., and McDonald, J. C. (1963): The direct removal of portal blood serotonin by the liver. *Surg. Gynecol. Obstet.*, 116:481–485.
290. Drapanas, T., MacDonald, J. C., and Stewart, J. D. (1962): Serotonin release following instillation of hypertonic glucose into the proximal intestine. *Ann. Surg.*, 156:528–536.
291. Dreiling, D. A. (1970): The lymphatics, pancreatic ascites and pancreatic disease: A new therapy for pancreatitis. *Am. J. Gastroenterol.*, 53:119–131.
292. Dreyfus, C. F., Bornstein, M. B., and Gershon, M. D. (1977): Uptake of serotonin by intrinsic neurons of the myenteric plexus grown in organotypic tissue culture. *Brain Res.*, 128:109–123.
293. Dreyfus, C. F., Sherman, D., and Gershon, M. D. (1977): Synthesis of serotonin by neurons of the myenteric plexus *in situ* and in organotypic culture. *Brain Res.*, 128:125–139.
294. Dreyfuss, F., and Zahavi, J. (1973): Adenosine diphosphate induced platelet aggregation in myocardial infarction and ischemic heart disease. *Atherosclerosis*, 17:107–120.
295. Drummond, A. H. (1976): Interaction of blood platelets with biogenic amines: Uptake, stimulation and receptor binding. In: *Platelets in Biology and Pathology*, edited by Gordon, pp. 203–239. Elsevier-North Holland Biomedical Press, New York.
296. Earley, C. J., and Leonard, B. E. (1978): Isolation and assay of noradrenaline, dopamine, 5-hydroxytryptamine and several metabolites from brain tissue using disposable Bio-Rad columns packed with Sephadex G-10. *J. Pharmacol. Methods*, 1:67–79.
297. Eccles, N. K., Grimmer, A. J., and Leathard, R. L. (1981): Action of ketanserin, a new 5-hydroxytryptamine antagonist on human isolated blood vessels. *Br. J. Pharmacol.*, 74:831P–832P.
298. Echizen, H., and Freed, C. R. (1982): Long-term infusion of L-5-hydroxytryptophan increases brain serotonin turnover and decreases blood pressure in normotensive rats. *J. Pharmacol. Exp. Ther.*, 220:579–584.
299. Edvinsson, L., Degueurce, A., Duverger, D., MacKenzie, E. T., and Scatton, B. (1983): Central serotonergic nerves project to the pial vessels of the brain. *Nature (Lond.)*, 306:55–57.
300. Edvinsson, L., Hardebo, J. E., and Owman, C. (1978): Pharmacological analysis of 5-hydroxytryptamine receptors in isolated intracranial and extracranial vessels of cat and man. *Circ. Res.*, 42:143–151.
301. Edvinsson, Högestätt, and Mackenzie (1983): 5-Hydroxytryptamine and the cerebral circulation. In: *Vascular Neuroeffector Mechanisms*, edited by J. A. Bevan et al., pp. 251–258. Raven Press, New York.
302. Eklund, B., Jogestrand, T., and Pernow, B. (1977): Effect of substance P on resistance and

capacitance vessels in the human forearm. In: *Substance P*, edited by U. von Euler and B. Pernow, pp. 275–277. Raven Press, New York.

303. Eklund, S., Fahrenkrug, J., Jodal, M., Lundgren, O., Schaffalitzky de Muckadell, O. B., and Sjöqvist, A. (1980): Vasoactive intestinal polypeptide, 5-hydroxytryptamine and reflex hyperemia in the small intestine of the cat. *J. Physiol. (Lond.)*, 302:549–557.
304. Ellis, E. F., Oelz, O., Roberts, L. J., Payne, N. A., Sweetman, B. J., Niess, A. S., and Oates, J. A. (1976): Coronary arterial smooth muscle contraction by a substance released from platelets: Evidence that it is thromboxane A_2. *Science*, 193:1135–1137.
305. Emerson, T. E., Jr., Meier, P. D., and Daugherty, R. M., Jr. (1973): Dependence of skeletal muscle vascular response to serotonin upon the level of vascular resistance. *Proc. Soc. Exp. Biol. Med.*, 142:1185–1188.
306. Emson, P., Gilbert, R. F. T., Mårtensson, H., and Nobin, A. (1984): Elevated concentrations of substance P and 5-HT in plasma in patients with carcinoid tumors. *Cancer (in press)*.
307. Engback, F., and Magnussen, I. (1978): Determination of 5-hydroxytryptophan in plasma by high performance liquid chromatography and fluorometric detection after phthaldialdehyde reaction. *Clin. Chem.*, 24:376–378.
308. Engel, G., Göthert, M., Müller-Schweinitzer, E., Schlicker, E., Sistonen, L., and Stadler, P. A. (1983): Evidence for common pharmacological properties of [^{3}H]5-hydroxytryptamine binding sites, presynaptic 5-hydroxytryptamine autoreceptors in CNS and inhibitory presynaptic 5-hydroxytryptamine receptors on sympathetic nerves. *Naunyn Schmiedebergs Arch. Pharmacol.*, 324:116–124.
309. Engel, G., Müller-Schweinitzer, E., and Palacios, J. M. (1984): 2-[^{125}Iodo]LSD, a new ligand for the characterization and localization of 5-HT_2 receptors. *Naunyn Schmiedebergs Arch. Pharmacol.*, 325:328–336.
310. Engelman, K., Lovenberg, W., and Sjoersdma, A. (1974): Inhibition of serotonin synthesis by parachlorophenylalanine in patients with the carcinoid syndrome. *N. Engl. J. Med.*, 277:1103–1108.
311. Ennis, C., Kemp, J. D., and Cox, B. (1981): Characterisation of inhibitory 5-hydroxytryptamine receptors that modulate dopamine release in the striatum. *J. Neurochem.*, 36:1515–1520.
312. Erde, S. M., Sherman, D., and Gershon, M. D. (1984): Morphology and serotonergic innervation of physiologically identified cells of the guinea pig myenteric plexus. *J. Neurosci.*, *(in press)*.
313. Erhardt, L. F. (1978): Coronary thrombosis—Cause or consequence. In: *Acute and Long-Term Medical Management of Myocardial Ischemia*, edited by L. Wilhelmsen and A. Hjalmarson, p. 70. Molndal, Sweden.
314. Erne, P., Bühler, F. R., Affolter, H., and Bürgisser, E. (1983): Excitatory and inhibitory modulation of intracellular free calcium in human platelets by hormones and drugs. *Eur. J. Pharmacol.*, 91:331–332.
315. Ernst, E., Matrai, A., Dormandy, J. A., and Flute, P. T. (1981): Is there a correlation between red cell filterability and blood viscosity measurements? *Clin. Haemorheol.*, 1:335–340.
316. Erspamer, V. (1954): Pharmacology of indole alkylamines. *Pharmacol. Rev.*, 6:425–487.
317. Erspamer, V. (1954): Quantitative estimation of 5-HT in gastrointestinal tract, spleen and blood of vertebrates. In: *Ciba Symposium on Hypertension*, edited by G. Wolstenholme and M. P. Cameron, pp. 78–84. Churchill, London.
318. Erspamer, V. (1966): Peripheral physiological and pharmacological actions of indole alkylamines. In: *Handbook of Experimental Pharmacology, Vol. 14: 5-Hydroxytryptamine and Related Indolealkylamines*, edited by V. Erspamer, pp. 245–359. Springer-Verlag, Berlin.
319. Erspamer, V., and Asero, B. (1952): Identification of enteramine, the specific hormone of the enterochromaffin cell system, as 5-hydroxytryptamine. *Nature (Lond.)*, 169:800–801.
320. Erspamer, V., and Testini, A. (1959): Observations of the release and turnover rate of 5-hydroxytryptamine in the gastrointestinal tract. *J. Pharm. Pharmacol.*, 11:618–623.
321. Erwin, V. G., and Deitrich, R. A. (1966): Brain aldehyde dehydrogenase. Localization, purification and properties. *J. Biol. Chem.*, 241:3533–3539.
322. Essman, W. B. (1978): Serotonin distribution in tissues and fluids. In: *Serotonin in Health and Disease, Vol. 1: Availability, Localization and Distribution*, edited by W. B. Essman, pp. 15–180. S.P. Medical and Scientific Books, New York, London.
323. Facer, P., Polak, J. M., Joffee, B. M., and Pearse, A. G. E. (1979): Immunocytochemical demonstration of 5-hydroxytryptamine in gastrointestinal endocrine cells. *Histochem. J.*, 11:117–121.
324. Fagard, R., Fiocchi, R., Lijnen, P., Staessen, J., Moerman, E., De Schaepdrijver, A., and

Amery, A. (1984): Hemodynamic and humoral responses to chronic ketanserin treatment in essential hypertension. *Br. Heart J.*, 51:149–156.

325. Fain, J. N., and Berridge, M. J. (1979): Relationship between hormonal activation of phosphatidylinositol hydrolysis, fluid secretion and calcium flux in the blowfly salivary gland. *Biochem. J.*, 178:45–58.
326. Falck, B., and Hellman, B. (1963): Evidence for the presence of biogenic amines in pancreatic islets. *Experientia*, 19:139–140.
327. Falck, B., Larson, B., von Mechlenberg, C., Rosengreen, E., and Svenaeus, K. (1964): On the presence of a second specific cell system in mammalian thyroid gland. *Acta Physiol. Scand.*, 62:491–492.
328. Falck, B., and Owman, C. (1968): 5-Hydroxytryptamine and related amines in endocrine cell systems. In: *Advances in Pharmacology*, edited by S. Garattini and P. A. Shore, pp. 211–233. Academic Press, New York.
329. Falk, A., Kaijser, B., Myrvold, H. E., and Haglund, U. (1980): Intestinal vascular and central hemodynamic responses in the cat following i.v. infusion of live *E. coli* bacteria. *Circ. Shock*, 7:239–250.
330. Falk, A., Myrvold, H. E., and Haglund, U. (1982): Cardiopulmonary function as related to intestinal mucosal lesions in experimental septic shock. *Circ. Shock*, 9:419–432.
331. Fara, J., and Haglund, U. (1973): Release of serotonin from the small intestine following hypotension in the cat. *Eur. Surg. Res.*, 5:152–161.
332. Farrow, J. T., and Van Vunakis, H. (1972): Binding of *d*-lysergic acid diethylamide to subcellular fractions from rat brain. *Nature*, 237:164–166.
333. Feldman, J. M. (1978): Serotonin metabolism in patients with carcinoid tumors. Incidence of 5-hydroxytryptophan secreting tumors. *Gastroenterology*, 75:1109–1114.
334. Felten, D. L., and Cummings, J. P. (1979): The raphe nuclei of the rabbit brainstem. *J. Comp. Neurol.*, 187:199–244.
335. Feniuk, W., Humphrey, P. P. A., and Watts, A. D. (1981): An analysis of the receptors involved in 5-hydroxytryptamine-induced smooth muscle relaxation. *Proc. Br. Pharm. Soc.*, 16–18 September, 1981, 850p–851p.
336. Feniuk, W., Humphrey, P. P. A., and Watts, A. D. (1983): 5-Hydroxytryptamine-induced relaxation of isolated mammalian smooth muscle. *Eur. J. Pharmacol.*, 96:71–78.
337. Fernstrom, J. D., and Wurtman, R. J. (1971): Brain serotonin content: Physiological dependence on plasma tryptophan levels. *Science*, 173:149–151.
338. Fillion, G. M. B., Lluch, S., and Uvnas, B. (1971): Release of noradrenaline from the dog heart *in situ* after intravenous and intracoronary administration of 5-hydroxytryptamine. *Acta Physiol. Scand.*, 83:115–123.
339. Fisher, S. K., Van Rooijen, L. A. A., and Agranoff, B. W. (1984): Renewed interest in the polyphosphoinositides. *TIBS* (February 1984):53–56.
340. Flink, J. R., Pitt, B. R., Hammond, G. L., and Gillis, C. N. (1982): Selective effect of microembolization on pulmonary removal of biogenic amines. *J. Appl. Physiol.*, 52:R421–R427.
341. Fock, S., and Mense, S. (1976): Excitatory effects of 5-hydroxytryptamine, histamine and potassium ions on muscular group IV afferent units: A comparison with bradykinin. *Brain Res.*, 105:459–469.
342. Folkow, B. (1982): Physiological aspects of primary hypertension. *Physiol. Rev.*, 62:347–504.
343. Folkow, B., and Neil, E. (1971): *Circulation*. Oxford University Press, New York.
344. Follansbee, W. P., Curtiss, E. I., Medsger, T. A., Steen, V. D., Uretsky, B. F., Owens, G. R., and Rodnan, G. P. (1984): Physiologic abnormalities of cardiac function in progressive systemic sclerosis with diffuse scleroderma. *N. Engl. J. Med.*, 310:142–148.
345. Forconi, S. (1983): Red cell deformability and ischaemia. In: *Red Cell Deformability and Filterability*, edited by J. A. Dormandy. Martinus Nijhoff, The Hague.
346. Forester, T. (1981): Adenosine or adenosine triphosphate? In: *Vasodilatation*, edited by P. M. Vanhoutte and I. Leusen, pp. 205–229. Raven Press, New York.
347. Forsberg, E. J., and Miller, R. J. (1983): Regulation of serotonin release from rabbit intestinal enterochromaffin cells. *J. Pharmacol. Exp. Ther.*, 227:755–766.
348. Forst, R., Rieger, H., and Schmid-Schönbein, H. (1979): Stimulation of human platelets under the influence of high shear stresses in tube flow. In: *Basic Aspects of Blood Trauma*, edited by H. Schmid-Schönbein and P. Teitel, pp. 46–62. Martinus Nijhoff Publishers, The Hague.
349. Forster, C., Whalley, E. T., Mohan, J., and Dutton, J. (1980): Vascular smooth muscle response

to fibrinogen degradation products and 5-hydroxytryptamine: Possible role in cerebral vasospasm in man. *Br. J. Clin. Pharmacol.*, 10:231–236.

350. Fozard, J. R. (1982): Mechanism of the hypotensive effect of ketanserin. *J. Cardiovasc. Pharmacol.*, 4:829–838.
351. Fozard, J. R., and Mobarok-Ali, A. T. M. (1978): Receptors for 5-hydroxytryptamine on the sympathetic nerves of the rabbit heart. *Naunyn Schmiedebergs Arch. Pharmacol.*, 301:223–235.
352. Frankhuyzen, A. L., and Mulder, A. H. (1982): Pharmacological characterization of presynaptic α-adrenoceptors modulating [^{3}H]noradrenaline and [^{3}H]5-hydroxytryptamine release from slices of the hippocampus of the rat. *Eur. J. Pharmacol.*, 81:97–106.
353. Franz, D. N., Madsen, P. W., Peterson, R. G., and Sangdee, C. (1982): Functional roles of monoaminergic pathways to sympathetic preganglionic neurons. *Clin. Exp. Hypertens. (A)*, 4(4–5):543–562.
354. Freedman, R., and Ianni, P. (1983): Role of cold and emotional stress in Raynaud's disease and scleroderma. *Br. Med. J.*, 287:1499–1502.
355. Freedman, R. J., Stemerman, M. B., Wenz, B., Moore, S., Gauldrie, J., Gent, R., Tiell, M. L., and Spaet, T. H. (1977): The effect of thrombocytopenia on experimental arteriosclerotic lesion formation in rabbits. I. Smooth muscle cell proliferation and re-endothelialization. *J. Clin. Invest.*, 60:1191–1201.
356. Friedhoff, L. T., Seibold, J. R., Kim, H. C., and Simester, K. S. (1984): Serotonin induced platelet aggregation in systemic sclerosis. *Clin. Exp. Rheumatol.*, 2:119–123.
357. Friedman, E., Lambert, G. A., and Buchweitz, E. (1978): 2,5-Dimethoxy-4-methylamphetamine (DOM)—A central component of its cardiovascular effects in rats; involvement of serotonin. *Eur. J. Pharmacol.*, 49:157–161.
358. Friedman, E. A., and Neff, R. K. (1977): *Pregnancy Hypertension: System Evaluation of Clinical Diagnostic Criteria: 2*. PSG Publishing Co., Littleton, Mass.
359. Friedman, P. A., Kappelman, A. H., and Kaufman, S. (1972): Partial purification and characterization of tryptophan hydroxylase from rabbit hindbrain. *J. Biol. Chem.*, 247:4165–4173.
360. Friedman, R., Barrett, R. J., and Sanders-Bush, E. (1984): Discriminative stimulus properties of quipazine: Mediation by serotonin$_2$ binding sites. *J. Pharmacol. Exp. Ther.*, 228:628–635.
361. Fries, J. F. (1969): Physiologic studies in systemic sclerosis (scleroderma). *Arch. Intern. Med.*, 123:22–25.
362. Fries, J. F., Lindgren, J. A., and Bull, J. M. (1973): Scleroderma-like lesions and the carcinoid syndrome. *Arch. Intern. Med.*, 131:550–553.
363. Fröhlich, J. C., Bloomgarden, Z. T., Oates, J. A., McGuigan, J. E., and Rabinowitz (1978): The carcinoid flush. Provocation by pentagastrin and inhibition by somatostatin. *N. Engl. J. Med.*, 299:1055–1057.
364. Fry, D. E., Pearlstein, L., Fulton, R. L., and Polk, H. C., Jr. (1980): Multiple system organ failure. The role of uncontrolled infection. *Arch. Surg.*, 115:136–140.
365. Fu, L. H. W., and Toda, N. (1983): Analysis of the contractile response to serotonin and tryptamine of isolated dog cerebral, femoral and mesenteric arteries. *Jpn. J. Pharmacol.*, 33:473–480.
366. Fujita, T. (1977): Concept of paraneurons. *Arch. Histol. Jpn. (Suppl.)*, 40:1–12.
367. Fuller, R. W. (1966): Serotonin oxidation by rat brain monoamine oxidase: Inhibition by 4-chloroamphetamine. *Life Sci.*, 5:2247–2252.
368. Furchgott, R. F. (1981): The requirement for endothelial cells in the relaxation of arteries by acetylcholine and some other vasodilators. *Trends Pharmacol. Sci.*, 2:173–176.
369. Furchgott, R. F. (1983): Role of the endothelium in responses of vascular smooth muscle. *Circ. Res.*, 53:557–72.
370. Furchgott, R. F., and Jothianandan, D. (1983): Relation of cyclic GMP levels to endothelium-dependent relaxation by acetylcholine in rabbit aorta. *Fed. Proc.*, 42:619.
371. Furchgott, R. F., and Zawadski, J. V. (1980): The obligatory role of endothelial cells in the relaxation of arterial smooth muscle by acetylcholine. *Nature (Lond.)*, 288:373–376.
372. Furchgott, R. F., Zawadski, J. V., and Cherry, P. D. (1981): Role of endothelium in vasodilator response to acetylcholine. In: *Vasodilatation*, edited by P. M. Vanhoutte and I. Leusen, pp. 49–66. Raven Press, New York.
373. Furness, J. B., and Costa, M. (1980): Types of nerves in the enteric nervous system. *Neuroscience*, 5:1–20.

374. Furness, J. B., and Costa, M. (1982): Neurons with 5-hydroxytryptamine-like immunoreactivity in the enteric nervous system: Their projections in the guinea pig small intestine. *Neuroscience*, 7:341–350.
375. Furst, D. E., Davis, J. A., Clements, P. J., Chopra, S. K., Theofilopoulos, A. N., and Chia, D. (1981): Abnormalities of pulmonary vascular dynamics and inflammation in early progressive systemic sclerosis. *Arthritis Rheum.*, 24:1403–1408.
376. Fuster, V., Bowie, E. J. W., Fass, D. N., Owen, C. A., and Brown, A. L. (1978): Resistance to arteriosclerosis in pigs with von Willebrand disease. *J. Clin. Invest.*, 61:722–730.
377. Gabbiani, G., Badonnel, M. C., and Majno, G. (1970): Intra-arterial injections of histamine, serotonin, or bradykinin: A topographic study of vascular leakage. *Proc. Soc. Exp. Biol. Med.*, 135:447–452.
378. Gaddum, J. H., Hebb, C. O., Silver, A., and Swan, A. A. B. (1953): 5-Hydroxytryptamine. Pharmacological action and destruction in perfused lungs. *Q. J. Exp. Physiol.*, 38:255–262.
379. Gaddum, J. H., and Picarelli, Z. P. (1957): Two kinds of tryptamine receptors. *Br. J. Pharmacol.*, 12:323–328.
380. Garattini, S., and Valzelli, L. (1965): *Serotonin*. Elsevier Publishing Co., Amsterdam, London, New York.
381. Gaudin-Chazal, G., Portalier, P., Puizillout, J. J., and Vigier, G. (1983): Simultaneous visualization of aortic and [^{3}H]-5-hydroxytryptamine-accumulating cell bodies in the nodose ganglion of the cat. *J. Physiol. (Lond.)*, 337:321–330.
382. Gautieri, R. F., and Ciuchta, H. P. (1961): Effect of certain drugs on perfused human placenta. I. *J. Pharm. Sci.*, 51:55.
383. Geaney, D. P., Schächter, M., Elliot, J. M., and Grahame-Smith, D. G. (1984): Characterisation of [^{3}H]lysergic acid diethylamide binding to a 5-hydroxytryptamine receptor on human platelet membranes. *Eur. J. Pharmacol.*, 97:87–94.
384. Gerber, J. G., Payne, N. A., Murphy, R. C., et al. (1981): Prostacyclin produced by the pregnant uterus in dog may act as a circulating vasodepressor substance. *J. Clin. Invest.*, 67:632.
385. Gershon, M. D. (1977): Biochemistry and physiology of serotonergic transmission. In: *Handbook of Physiology, Section I: The Nervous System*, edited by J. M. Brookhart and V. B. Mountcastle, *Vol. I*, pp. 573–623. American Physiological Society, Bethesda, MD.
386. Gershon, M. D. (1982): Enteric serotonergic neurons. In: *Biology of Serotonergic Transmission*, edited by N. N. Osborne, pp. 363–399. John Wiley and Sons, Ltd., New York.
387. Gershon, M. D. (1983): Serotonin storage outside the central nervous system. Abstracts of the Satellite Symposium to the First European Meeting on Hypertension, "Serotonin and Cardiovascular Diseases," Milan, 1983.
388. Gershon, M. D., and Altman, R. F. (1971): An analysis of the uptake of 5-hydroxytryptamine by the myenteric plexus of the small intestine of the guinea pig. *J. Pharmacol. Exp. Ther.*, 179:29–41.
389. Gershon, R. K., Askenase, P. W., and Gershon, M. D. (1975): Requirement for vasoactive amines for production of delayed-type hypersensitivity skin reactions. *J. Exp. Med.*, 142:732–747.
390. Gershon, M. D., Drakontides, A. B., and Ross, L. L. (1965): Serotonin: Synthesis and release from the myenteric plexus of the mouse intestine. *Science*, 149:197–199.
391. Gershon, M. D., and Dreyfus, C. F. (1980): Stimulation of tryptophan uptake into enteric neurons by 5-hydroxytryptamine: A novel form of neuromodulation. *Brain Res.*, 184:229–233.
392. Gershon, M. D., Dreyfus, C. F., Pickel, V. M., Joh, T. H., and Reis, D. J. (1977): Serotonergic neuron in the peripheral nervous system: Identification in gut by immunohistochemical localization of tryptophan hydroxylase. *Proc. Natl. Acad. Sci. U.S.A.*, 74:3086–3089.
393. Gershon, M. D., and Erde, S. M. (1981): The nervous system of the gut. *Gastroenterology*, 80:1571–1594.
394. Gershon, M. D., Liu, K. P., Karpiak, S. E., and Tamir, H. (1983): Storage of serotonin *in vivo* as complex with serotonin-binding protein in central and peripheral serotonergic neurons. *J. Neurosci.*, 3:1901–1911.
395. Gershon, M. D., and Nunez, E. A. (1976): Uptake of biogenic amines by thyroid glands. *Mol. Cell. Endocrinol.*, 5:169–180.
396. Gershon, M. D., Robinson, R. G., and Ross, L. L. (1976): Serotonin accumulation in the guinea pig's myenteric plexus: Ion dependence, structure activity relationship and the effect of drugs. *J. Pharmacol. Exp. Ther.*, 198:548–561.

397. Gershon, M. D., and Ross, L. L. (1962): Studies on the relationship of 5-hydroxytryptamine and the enterochromaffin cell to anaphylactic shock in mice. *J. Exp. Med.*, 115:367–382.
398. Gershon, M. D., and Ross, L. L. (1966): Location of sites of 5-hydroxytryptamine storage and metabolism by radioautography. *J. Physiol. (Lond.)*, 186:451–476; 477–492.
399. Gershon, M. D., and Sherman, D. (1982): Selective demonstration of serotonergic neurons and terminals in electron micrographs: Loading with 5,7-dihydroxytryptamine and fixation with $NaMnO_4$. *J. Histochem. Cytochem.*, 30:769–773.
400. Gershon, M. D., Sherman, D. L., and Dreyfus, C. F. (1980): Effects of indolic neurotoxins on enteric serotonergic neurons. *J. Comp. Neurol.*, 190:581–596.
401. Gershon, M. D., Takaki, M., Tamir, H., and Branchek, T. (1984): The enteric neural receptor for 5-hydroxytryptamine. *Experientia (in press).*
402. Gershon, M. D., and Tamir, H. (1981): Release of endogenous 5-hydroxytryptamine from resting and stimulated enteric neurons. *Neuroscience*, 6:2277–2286.
403. Giarcovich-Martinez, S., Fernandez, M., Chemerinski, E., and Enero, M. A. (1983): Central serotonergic activity after neurogenic hypertension. *Eur. J. Pharmacol.*, 86:337–345.
404. Gilbey, M. P., Coote, J. H., Macleod, V. H., and Peterson, D. F. (1981): Inhibition of sympathetic activity by stimulating in the raphe nuclei and the role of 5-hydroxytryptamine in this effect. *Brain Res.*, 226:131–142.
405. Gillis, C. N. (1980): Metabolism of vasoactive hormones by pulmonary vascular endothelium: Possible functional significance. In: *Vascular Neuroeffector Mechanisms*, edited by J. A. Bevan, T. Godfraind, R. A. Maxwell, and P. M. Vanhoutte, pp. 304–315. Raven Press, New York.
406. Gillis, C. N., and Catravas, J. D. (1982): Altered removal of vasoactive substances in the injured lung: Detection of lung microvascular injury. *Ann. N.Y. Acad. Sci.*, 384:458–474.
407. Gillis, C. N., Cronau, L. H., and Hammond, G. L. (1974): Removal of 5-hydroxytryptamine and norepinephrine from the pulmonary vascular space of man: Influence of cardiopulmonary bypass and pulmonary arterial pressure on these processes. *Surgery*, 76:1608–1616.
408. Gillis, C. N., Cronau, L. H., Mandel, S., and Hammond, G. L. (1979): Indicator dilution measurement of 5-hydroxytryptamine clearance by human lung. *J. Appl. Physiol.*, 46:1178–1183.
409. Gillis, C. N., and Pitt, B. R. (1982): The fate of circulating amines within the pulmonary circulation. *Annu. Rev. Physiol.*, 44:269–281.
410. Glennon, R. A., Young, R., and Rosecrans, J. A. (1983): Antagonism of the effects of the hallucinogen DOM and the purported 5-HT agonist quipazine by 5-HT_2 antagonists. *Eur. J. Pharmacol.*, 91:197–206.
411. Goetz, U., Da Prada, M., and Pletscher, M. (1971): Adenine-guanine and uridine-5′-phosphonucleotides in blood platelets and storage organelles of various species. *J. Pharmacol. Exp. Ther.*, 178:210–215.
412. Gold, S., Gershon, M. D., and Tamir, H. (1982): Comparison of serotonin binding proteins of neuronal and enterochromaffin cells. *Neurosci. Abst.*, 8:408.
413. Gold, W. M., Kessler, G. F., and Yu, D. Y. C. (1972): Role of vagus nerves in experimental asthma in allergic dogs. *J. Appl. Physiol.*, 33:719–725.
414. Goldberg, A. D., Raftery, E. B., and Green, H. L. (1976): The Oxford continuous blood-pressure recorder, technical and clinical evaluation. *Postgrad. Med. J.*, 52(*Suppl.* 7):375–378.
415. Goldsmith, J. C., and Needleman, S. W. (1982): A comparative study of thromboxane and prostacyclin release from *ex vivo* and cultured bovine vascular endothelium. *Prostaglandins*, 24:173–178.
416. Goodman, R. P., Killam, A. P., Brash, A. R., et al. (1982): Prostacyclin production during normal, preeclamptic, and some other complicated pregnancies. *Am. J. Obstet. Gynecol.*, 142:817–822.
417. Goodwin, J. D. (1975): Carcinoid tumors. An analysis of 2837 cases. *Cancer*, 36:560–569.
418. Gordon, T., Castelli, W. P., Hjortland, M. C., Kannel, W. E., and Dawber, T. R. (1977): High density lipoproteins as a protective factor against coronary heart disease. The Framingham study. *Am. J. Med.*, 62:707–714.
419. Gothert, M., Huth, H., and Schlicker, E. (1981): Characterization of receptor subtype involved in alpha-adrenoceptor-mediated modulation of serotonin release from rat brain cortex slices. *Naunyn Schmiedebergs Arch. Pharmacol.*, 317:199–203.
420. Gozlan, H., El Mestikawy, S., Pichat, L., Glowinski, J., and Hamon, M. (1983): Identification of presynaptic serotonin autoreceptors using a new ligand: ^{3}H-PAT. *Nature*, 305:140–142.
421. Grafstein, B. (1977): Axonal transport: The intracellular traffic of the neuron. In: *Handbook of*

Physiology, edited by J. M. Brookhart, V. B. Mountcastle, E. R. Kandel, and S. R. Geiger, *Vol. I*, pp. 691–717. American Physiological Society, Bethesda, MD.
422. Grahame-Smith, D. G. (1964): Tryptophan hydroxylation in brain. *Biochem. Biophys. Res. Commun.*, 16:586–592.
423. Grahame-Smith, D. G. (1972): *The Carcinoid Syndrome*. W. Heinemann Medical Books, London.
424. Green, A. R., Koslow, S. H., and Costa, E. (1973): Identification and quantitation of a new indolethylamine in rat hypothalamus. *Brain Res.*, 51:371–374.
425. Green, H., and Sawyer, J. L. (1966): Demonstration, characterization, and assay procedure of tryptophan hydroxylase in rat brain. *Anal. Biochem.*, 15:53–64.
426. Griffith, S. G., Lincoln, J., and Burnstock, G. (1982): Serotonin as a neurotransmitter in cerebral arteries. *Brain Res.*, 247:388–392.
427. Grimelius, L. (1968): The argyrophil reaction in islet cells of adult human pancreas studied with a new silver nitrate procedure. *Acta Soc. Med. Upsal.*, 271–294.
428. Grönstad, K., Dahlström, A., Zinner, M. J., Jaffe, B. M., and Ahlman H. (1984): Regulatory mechanisms in the endoluminal release of serotonin and substance P from the feline jejunum *(submitted for publication)*.
429. Grönstad, K., Yeo, K., Ahlman, H., Jaffe, B. M., and Zinner, M. J. (1983): The effect of endoluminal SP on jejunal blood flow. In: *Substance P*, edited by P. Skrabanek and D. Powell, pp. 133–134. Boole Press, Dublin.
430. Guerrini, M., Laghi Pasini, F., Pieragalli, D., Acciavatti, A., Pasqui, A. L., Ceccatelli, L., Del Bigo, C., Galigani, C., Forconi, S., and Di Perri, T. (1983): Whole blood filterability versus red blood cell (RBC). II. *In vitro* influence of platelet activation. *Clin. Haemorheol.*, 3:304 (abstract).
431. Gujrati, V. R., Shanker, K., and Bhargava, K. P. (1982): 5-Hydroxytryptamine in toxemia of pregnancy. Abstract, FIGO, San Francisco.
432. Gwee, M. C. E., and Yeoh, T. S. (1968): The release of 5-hydroxytryptamine from rabbit small intestine *in vitro*. *J. Physiol. (Lond.)*, 194:817–825.
433. Haddy, F. J., Gordon, M. S., and Emanuel, D. A. (1959): The influence of tone upon responses of small and large vessels to serotonin. *Circ. Res.*, 7:123–130.
434. Haeusler, G., and Finch, L. (1972): Vascular reactivity to 5-hydroxytryptamine and hypertension in the rat. *Naunyn Schmiedebergs Arch. Pharmacol.*, 272:101–116.
435. Haft, J. E., and Fani, K. (1973): Intravascular platelet aggregation in the heart induced by stress. *Circulation*, 47:353–357.
436. Haglind, E., Falk, A., and Haglund, U. (1983): The effects of 5-HT blockade in graded intestinal vascular obstruction in the rat. *Acta Chir. Scand.*, 149:273–277.
437. Haglund, U. (1973): Vascular reactions in the small intestine of the cat during hemorrhage. *Acta Physiol. Scand.*, 89:129–141.
438. Haglund, U. (1983): Shock toxins. In: *Handbook of Shock and Trauma, Vol 1: Basic Science*, edited by B. M. Altura, A. M. Lefer, and W. Schumer, pp. 377–390. Raven Press, New York.
439. Haglund, U., Hultén, L., Lundgren, O., and Åhrén, C. (1975): Mucosal lesions in the human small intestine in shock. *Gut*, 16:979–984.
440. Haglund, U., Jodal, M., and Lundgren, O. (1984): The small bowel in arterial hypotension and shock. In: *Physiology of the Intestinal Circulation*, edited by A. P. Shepherd and D. N. Granger. Raven Press, New York.
441. Haglund, U., and Lundgren, O. (1973): The effects of vasoconstrictor fibre stimulation on consecutive vascular sections of cat small intestine during haemorrhage hypotension. *Acta Physiol. Scand.*, 88:95–108.
442. Hahn, H. L., Wilson, A. G., Graf, P. D., Fischer, S. P., and Nadel, J. A. (1978): Interactions between serotonin and efferent vagus nerves in dog lungs. *J. Appl. Physiol.*, 44:144–149.
443. Halevy, S., Spector, S., and Altura, B. M. (1980): Serum serotonin levels determined by radioimmunoassay in normal male and female rats after intestinal ischaemia shock. *Biochem. Med.*, 23:236–238.
444. Hall, E. D. (1981): Non-serotonergic depression of spinal monosynaptic reflex transmission by 5-hydroxytryptophan. *Neuropharmacology*, 20:109–114.
445. Halpern, A., Kuhn, P. H., Shaftel, H. E., Samuels, S. S., Shaftel, N., Selman, D., and Birch, H. G. (1960): Raynaud's disease, Raynaud's phenomenon, and serotonin. *Angiology*, 11:151–167.
446. Hammond, D. L., Yaksh, T. L., and Tyce, G. M. (1981): Presence of conjugated 5-hydroxytryptamine in *in vivo* superfusates of rat spinal cord. *J. Neurochem.*, 37:1068–1071.

447. Hammond, D. L., Yaksh, T. L., and Tyce, G. M. (1984): Conjugates of dopamine and serotonin in ventriculocisternal perfusates of the cat. *J. Neurochem.*, 42:1752–1757.
448. Hammond, D. L., Tyce, G. M., and Yaksh, T. L. (1984): Efflux of 5-hydroxytryptamine and noradrenaline into spinal cord superfusates during stimulation of the rat medulla. *J. Physiol. (in press).*
449. Hamon, M., Bourgoin, S., Artand, F., and Glowinski, J. (1979): The role of intraneuronal 5HT and of tryptophan hydroxylase activation in the control of 5HT synthesis in rat brain slices incubated in K^+-enriched medium. *J. Neurochem.*, 33:1031–1042.
450. Hampton, J. R., and Gorlin, R. (1972): Platelet studies in patients with coronary artery disease and their relatives. *Br. Heart J.*, 34:465–471.
451. Hardebo, J. E., and Owman, C. (1980): Characterization of the *in vitro* uptake of monoamines into brain microvessels. *Acta Physiol. Scand.*, 108:223–229.
452. Hardisty, R. M., and Stacey, R. S. (1955): 5-Hydroxytryptamine in normal human platelet. *J. Physiol. (Lond.)*, 130:711–720.
453. Harker, L. A., Ross, R., Slichter, S. J., and Scott, R. C. (1976): Homocystine-induced arteriosclerosis. The role of endothelial cell injury and platelet response in its genesis. *J. Clin. Invest.*, 58:731–741.
454. Harker, L. A., and Slichter, S. J. (1974): Arterial and venous thromboembolism: Kinetic characterization and evaluation of therapy. *Thromb. Diathes. Haemorrh.*, 31:188–203.
455. Harlan, R. W., Nadir, B., Harker, L., and Hildebrandt, J. (1983): Thromboxane A_2 mediates lung vasoconstriction but not permeability after endotoxin. *J. Clin. Invest.*, 72:911–918.
456. Harris, A. L. (1981): Chemotherapy for the carcinoid syndrome. *Cancer Chemother. Pharmacol.*, 5:133–138.
457. Hartig, P. R., Kadan, M. J., Evans, M. J., and Krohn, A. M. (1983): ^{125}I-LSD: A high sensitivity ligand for serotonin receptors. *Eur. J. Pharmacol.*, 89:321–322.
458. Hartman, E. (1974): L-Tryptophan: A possible natural hypnotic substance. *JAMA*, 230:1680–1681.
459. Haslam, R. J., Davidson, M. L., Davies, T., Lynham, J. A., and McClenaghan, M. D. (1978): Regulation of blood platelet function by cyclic nucleotides. *Adv. Cyclic Nucleotide Res.*, 9:552–553.
460. Haverback, B. J., Bogdanski, D., and Hogben, C. A. M. (1958): Inhibition of gastric acid secretion in the dog by the precursor of serotonin, 5-hydroxytryptophan. *Gastroenterology*, 34:188–195.
461. Head, G. A., and Korner, P. I. (1982): Cardiovascular functions of brain serotonergic neurons in the rabbit as analyzed from the acute and chronic effects of 5,6-dihydroxytryptamine. *J. Cardiovasc. Pharmacol.*, 4:398–408.
462. Head, G. A., Korner, P. I., Lewis, S. L., and Badoer, E. (1983): Contribution of noradrenergic and serotonergic neurons to the circulatory effects of centrally acting clonidine and α-methyldopa in rabbits. *J. Cardiovasc. Pharmacol.*, 5:945–953.
463. Hechtman, D., and Jageneau, A. (1984): Inhibition of cold-induced vasoconstriction with ketanserin. *Ann. Surg. (in press).*
464. Hechtman, H. B., Huval, W. V., Mathieson, M. A., Stemp, L. I., Valeri, C. R., and Shepro, D. (1983): Prostaglandin and thromboxane mediation of cardiopulmonary failure. *Surg. Clin. North Am.*, 63:263–283.
465. Hechtman, H. B., Lonergan, E., and Shepro, D. (1978): Platelet and leukocyte lung interactions in patients with respiratory failure. *Surgery*, 83:155–163.
466. Hechtman, H. B., Lonergan, E. A., Staunton, P. B., Dennis, R. C., and Shepro, D. (1978): Pulmonary entrapment of platelets during acute respiratory failure. *Surgery*, 83:277–283.
467. Hechtman, H. B., and Shepro, D. (1982): Lung metabolism and systemic organ function. *Circ. Shock*, 9:457–467.
468. Hedner, T., Persson, B., and Berglund, G. (1983): Ketanserin, a novel 5-hydroxytryptamine antagonist: Monotherapy in essential hypertension. *Br. J. Clin. Pharmacol.*, 16:121–125.
469. Heistad, D. D., Armstrong, M. L., Marcus, M. L., Piegors, D. J., and Mark, A. L. (1984): Augmented responses to vasoconstrictor stimuli in hypercholesterolemic and atherosclerotic monkeys. *Circ. Res.*, 54:711–718.
470. Heldin, C. H., Westermark, B., and Wasteson, A. (1979): Platelet derived growth factor: Purification and partial characterization. *Proc. Natl. Acad. Sci. U.S.A.*, 76:3722–3726.

471. Hellem, A. J. (1960): The adhesiveness of human blood platelets *in vitro*. *Scand. J. Clin. Lab. Invest.*, 12(*Suppl.* 51):1–117.
472. Henry, J. L., and Calaresu, F. R. (1974): Excitatory and inhibitory inputs from medullary nuclei projecting to spinal cardioacceleratory neurons in the cat. *Exp. Brain Res.*, 20:485–504.
473. Henry, P. D., and Yokoyama, M. (1980): Supersensitivity of atherosclerotic rabbit aorta to ergonovine. Mediation by a serotonergic mechanism. *J. Clin. Invest.*, 66:306–313.
474. Hertz, L., and Tamir, H. (1981): Some properties of an astrocytic protein fraction that binds serotonin. *J. Neurochem.*, 37:1331–1334.
475. Heuser, J. E., and Reese, T. S. (1981): Structural changes after transmitter release at the frog neuromuscular junction. *J. Cell Biol.*, 88:564–580.
476. Heykants, J., Roestenborgs, R., Scheijorond, H., and Symoeus, J. (1981): Preliminary study of the pharmacokinetics of seroquinaline in healthy subjects following intravenous intramuscular and oral administration. Janssen Clinical Research Report on R41468, No. 33, January 1981.
477. Hidaka, H., Nagatsu, T., Takeya, K., Matsumoto, S., and Yagi, K. (1969): Inactivation of serotonin by sulfotransferase system. *J. Pharmacol. Exp. Ther.*, 166:272–279.
478. Hidaka, H., Nagatsu, T., and Yagi, K. (1969): Occurrence of a serotonin sulphotransferase in the brain. *J. Neurochem.*, 16:783–785.
479. Hirst, G. D. S., and Silinsky, E. M. (1975): Some effects of 5-hydroxytryptamine, dopamine and noradrenaline on neurones in the submucous plexus of guinea-pig small intestine. *J. Physiol. (Lond.)*, 251:817–832.
480. Hjorth, S., Carlsson, A., Lindberg, P., Sanchez, D., Wikström, H., Arvidsson, L.-E., Hacksell, U., and Nilsson, J. L. G. (1982): 8-Hydroxy-2-(di-*n*-propylamino)tetralin, 8-OH-DPAT, a potent and selective simplified ergot congener with central 5-HT-receptor stimulating activity. *J. Neural Transm.*, 55:169–188.
481. Hodge, C. J., Woods, C. I., and Delatizky, J. (1980): Noradrenalin, serotonin, and the dorsal horn. *J. Neurosurg.*, 52:674–685.
482. Hofmann, S. L., and Majerus, P. W. (1982): Identification and properties of two distinct phosphatidylinositol-specific phospholipase C enzymes from sheep seminal vesicular glands. *J. Biol. Chem.*, 257:6461–6469.
483. Hökfelt, T., Kellerth, O., Nilsson, G., and Pernow, B. (1975): Substance P: Localization in the central nervous system and in some primary sensory neurons. *Science*, 190:889–890.
484. Holmsen, H. (1975): Biochemistry of the platelet release reaction. In: *Biochemistry and Pharmacology of Platelets, Ciba Foundation Symposium 35*, pp. 175–205. Elsevier-North Holland Biomedical Press, Amsterdam.
485. Holmsen, H. (1977): Prostaglandin endoperoxide—Thromboxane synthesis and dense granule secretion as positive feedback loops in the propagation of platelet responses during "the basic platelet reaction." *Thromb. Haemost.*, 38:1030–1041.
486. Holmsen, H. (1982): Biochemistry of the platelet: Energy metabolism. In: *Hemostasis and Thrombosis*, edited by R. W. Colman, J. Hirsh, V. J. Marder, and E. W. Salzman, pp. 431–443. J. B. Lippincott Co., Philadelphia.
487. Holmsen, H., and Dangelmaier, C. A. (1981): Evidence that the platelet plasma membrane is impermeable to calcium and magnesium complexes of A23187. A23187-induced secretion is inhibited by Mg^{2+} and Ca^{2+} and requires aggregation and active cyclooxygenase. *J. Biol. Chem.*, 256:10449–10452.
488. Holmsen, H., Dangelmaier, C. A., and Holmsen, H. K. (1981): Thrombin-induced platelet responses differ in requirement for receptor occupancy: Evidence for tight coupling of occupancy and compartmentalized phosphatidic acid formation. *J. Biol. Chem.*, 256:9393–9396.
489. Holmsen, H., and Karpatkin, S. (1983): Metabolism of platelets. In: *Hematology*, edited by W. J. Williams, E. Beutler, A. J. Erslev, and M. A. Lichtman, pp. 1149–1176. McGraw-Hill Book Co., New York.
490. Holzer, P., and Skofitsch, G. (1984): Release of 5-hydroxytryptamine from the myenteric plexus of the guinea pig isolated small intestine. *Br. J. Pharmacol.*, 81:381–386.
491. Holzwarth, M., Sawetawan, C., and Brownfield, M. S. (1984): Serotonin immunoreactivity in the adrenal medulla: Distribution and response to pharmacological manipulation. *Brain Res. Bull. (in press)*.
492. Hope, W., Martin, T. J., Chesterman, C. N., and Morgan, F. J. (1979): Human β-thromboglobulin inhibits PGI_2 production and binds to a specific site in bovine aortic endothelial cells. *Nature (Lond.)*, 282:210–212.

493. Hornstra, G., Christ-Hazelhof, E., and Haddeman, E. (1981): Fish oil feeding lowers thromboxane and prostacyclin production by rat platelets and aorta and does not result in the formation of prostaglandin I_3. *Prostaglandins*, 21:727–738.
494. Hornstra, G., Haddeman, E., and Don, J. A. (1979): Blood platelets do not provide endoperoxides for vascular prostaglandin production. *Nature (Lond.)*, 279:66–68.
495. Hornstra, G., and Hemker, H. C. (1979): Clot-promoting effects of platelet-vessel wall interactions: Influence of dietary fats in relation to arterial thrombus formation in rats. *Haemostasis*, 8:211–226.
496. Horst, W. D., and Jester, T. (1972): Influence of serotonin on adrenergic mechanisms. *Biochem Pharmacol*, 21:333–338.
497. Houston, D. S., Shepherd, J. T., and Vanhoutte, P. M. (1984): Platelets cause endothelium-dependent relaxations in some but not all isolated canine vessels. *Fed. Proc.*, 43:1084.
498. Houston, D. S., and Vanhoutte, P. M. (1984): Platelet activating factor induces endothelium-independent relaxation of isolated canine blood vessels. *Physiologist (in press)*.
499. Howe, P. R. C., Kuhn, D. M., Minson, J. B., Stead, B. H., and Chalmers, J. P. (1983): Evidence for a bulbospinal serotonergic pressor pathway in the rat brain. *Brain Res.*, 270:29–36.
500. Huebschmann, P. (1910): Sur le carcinome primitif de l'appendice vermiculaire. *Rev. Med. Suiss. Rom. Geneve*, 30:317–332.
501. Humphrey, J. H., and Jacques, R. (1954): The histamine and serotonin content of the platelets and polymorphonuclear leucocytes of various species. *J. Physiol. (Lond.)*, 124:305–310.
502. Humphrey, J. H., and Toh, C. C. (1954): Absorption of serotonin (5-hydroxytryptamine) and histamin by dog platelets. *J. Physiol. (Lond.)*, 124:300–304.
503. Humphrey, P. P. A., Feniuk, W., and Watts, A. D. (1983): Prejunctional effects of 5-hydroxytryptamine on noradrenergic nerves in the cardiovascular system. *Fed. Proc.*, 42:218–222.
504. Humphrey, W. R., and Aiken, J. W. (1982): Antithrombotic and hypotensive effects of the serotonin-2 receptor antagonist ketanserin in anesthetized dogs. *Pharmacologist*, 24:196.
505. Hussain, M. N., and Sole, M. J. (1981): A simple, specific radioenzymatic assay for picogram quantities of serotonin or acetylserotonin in biological fluids and tissues. *Anal. Biochem.*, 111:105–110.
506. Hutton, R. A., Mikhailidis, D. P., Bernstein, R. M., Jeremy, J. Y., Hughes, G. R. V., and Dandona, P. (1984): Assessment of platelet function in patients with Raynaud's syndrome. *J. Clin. Pathol.*, 37:182–187.
507. Huval, W. V., Dunham, B. M., Lelcuk, S., Valeri, C. R., Shepro, D., and Hechtman, H. B. (1983): Thromboxane mediation of cardiovascular dysfunction following aspiration. *Surgery*, 94:259–266.
508. Huval, W. V., Lelcuk, S., Allen, P. D., Mannick, J. A., Shepro, D., and Hechtman, H. B. (1984): Determinants of cardiovascular stability during abdominal aortic aneurysmectomy (AAA). *Ann. Surg.*, 199:216–222.
509. Huval, W. V., Lelcuk, S., Hechtman, H. B., and Demling, R. H. (1983): Serotonin content of lung lymph and plasma after endotoxemia in sheep. *Fed. Proc.*, 742:490.
510. Huval, W. V., Lelcuk, S., Shepro, D., and Hechtman, H. B. (1984): Role of serotonin in patients with acute respiratory failure. *Ann. Surg.*, 200:166–172.
511. Huval, W. V., Mathieson, M. A., Stemp, L. I., Dunham, B. M., Jones, A. G., Shepro, D., and Hechtman, H. B. (1983): Therapeutic benefits of 5-hydroxytryptamine inhibition following pulmonary embolism. *Ann. Surg.*, 197:220–225.
512. Hylden, J. L. K., and Wilcox, G. L. (1983): Intrathecal serotonin in mice: Analgesia and inhibition of a spinal action of substance P. *Life Sci.*, 33:789–795.
513. Hyman, A. (1969): The direct effects of vasoactive agents on pulmonary veins. Studies of responses to acetylcholine, serotonin, histamine, and isoproterenol in intact dogs. *J. Pharmacol. Exp. Ther.*, 168:96–105.
514. Hyttel, J., Fredericson Overø, K., and Arnt, J. (1984): Biochemical effects and drug levels in rats after long-term treatment with the specific 5-HT-uptake inhibitor, citalopram. *Psychopharmacology*, 83:20–27.
515. Islam, M. D., Melville, G. N., and Ulmer, W. T. (1974): Role of atropine in antagonizing the effect of 5-hydroxytryptamine (5HT) on bronchial and pulmonary vascular systems. *Respiration*, 31:47–59.
516. Iwasawa, Y., and Gillis, C. N. (1974): Pharmacological analysis of norepinephrine and 5-hydroxy-

tryptamine removal from the pulmonary circulation: Differentiation of uptake sites for each amine. *J. Pharmacol. Exp. Ther.*, 188:386–393.

517. Jackson Roberts, L., II, Bloomgarden, Z. T., Marney, S. R., Rabin, D., and Oates, J. A. (1983): Histamine release from a gastric carcinoid. Provocation by pentagastrin and inhibition by somatostatin. *Gastroenterology*, 84:272–275.
518. Jacobs, B. L. (1976): An animal behavior model for studying central serotonergic synapses. *Life Sci.*, 19:777–786.
519. Jacobs, M. J. H. M., Slaaf, D. W., Reneman, R. J., and Lemmens, H. A. J. (1984): Haemorheological and microcirculatory measurements in patients with Raynaud's phenomenon, before and after treatment with ketanserin *(submitted for publication)*.
520. Jaffe, B. M., Kopen, D. F., and Lazan, D. W. (1977): Endogenous serotonin in the control of gastric acid secretion. *Surgery*, 82:156–163.
521. Jageneau, A. H. M., Hörig, C., Loots, W., and Symoens, J. (1982): Plethysmographic registration of volume changes in a hand vein. Effects of serotonin and a specific antagonist. *Angiology*, 31:828–832.
522. Jamieson, G. G., Ludbrook, J., and Wilson, A. (1971): Cold hypersensitivity in Raynaud's phenomenon. *Circulation.*, 44:254–264.
523. Janeway, T. C., Richardson, H. B., and Park, E. A. (1918): Experiments on the vasoconstrictor action of blood serum. *Arch. Int. Med.*, 21:565–603.
524. Janoff, A., Nagler, A. L., Baez, S., and Zweifach, B. W. (1961): Pathogenesis of experimental shock. III. A lethal factor in the blood of rabbits following occlusion of the superior mesenteric artery. *J. Exp. Med.*, 114:205.
525. Janssen, P. A. J. (1982): The pharmacology of specific; pure and potent serotonin, $5HT_2$ or S_2-antagonists. In: *Advances In Pharmacology and Therapeutics, Vol. 4*, edited by H. Yoshida, Y. Hagihara, and S. Ebashi, pp. 21–33. Pergamon Press, Oxford.
526. Janssen, P. A. J. (1983): $5\text{-}HT_2$ receptor blockade to study serotonin-induced pathology. *Trends Pharmacol. Sci.*, 4:198–206.
527. Janssen Pharmaceutica (1980): Investigational New Drug Brochure on R41468, 4th edition, June 1980.
528. Janssen Pharmaceutica. (1983): Ketanserin: A novel serotonin S2 receptor blocking agent. Investigational New Drug Brochure 1983, 7th edition.
529. Janssens, W. J., Beyens, G. G., and Verhaege, R. H. (1983): Dependency of $alpha_1$- and $alpha_2$-adrenergic contractions of canine saphenous veins on intra- and extracellular activation of calcium. *Arch. Int. Pharmacodyn. Ther.*, 263:320–321.
530. Janssens, W. J., and Vanhoutte, P. M. (1978): Instantaneous changes of alpha-adrenoceptor affinity caused by moderate cooling in canine cutaneous veins. *Am. J. Physiol.*, 234:H330–H337.
531. Jayson, M. I. V., Tomlinson, I. W., Gush, R., and Salisbury, R. (1983): Acute effects of a selective 5-HT receptor blocking agent in a systemic sclerosis and primary Raynaud phenomenon. *Arthritis Rheum.*, 26:S26.
532. Jenner, W. N., and Rose, F. A. (1973): Studies on the sulfation of 3,4-dihydroxyphenylethylamine (dopamine) and related compounds by rat tissues. *Biochem. J.*, 135:109–114.
533. Jequier, E. W., Lovenberg, W., and Sjoerdsma, A. (1967): Tryptophan hydroxylase inhibition: The mechanism by which *p*-chlorophenylalanine depletes rat brain serotonin. *Mol. Pharmacol.*, 3:274–278.
534. Joh, T. H., Shikimi, T., Pickel, V. M., and Reis, D. J. (1975): Brain tryptophan hydroxylase: Purification of, production of antibodies to, and cellular and ultrastructural localization in serotonergic neurons of rat midbrain. *Proc. Natl. Acad. Sci. U.S.A.*, 72:3575–3579.
535. Johansson, B., and Somlyo, A. P. (1980): Electrophysiology and excitation-contraction coupling. In: *Handbook of Physiology, Vascular Smooth Muscle*, edited by D. F. Bohr, A. P. Somlyo, and H. V. Sparks, pp. 301–321. The American Physiological Society.
536. Johnson, L. A., Lavin, P., Moertel, C. G., Weiland, L., Dayal Y., Doos, W. G., Geller, S. A., Cooper, H. S., Nome, F., Massé, S., Simson, I. W., Sumner, H., Fölsch, E., and Engström P. (1983): Carcinoids: The association of histologic growth pattern and survival. *Cancer*, 51:882–889.
537. Johnsson, R., Frick, M. H., Vallem, Vuopio, P., and Tarssanen, L. (1981): Red cell flexibility and oxygen affinity in patients with angina pectoris and normal coronary arteries. *Ann. Clin. Res.*, 13:77–80.

538. Johnston, J. P. (1968): Some observations upon a new inhibitor of monoamine oxidase in brain tissue. *Biochem. Pharmacol.*, 17:1285–1297.

539. Joist, J. H., Kendall Baker, R., and Schonfield, G. (1979): Increased *in vivo* and *in vitro* platelet function in type II- and type IV-hyperlipoproteinaemia. *Thromb. Res.*, 15:95–108.

540. Jonakait, J. M., Gintzler, A. R., and Gershon, M. D. (1979): Isolation of axonal varicosities (autonomic synaptosomes) from the enteric nervous system. *J. Neurochem.*, 32:1387–1400.

541. Jonakait, J. M., Tamir, H., Gintzler, A. R., and Gershon, M. D. (1979): Release of serotonin and its binding protein by enteric neurons. *Brain Res.*, 174:55–69.

542. Jonakait, J. M., Tamir, H., Rapport, M. M., and Gershon, M. D. (1977): Detection of a soluble serotonin binding protein in mammalian myenteric plexus and other peripheral sites of serotonin storage. *J. Neurochem.*, 28:277–284.

543. Jones, A. W., Heidlage, J. F., Meyer, R., Day, B., and Freeland, A. (1981): Non-specific supersensitivity of aortic ^{42}K effluxes during DOCA hypertension in the rat, and the effects of antihypertensive therapy. In: *New Trends in Arterial Hypertension*, edited by M. Worcel et al., pp. 163–173. Elsevier/North Holland Biomedical Press, Amsterdam.

544. Jones, J. B., and Rowesel, A. (1973): Fetal 5-hydroxytryptamine levels in late pregnancy. *Br. J. Obstet. Gynecol.*, 80:687.

545. Jule, Y. (1980): Nerve-mediated descending inhibition in the proximal colon of the rabbit. *J. Physiol. (Lond.)*, 159:361–368.

546. Junod, A. F. (1975): Metabolism, production and release of hormones and mediators in the lung. *Am. Rev. Resp. Dis.*, 112:93–108.

547. Junod, A. F., and Ody, C. (1977): Amine uptake and metabolism by endothelium of pig pulmonary artery and aorta. *Am. J. Physiol.*, 232:C88–C94.

548. Kadzielawa, K. (1983): Antagonism of the excitatory effects of 5-hydroxytryptamine on sympathetic preganglionic neurones and neurones activated by visceral afferents. *Neuropharmacology*, 22:19–27.

549. Kahaleh, M. B., Osborn, I., and LeRoy, E. C. (1981): Increased factor VIII/von Willebrand factor antigen and von Willebrand factor activity in scleroderma and in Raynaud's phenomenon. *Ann. Intern. Med.*, 94:482–484.

550. Kahaleh, M. B., Osborn, I., and LeRoy, E. C. (1982): Elevated levels of circulating platelet aggregates and beta-thromboglobulin in scleroderma. *Ann. Intern. Med.*, 96:610–613.

551. Kalkman, H. O., Boddeke, H. W. G. M., Doods, H. N., Timmermans, P. B. M. W. M., Van Zwieten, P. A. (1983): Hypotensive activity of serotonin receptor agonists in rats is related to their affinity for 5-HT_1 receptors. *Eur. J. Pharmacol.*, 91:155–156.

552. Kalkman, H. O., Harms, Y. M., van Gelderen, E. M., Batink, H. D., Timmermans, P. B. M. W. M., and Van Zwieten, P. A. (1983): Hypotensive activity of serotonin antagonists; correlation with α_1-adrenoceptor and serotonin receptor blockade. *Life Sci.*, 32:1499–1505.

553. Kalkman, H. O., Timmermans, P. B. W. M., and Van Zwieten, P. A. (1982): Characterization of the antihypertensive properties of ketanserin (R41 468) in rats. *J. Pharmacol. Exp. Ther.*, 222:227–231.

554. Kallenberg, C. G. M., Pastoor, G. W., Wouda, A. A., and The, T. H. (1982): Antinuclear antibodies in patients with Raynaud's phenomenon: Clinical significance of anticentromere antibodies. *Ann. Rheum. Dis.*, 41:382–387.

555. Kallenberg, C. G. M., Vellenga, E., Wouda, A. A., and The, T. H. (1982): Platelet activation, fibrinolytic activity and circulating immune complexes in Raynaud's phenomenon. *J. Rheumatol.*, 9:878–884.

556. Kaplan, K. L. (1980): Beta-thromboglobulin. *Prog. Hemost. Thromb.*, 5:153–178.

557. Kaplan, K. L., Jaffe, B. M., and Peskin, G. W. (1972): A new provocative test for the diagnosis of carcinoid syndrome 123:173–179.

558. Kaplan, K. L., and Owen, J. (1981): Plasma levels of β-thromboglobulin and platelet factor 4 as indices of platelet activation *in vivo*. *Blood*, 57:199–202.

559. Karpatkin, S., and Holmsen, H. (1983): Composition of platelets. In: *Hematology*, edited by W. J. Williams, E. Beutler, A. J. Erslev, and M. A. Lichtman, pp. 1136–1149. McGraw-Hill Book Co., New York.

560. Katusic, Z. S., Shepherd, J. T., and Vanhoutte, P. M. (1984): Vasopressin causes endothelium-dependent relaxations of the canine basilar artery. *Circ. Res.*, 55:575–579.

561. Katz, I. R. (1981): Interaction between the oxygen and tryptophan dependence of synaptosome tryptophan hydroxylase. *J. Neurochem.*, 37:447–451.

562. Kaumann, A. J. (1983): A classification of heart serotonin receptors. *Naunyn Schmiedebergs Arch. Pharmacol.*, 322(Suppl.):R42 (abstract).
563. Keenan, E. J., and Porter, J. M. (1983): Alpha-adrenergic receptors in platelets from patients with Raynaud's syndrome. *Surgery.*, 94:204–209.
564. Kellar, K. J., Cascio, C. S., Butler, J. A., and Kurtzke, R. N. (1981): Differential effects of electroconvulsive shock and antidepressant drugs on serotonin$_2$ receptors in rat brain. *Eur. J. Pharmacol.*, 69:515–518.
565. Kellum, J. M., and Jaffe, B. M. (1976): Release of immunoreactive serotonin following acid perfusion of the duodenum. *Ann. Surg.*, 184:633–638.
566. Kellum, J. M., Jr., and Jaffe, B. M. (1976): Validation and application of a radioimmunoassay for serotonin. *Gastroenterology*, 70:516–522.
567. Kellum, J., McCabe, M., Schneider, J., and Donowitz, M. (1983): Neuronal control of acid-induced serotonin release. *Am. J. Physiol. (Gastrointest. Liver Physiol.)*, 8:G284–G831.
568. Kishimoto, Y., Takahashi, N., and Egami, F. (1961): Synthesis and properties of serotonin *O*-sulfate. *J. Biochem.*, 49:436–440.
569. Kiskoff, B., and Moore, E. W. (1977): Effect of serotonin on water and electrolyte transport in the *in vivo* rabbit small intestine. *Gastroenterology*, 71:1033–1038.
570. Kissinger, P. T., Bruntlett, C. S., and Shoup, R. E. (1981): Neurochemical applications of liquid chromatography with electrochemical detection. *Life Sci.*, 28:455–465.
571. Kjeldsen, S. E., Gjesdal, K., Eide, I., Aakesson, I., Amundsen, R., Foss, O. P., and Leren, P. (1983): Increased beta-thromboglobulin in essential hypertension; interactions between arterial plasma adrenaline, platelet function and blood lipids. *Acta Med. Scand.*, 213:359–373.
572. Klapdor, R. (1980): Effects of somatostatin in bronchial constriction in a patient with carcinoid syndrome. *N. Engl. J. Med.*, 303:464.
573. Klein, D. C., and Weller, J. L. (1970): Indole metabolism in the pineal gland: An arcadian rhythm in *N*-acetyltransferase. *Science*, 169:1093–1095.
574. Kling, E., Penttila, O., and Tissari, A. (1964): The content of 5-hydroxytryptamine in human uterine and placental tissue, and their 5-hydroxytryptamine decarboxylase and monoamine oxidase activities in normal and toxemic pregnancies. *Acta Obstet. Gynecol. Scand.*, 43:107.
575. Kloprogge, E., DeHaas, G. H., Gorter, G., and Akkerman, J. W. N. (1983): Stimulus response coupling in human platelets. Evidence against a role of PAF-acether in the "third" pathway. *Thromb. Res.*, 30:107–114.
576. Knoll, J., and Magyar, K. (1972): Some puzzling pharmacological effects of monoamine oxidase inhibitors. *Adv. Biochem. Psychopharmacol.*, 5:393–408.
577. Kobayashi, S., Fujita, T., and Sasagawa, T. (1970): The endocrine cells of human duodenal mucosa. An electron microscope study. *Arch. Histol. Jpn.*, 31:477–494.
578. Kobayashi, S., and Sasagawa, T. (1976): Morphological aspects of the secretion of gastroenteric hormones. In: *Endocrine Gut and Pancreas*, edited by T. Fujita, pp. 255–271. Elsevier, Amsterdam.
579. Kobold, E. E., and Thal, A. P. (1963): Quantitation and identification of vasoactive substances liberated during various types of experimental and clinical intestinal ischemia. *Surg. Gynecol. Obstet.*, 117:315–322.
580. Koch, D. D., and Kissinger, P. (1979): Determination of tryptophan and several of its metabolites in physiological samples by reversed-phase liquid chromatography with electrochemical detection. *J. Chromatogr.*, 164:441–455.
581. Koch, D. D., and Kissinger, P. T. (1980): Determination of serotonin in serum and plasma by liquid chromatography with precolumn sample enrichment and electrochemical detection. *Anal. Chem.*, 52:27–29.
582. Koevary, S. B., Azmitia, E., and McEvoy, R. C. (1983): Rat pancreatic serotonergic nerves: Morphologic, pharmacologic and physiologic studies. *Brain Res.*, 265:328–332.
583. Kontos, H. A., and Wasserman, A. J. (1969): Effect of reserpine in Raynaud's phenomenon. *Circulation*, 39:259–266.
584. Konturek, S. J., Tasler, J., Pawlik, W., and Cieszkowski, M. (1980): Action of substance P and its C-terminal hexapeptide on gastrointestinal secretion and circulation. *Regul. Pept. (Suppl.)*, 1:S124.
585. Kopin, I. J., Pare, C. M. B., Axelrod, J., and Weissbach, H. (1961): The fate of melatonin in animals. *J. Biol. Chem.*, 236:3072–3075.

586. Kordenat, R. K., and Kezdi, P. (1979): Serotonin blockade during experimental coronary thrombosis. *Am. Heart J.*, 97:329–339.
587. Koren, Z., Pfeifer, Y., and Sulmon, F. G. (1965): Serotonin content of human placenta and fetus during normal pregnancy. *Am. J. Obstet. Gynecol.*, 93:411.
588. Korf, J. (1969): The determination of 5-hydroxytryptamine in human urine. *Clin. Chim. Acta*, 23:483–487.
589. Korner, P. I. (1980): Central nervous control of autonomic cardiovascular function. In: *Handbook of Physiology—The Cardiovascular System I*, edited by R. M. Berne, pp. 691–739. American Physiological Society, Bethesda, MD.
590. Korner, P. I., and Angus, J. A. (1981): Central nervous control of blood pressure in relation to antihypertensive drug treatment. *Pharmacol. Ther.*, 13:321–356.
591. Korner, P. I., and Head, G. A. (1981): Effects of noradrenergic and serotonergic neurons on blood pressure, heart rate and baroreceptor-heart rate reflex of the conscious rabbit. *J. Auton. Nerv. Syst.*, 3:511–523.
592. Kovalchik, M. T., Guggenheim, S. J., Silverman, M. D., Robertson, J. S., and Steigerwald, J. C. (1978): The kidney in progressive systemic sclerosis. A prospective study. *Ann. Intern. Med.*, 89:881–887.
593. Kovats, T. G. (1967): Endotoxin susceptibility and endotoxin hypersensitivity. *Monograph of the Medical University of Szeged*, 4:1–101.
594. Kowlessar, O. D., Law, D. H., and Sleisenger M. H. (1959): Malabsorption syndrome associated with metastatic carcinoid tumor. *Am. J. Med.*, 27:673–677.
595. Krstic, M. K., and Djurkovic, D. (1980): Analysis of cardiovascular responses to central administration of 5-hydroxytryptamine in rats. *Neuropharmacology*, 19:455–463.
596. Krstic, M. K., and Djurkovic, D. (1981): Comparison of the cardiovascular responses to intracerebroventricular administration of tryptamine, 5-hydroxytryptamine, tryptophan and 5-hydroxytryptophan in rats. *Arch. Int. Physiol. Biochim.*, 89:385–391.
597. Krupp, P., and Krupp, I. (1960): Serotonin and toxemia of pregnancy. *Obstet. Gynecol.*, 15:237.
598. Ku, D. D. (1982): Coronary vascular reactivity after acute myocardial ischemia. *Science*, 218:576–578.
599. Kubo, T., and Su, C. (1983): Effects of serotonin and some other neurohumoral agents on adrenergic neurotransmission in spontaneously hypertensive rat vasculature. *Clin. Exp. Hypertens.* (A), 5:1501–1510.
600. Kuhar, M. J., Aghajanian, G. K., and Roth, R. H. (1972): Tryptophan hydroxylase activity and synaptosomal uptake of serotonin in discrete brain regions after midbrain raphe lesions: Correlations with serotonin levels and histochemical fluorescence. *Brain Res.*, 44:165–176.
601. Kuhar, M. J., Roth, R. H., and Aghajanian, G. K. (1972): Synaptosomes from forebrains of rats with midbrain raphe lesions: Selective reduction of serotonin uptake. *J. Pharmacol. Exp. Ther.*, 181:36–45.
602. Kuhn, D. M., and Lovenberg, W. (1983): Assay of serotonin, related metabolites, and tryptophan hydroxylase. In: *Methods in Biogenic Amine Research*, edited by S. Parvez, T. Nagatsu, I. Nagatsu, and H. Parvez, pp. 515–548. Elsevier Science Publishers, New York.
603. Kuhn, D. M., Rosenberg, R. C., and Lovenberg, W. (1979): Determination of some molecular parameters of tryptophan hydroxylase from rat midbrain and murine mast cells. *J. Neurochem.*, 33:15–21.
604. Kuhn, D. M., Ruskin, B., and Lovenberg, W. (1980): Tryptophan hydroxylase. The role of oxygen, iron, and sulfhydryl groups as determinants of stability and catalytic activity. *J. Biol. Chem.*, 255:4137–4143.
605. Kuhn, D. M., Wolf, W. A., and Lovenberg, W. (1980): Pressor effects of electrical stimulation of the dorsal and median raphe nuclei in anaesthetized rats. *J. Pharmacol. Exp. Ther.*, 214:403–409.
606. Kuhn, D. M., Wolf, W. A., and Lovenberg W. (1980): Review of the role of the central serotonergic neuronal system in blood pressure regulation. *Hypertension*, 2:243–255.
607. Kuida, H., Hinshaw, L. B., Gilbert, R. P., and Visscher, M. B. (1958): Effect of gram-negative endotoxin on pulmonary circulation. *Am. J. Physiol.*, 192:335–344.
608. Kusajima, K., Ozdemir, I. A., Webb, W. R., Wax, S. D., and Parker, F. B., Jr. (1974): Role of serotonin and serotonin antagonist on pulmonary hemodynamics and microcirculation in hemorrhagic shock. *J. Thorac. Cardiovasc. Surg.*, 67:908–914.
609. Kutti, J., and Weinfeld, A. (1971): Platelet survival in man. *Scand. J. Haematol.*, 8:336–346.

610. Kwaan, H. C., Colwell, J. A., Cruz, S., Suwanela, N., and Dobbie, J. G. (1972): Increased platelet aggregation in diabetes mellitus. *J. Lab. Clin. Med.*, 80:236–246.
611. Lackovic, Z., Parenti, M., and Neff, N. H. (1981): Simultaneous determination of femtomole quantities of 5-hydroxytryptophan, serotonin, and 5-hydroxyindoleacetic acid in brain using HPLC with electrochemical detection. *Eur. J. Pharmacol.*, 69:347–350.
612. Laghi Pasini, F., Guerrini, M., Pasqui, A. L., Ceccatelli, L., Acciavatti, A., Pieragalli, D., Del Bigo, C., Galigani, C., Forconi, S., and Di Perri, T. (1983): Whole blood filterability versus red cell filterability. I. *In vitro* influence of PMA-induced polymorphonuclear leucocyte activation. *Clin. Haemorheol.*, 3:303.
613. Lambert, G. A., Friedman, E., Buchweitz, E., and Gershon, S. (1978): Involvement of 5-hydroxytryptamine in the central control of respiration, blood pressure and heart rate in the anesthetized rat. *Neuropharmacology*, 17:807–813.
614. Lampugnani, M. G., De Gaetano, G., and Rossi, E. C. (1982): Functional distinction between serotonin uptake and serotonin induced shape change receptors in rat platelets. *Biochim. Biophys. Acta*, 693:22–26.
615. Langley, (1878): On the physiology of the salivary secretion. *J. Physiol. (Lond.)*, 1:339–369.
616. Larsson, L.-I. (1980): Gastrointestinal cells producing endocrine, neurocrine and paracrine messengers. *Clin. Gastroenterol.*, 9:485–516.
617. Laubscher, A., and Pletscher, A. (1979): Shape change and uptake of 5-hydroxytryptamine in human blood platelets: Action of neuropsychotropic drugs. *Life Sci.*, 24:1833–1840.
618. Lauder, J. M., Petrusz, P., Wallace, J. A., Di Nome, A., Wilkie, M. B., and McCarthy, K. (1982): Combined serotonin immunohistochemistry and ^{3}H-thymidine autoradiography. *J. Histochem. Cytochem.*, 30:788–793.
619. Lawler, J. W., Slayter, H. S. L., and Coligan, J. E. (1978): Isolation and characterization of high molecular weight glycoprotein from human platelets. *J. Biol. Chem.*, 253:8609–8616.
620. LeDouarin, N. (1982): In: *Neuronal Crest.* Cambridge University Press, Cambridge.
621. Leempoels, J., De Cock, W., De Cree, J., and Verhaegen, H. (1983): The effect of ketanserin on 5-HT-induced irreversible platelet aggregation in patients with cardiovascular disease. First European Meeting on Hypertension, May 29–June 1, Milan (abstract).
622. Leeper, L. C., Weissbach, H., and Udenfriend, S. (1968): Studies on the metabolism of norepinephrine, epinephrine, and their *O*-methyl analogs by partially purified enzyme preparations. *Arch. Biochem. Biophys.*, 77:417–423.
623. Lefer, A. M. (1978): Properties of cardioinhibitory factors produced in shock. *Fed. Proc.*, 37:2734–2740.
624. Legay, C., Faudon, M., Hery, F., and Ternaux, J. P. (1983): 5-HT metabolism in the intestinal wall of the rat. I. The mucosa. *Neurochem. Int.*, 5:721–727.
625. Legay, C., Faudon, M., Hery, F., and Ternaux, J. P. (1983): 5-HT metabolism in the intestinal wall of the rat. II. The nerve plexuses—interactions between 5-HT containing cells. *Neurochem. Int.*, 5:571–577.
626. Lehmann, J. (1982): Tryptophan deficiency stupor—A new psychiatric syndrome. *Acta Psychiatr. Scand. (Suppl.)* 300(65).
627. Lembeck, F. (1953): 5-Hydroxytryptamine in a carcinoid tumor. *Nature (Lond.)*, 172:910–911.
628. Leny, R. I., Wiener, L., Walinsky, P., Lefer, A. M., Silver, M. J., and Smith, J. B. (1980): Thromboxane release during pacing-induced angina pectoris: Possible vasoconstrictor influence on the coronary vasculature. *Circulation*, 61:1165–1171.
629. LeRoy, E. C. (1981): Scleroderma (systemic sclerosis). In: *Textbook of Rheumatology*, edited by W. N. Kelley, E. D. Harris, S. Ruddy, and C. B. Sledge, pp. 1211–1230. W. B. Saunders Co., Philadelphia.
630. Levine, R. J. (1974): Serotonin and the carcinoid syndrome: Histamine and mastocytosis. In: *Duncan's Diseases of Metabolism: Endocrinology*, 7th edition, edited by P. K. Bondy and L. E. Rosenberg, pp. 1651–1684. W. B. Saunders Co., Philadelphia.
631. Levine, R. J., and Sjoerdsma, A. (1963): Pressor amines and the carcinoid flush. *Ann. Intern. Med.*, 58:818–828.
632. Lewis, P. J., Boyling, P., Friedman, L. A., et al. (1980): Prostacyclin in pregnancy. *Br. Med. J.*, 280:1581.
633. Leysen, J. E. (1979): Different neuroleptic receptors in various rat brain areas. In: *Catecholamines: Basic and Clinical Frontiers*, edited by F. Usdin, I. J. Kopin, and J. Barchas, pp. 556–558. Pergamon Press, New York.

634. Leysen, J. E. (1981): Serotoninergic receptors in brain tissue: Properties and identification of various [^{3}H]ligand binding sites *in vitro*. *J. Physiol. (Lond.)*, 77:351–362.
635. Leysen, J. E. (1983): Rationale and importance of binding studies in respect to characterization of serotonin receptors. In: *Vascular Neuroeffector Mechanisms*, edited by J. A. Bevan et al., pp. 259–264. Raven Press, New York.
636. Leysen, J. E. (1984): Problems in *in vitro* receptor binding studies and identification and role of serotonin receptor sites. *Neuropharmacology*, 23:247–254.
637. Leysen, J. E., Awouters, F., Kennis, L., Laduron, P. M., Vandenberk, J., and Janssen, P. A. J. (1981): Receptor binding profile of R41,468, a novel antagonist at 5HT$_2$ receptors. *Life Sci.*, 28:1015–1022.
638. Leysen, J. E., de Chaffoy de Courcelles, D., De Clerck, F., Niemegeers, C. J. E., and Van Nueten, J. M. (1984): Serotonin-S$_2$ receptor binding sites and functional correlates. *Neuropharmacology (in press)*.
639. Leysen, J. E., Geerts, R., Gommeren, W., Verwimp, M., and Van Gompel, P. (1982): Regional distribution of serotonin-2 receptor binding sites in the brain and effects of neuronal lesions. *Arch. Int. Pharmacodyn. Ther.*, 256:301–305.
640. Leysen, J. E., Gommeren, W., and De Clerck, F. (1983): Demonstration of S$_2$-receptor binding sites on cat blood platelets using [^{3}H]ketanserin. *Eur. J. Pharmacol.*, 88:125–130.
641. Leysen, J. E., and Laduron, P. M. (1977): A serotonergic component of neuroleptic receptors. *Arch. Int. Pharmacodyn. Ther.*, 230:337–339.
642. Leysen, J. E., Niemegeers, C. J. E., Tollenaere, J. P., and Laduron, P. M. (1978): Serotonergic component of neuroleptic receptors. *Nature (Lond.)*, 272:168–171.
643. Leysen, J. E., Niemegeers, C. J. E., Van Nueten, J. M., and Laduron, P. M. (1982): [^{3}H]Ketanserin (R 41 468), a selective ^{3}H-ligand for serotonin$_2$ receptor binding sites. Binding properties, brain distribution and functional role. *Mol. Pharmacol.*, 21:301–314.
644. Leysen, J. E., and Tollenaere, J. P. (1982): Biochemical models for serotonin receptors. *Annu. Rep. Med. Chem.*, 17:1–10.
645. Leysen, J. E., Van Gompel, P., Verwimp, M., and Niemegeers, C. J. E. (1983): Role and localization of serotonin$_2$ (S$_2$)-receptor-binding sites: Effects of neuronal lesions. In: *CNS Receptors. From Molecular Pharmacology to Behavior*, edited by P. Mandel and F. V. DeFeudis, pp. 373–383. Raven Press, New York.
646. Lidberg, P., Dahlström, A., Lundberg, J. M., and Ahlman, H. (1983): Different modes of action of substance P in the motor control of the feline stomach and pylorus. *Regul. Pept.*, 7:41–52.
647. Liljedahl, S.-O., Mattson, B., and Pernow, B. (1958): Effect of substance P on intestinal motility in man. *Scand. J. Clin. Lab. Invest.*, 10:16–25.
648. Lin, M. T., and Chern, S. I. (1979): Effect of brain 5-hydroxytryptamine alterations on reflex bradycardia in rats. *Am. J. Physiol.*, 236:R302–R306.
649. Lindblad, L. E., Shepherd, J. T., and Vanhoutte, P. M. (1984): Cooling augments platelet-induced contraction of peripheral arteries of the dog. *Proc. Soc. Exp. Biol. Med.*, 176:119–122.
650. Liu, H. M. (1983): Effects of raphe stimulation on cardiovascular response to noxious input in cats. *J. Formosan Med. Assoc.*, 82:964–966.
651. Liu, K. P., Tamir, H., Karpiak, S. E., and Gershon, M. D. (1982): Serotonin is stored as a complex with serotonin binding protein in situ. *Neurosci. Abst.*, 8:489.
652. Ljungqvist, U. (1973): Platelets in shock and trauma. *J. Surg. Res.*, 15:132–162.
653. Loewy, A. D. (1981): Raphe pallidus and raphe obscurus projections to the intermediolateral cell column in the rat. *Brain Res.*, 222:129–133.
654. Loewy, A. D., and Burton, H. (1978): Nuclei of the solitary tract: Efferent projections to the lower brain stem and spinal cord of the cat. *J. Comp. Neurol.*, 181:421–450.
655. Loewy, A. D., and McKellar, S. (1981): Serotonergic projections from the ventral medulla to the intermediolateral cell column in the rat. *Brain Res.*, 211:146–152.
656. Loewy, A. D., and Neil, J. J. (1981): The role of descending monoaminergic systems in central control of blood pressure. *Fed. Proc.*, 40:2778–2785.
657. Long, R. G., Peter, J. R., Bloom, S. R., Brown, M. R., Vale, W., Rivier, J. E., and Grahame-Smith, D. G. (1981): Somatostatin, gastrointestinal peptides and the carcinoid syndrome. *Gut*, 22:549–553.
658. Longstaff, J., Gush, E. H., Williams, S. E., and Jayson, M. (1984): Effects of ketanserin, a novel serotonin antagonist, on haemorheology and blood flow in patients with Raynaud's phenomenon. Abstracts of the Satellite Symposium to the 10th Scientific Meeting of the International

Society of Hypertension, "Serotonergic Mechanisms in the Cardiovascular System," Zurich, 1984, p. 49.

659. Loose, R. W., Patterson, W. G., and Crimgeour, J. B. (1970): The concentration of 5-hydroxyindoleacetic acid in fetal urine in normal and toxemic pregnancies. *Biol. Neonate*, 16:323.

660. Lorenz, R. R., and Vanhoutte, P. M. (1984): Aggregating platelets and adrenergic neurotransmission in isolated canine blood vessels. *Fed. Proc.*, 43:430.

661. Lovenberg, W., Jequier, E., and Sjoerdsma, A. (1967): Tryptophan hydroxylation in pineal gland, brainstem and carcinoid tumor. *Science*, 155:217–219.

662. Lovenberg, W., Levine, R. J., and Sjoerdsma, A. (1965): A tryptophan hydroxylase in cell-free extracts of malignant mouse mast cells. *Biochem. Pharmacol.*, 14:887–889.

663. Lowe, G. D. O., Reavey, M. M., Johnston, R. V., Forbes, C. D., and Prentice, C. R. M. (1979): Increased platelet aggregates in vascular and non-vascular illness. Correlation with plasma fibrinogen and effect of Ancrod. *Thromb. Res.*, 14:377–386.

664. Lubarsch, O. (1888): Über den primaren Krebs des Ileum nebst Bemerkungen über das gleichzeitige Vorkommen von Krebs und Tuberculose. *Virchows Archiv. (Pathol. Anat.)*, 111:280–317.

665. Ludlam, C. A. (1979): Evidence for the platelet specificity of β-thromboglobulin and studies on its plasma concentration in healthy individuals. *Br. J. Haematol.*, 41:271–278.

666. Ludlam, C., Bolton, A., Moore, S., and Cash, J. (1975): New rapid method for diagnosis of deep venous thrombosis. *Lancet*, 2:259–260.

667. Ludlam, C. A., Moore, S., Bolton, A. E., Pepper, D. S., and Cash, J. D. (1975): The release of a human platelet specific protein measured by radioimmunoassay. *Thromb. Res.*, 6:543–548.

668. Lundberg, J. M., and Saria, A. (1982): Vagal SP nerves involved in control of vascular permeability and smooth muscle tone in the trachea and bronchi. *Br. J. Pharmacol.*, 77:441P.

669. Luscombe, G., Jenner, P., and Marsden, C. D. (1981): Pharmacological analysis of the myoclonus induced by 5-hydroxytryptophan in the guinea pig suggests the presence of multiple 5-hydroxytryptamine receptors in the brain. *Neuropharmacology*, 20:819–831.

670. MacDonald, R. A., Robins, S. C., and Mallory, G. K. (1958): Dermal fibrosis following subcutaneous injections of serotonin creatinine sulphate. *Proc. Soc. Exp. Biol.*, 97:334–337.

671. Mack, A. J., and Todd, J. K. (1969): A study of human gallbladder muscle *in vitro*. *Gut*, 9:548–549.

672. Madan, B. R., Khanna, N. K., Godhwani, J. L., and Pendse, V. K. (1970): Changes in the 5-hydroxytryptamine content of the heart during ectopic ventricular arrhythmias and consequent to its reversion by quinidine. *Indian J. Med. Res.*, 58:130–134.

673. Maggi, C. A., Manzini, S., and Meli, A. (1983): Contribution of cellular and extracellular Ca^{2+} during 5-hydroxytryptamine-induced contractions of rabbit ear artery. *Eur. J. Pharmacol.*, 94:251–260.

674. Majno, G., and Palade, G. E. (1961): Studies on inflammation. I. The effect of histamine and serotonin on vascular permeability: An electron microscopic study. *J. Biophys. Biochem. Cytol.*, 11:571.

675. Majno, G., Palade, G. E., and Schoefl, G. I. (1961): Studies on inflammation. II. The site of action of histamine and serotonin along the vascular tree: A topographic study. *J. Biophys. Biochem. Cytol.*, 11:607–626.

676. Majno, G., Shea, S. M., and Leventhal, M. (1969): Endothelial contractions induced by histamine-type mediators. *J. Cell Biol.*, 42:647–672.

677. Makabali, G. L., Mandal, A. K., and Morris, J. A. (1983): An assessment of the participatory role of prostaglandins and serotonin in the pathophysiology of endotoxic shock. *Am. J. Obstet. Gynecol.*, 145:439–445.

678. Makila, U., Jouppila, P., Kirkinen, P., et al. (1983): Relation between umbilical prostacyclin production and blood flow in the fetus. *Lancet*, 1:728.

679. Maksad, A. K., Chung-Ja, C., Stuart, R-C., Brosco, F. A., and Clowes, G. H. A., Jr. (1979): Myocardial depression in septic shock: Physiologic and metabolic effects of a plasma factor on an isolated heart. *Circ. Shock (Suppl.)*, 1:35–42.

680. Malmgren, R., Olsson, P., and Unge, G. (1981): The effect of various inhibitors on the 5HT binding and uptake by human platelets. *Thromb. Res.*, 22:469–480.

681. Malofiejew, M. (1971): The biological and pharmacological properties of some fibrinogen degradation products. *Scand. J. Haematol. (Suppl.)*, 13:303–308.

682. Mandel, A. J., and Knapp, S. (1977): Regulation of serotonin biosynthesis in brain: Role of the high affinity uptake of tryptophan into serotonergic neurons. *Fed. Proc.*, 36:2142–2148.

683. Man in't Veld, A. J., Wenting, G. J., Boomsma, F., Verhoeven, R. P., and Schalekamp, M. A. D. H. (1980): Sympathetic and parasympathetic components of reflex cardiostimulation during vasodilator treatment of hypertension. *Br. J. Clin. Pharmacol.*, 9:547–551.
684. Mantyh, P. W., and Peschanski, M. (1982): Spinal projections from the periaqueductal grey and dorsal raphe in the rat, cat and monkey. *Neuroscience*, 7:2769–2776.
685. Marcinkiewicz, M., Vergé, D., Gozlan, H., Pichat, L., and Hamon, M. (1984): Autoradiographic evidence for the heterogeneity of 5-HT_1 sites in the rat brain. *Brain Res.*, 291:159–163.
686. Marcus, A. J., Weksler, B. B., Jaffe, E. A., and Broekman, M. J. (1980): Synthesis of prostacyclin from platelet-derived endoperoxide by cultured human endothelial cells. *J. Clin. Invest.*, 66:979–986.
687. Marguerie, G. A., Ardaillo, N., Cherel, G., and Plow, E. F. (1982): The binding of fibrinogen to its platelet receptor. Involvement of the D domain. *J. Biol. Chem.*, 257:11872–11875.
688. Maricq, H. R., Spencer-Green, G., and LeRoy, E. C. (1976): Skin capillary abnormalities as indicators of organ involvement in scleroderma (systemic sclerosis), Raynaud's syndrome and dermatomyositis. *Am. J. Med.*, 61:862–870.
689. Maricq, H. R., Weinberger, A. B., and LeRoy, E. C. (1982): Early detection of scleroderma-spectrum disorders by *in vivo* capillary microscopy: A prospective study of patients with Raynaud's phenomenon. *J. Rheumatol.*, 9:289–291.
690. Markgraf, W. H., and Dunn, T. M. (1964): Appendiceal carcinoid with carcinoid syndrome. *Am. J. Surg.*, 107:730–732.
691. Mårtensson, H., Nobin, A., and Sundler, F. (1984): Endocrine tumors of the ileum—Cytochemical and clinical aspects. *Gut (in press).*
692. Martin, L. L., and Sanders-Bush, E. (1982): Comparison of the pharmacological characteristics of 5 HT_1 and 5 HT_2 binding sites with those of serotonin autoreceptors which modulate serotonin release. *Naunyn Schmiedebergs Arch. Pharmacol.*, 321:165–170.
693. Martin, T. J., Smith, I. L., Nolan, R. D., and Dusting, G. J. (1983): Prostanoids in platelet vascular interactions. *Am. J. Cardiol.*, 52:22a.
694. Marwood, J. F., and Stokes, G. S. (1983): Vascular smooth muscle serotonin and α-adrenoceptors. *Clin. Exp. Pharmacol. Physiol.*, 10:265–267.
695. Marwood, J. F., and Stokes, G. S. (1984): Serotonin (5-HT) and its antagonists: Involvement in the cardiovascular system. *Clin. Exp. Pharmacol. Physiol. (in press).*
696. Marwood, J. F., and Stokes, G. S. (1984): Studies on the mechanism of action of the hypotensive effect of ketanserin. *Clin. Exp. Pharmacol. Physiol.*, 11:125–132.
697. Maseri, A., L'Abbate, A., Baroldi, G., Chiercha, S., Marzilli, M., Ballestra, A. M., Sereri, S., Parodi, O., Biagini, A., Distante, A., Pesola, A. (1978): Coronary vasospasm as a possible cause of myocardial infarction. *N. Engl. J. Med.*, 299:1271–1277.
698. Massini, P., and Naf, U. (1980): Ca^{2+} ionophores and the activation of human blood platelets. The effect of ionomycin, beauvericin, lysocellin, virginamycin S lasalocid-derivatives and McN 4308. *Biochim. Biophys. Acta*, 598:575–587.
699. Masson, P. (1928): Carcinoids (argentaffin cell tumors) and nerve hyperplasia of appendicular mucosa. *Am. J. Pathol.*, 4:181–211.
700. Matsumo, M., Kusunoki, M., Uyama, O., Fujisawa, A., Matsuyama, T., Yoneda, S., Kimura, K., and Abe, H. (1981): Platelet aggregation induced by arachidonic acid and thromboxane generation in patients with hypertension or cerebrovascular disease. *Prostaglandins Med.*, 7:553–562.
701. McAllen, R. M., Neil, J. J., and Loewy, A. D. (1982): Effects of kainic acid applied to the ventral surface of the medulla oblongata on vasomotor tone, the baroreceptor reflex and hypothalamic autonomic responses. *Brain Res.*, 238:65–76.
702. McCall, R. B. (1984): Evidence for serotonergically mediated sympatho-excitatory response to stimulation of medullary raphe nuclei. *Brain Res. (in press).*
703. McCall, R. B., Gebber, G. L., and Barman, S. M. (1977): Spinal interneurons in the baroreceptor reflex arc. *Am. J. Physiol.*, 232:H657–H665.
704. McCall, R. B., and Humphrey, S. J. (1982): Involvement of serotonin in the central regulation of blood pressure: Evidence for a facilitating effect on sympathetic nerve activity. *J. Pharmacol. Exp. Ther.*, 222:94–102.
705. McCall, R. B., and Schuette, M. R. (1984): Evidence for an alpha-1 receptor-mediated central sympathoinhibitory action of ketanserin. *J. Pharmacol. Exp. Ther.*, 228:704–710.

706. McCartney, C. P. (1964): Pathologic anatomy of acute hypertension of pregnancy. *Circulation (Suppl.* 2), 30:37–43.
707. McCubbin, J. W., Kaneko, Y., and Page, I. H. (1962): Inhibition of neurogenic vasoconstriction by serotonin: Vasodilator action of serotonin. *Circ. Res.*, 11:74–83.
708. McGoon, M. D., and Vanhoutte, P. M. (1984): Aggregating platelets contract isolated canine pulmonary arteries by releasing 5-hydroxytryptamine. *J. Clin. Invest.*, 74:828–833.
709. McGrath, M. A. (1977): 5-Hydroxytryptamine and neurotransmitter release in canine blood vessels. *Circ. Res.*, 41:428–435.
710. McGrath, M. A., and Shepherd, J. T. (1977): Histamine and 5-hydroxytryptamine-inhibition of transmitter release mediated by H_2 and 5-hydroxytryptamine receptors. *Fed. Proc.*, 37:195–198.
711. McGregor, D. D., and Smirk, F. H. (1970): Vascular responses to 5-hydroxytryptamine in genetic and renal hypertensive rats. *Am. J. Physiol.*, 219:687–690.
712. McIsaac, W. M., and Page, I. H. (1958): New metabolites of serotonin in carcinoid urine. *Science*, 128:537.
713. McLean, E. K. (1970): The toxic actions of pyrrolizidine (Senecio) alkaloid. *Pharmacol. Rev.*, 22:429–518.
714. McNeil, J. R., Stark, R. D., and Greenway, C. V. (1970): Intestinal vasoconstriction after hemorrhage: Roles of vasopressin and angiotensin. *Am. J. Physiol.*, 219:1342–1347.
715. Mecca, T. E., and Webb, R. C. (1984): Serotonin and vasodilatation. In: *Vasodilator Mechanisms*, edited by P. M. Vanhoutte and S. F. Vatner. Raven Press, New York *(in press)*.
716. Mecca, T. E., and Webb, R. C. (1984): Vascular responses to serotonin in steroid hypertensive rats. *Hypertension (in press)*.
717. Medcalf, R. L., Iwanov, V., and Moulds, R. F. (1983): Responses of human arterial muscle to stable vasoactive substances released from human platelets by collagen and heat aggregated IgG. *Clin. Exp. Pharmacol. Physiol.*, 10:567–575.
718. Meek, J. L. (1980): Nomogram for adjusting mobile phase composition in reverse phase high pressure liquid chromatograph. *Anal. Chem.*, 52:1370–1371.
719. Mefford, I. N. (1981): Application of high performance liquid chromatography with electrochemical detection to neurochemical analysis: Measurement of catecholamines, serotonin, and metabolites, in rat brain. *J. Neurosci. Methods*, 3:207–224.
720. Mefford, I. N., and Barchas, I. D. (1980): Determination of tryptophan and metabolites in rat brain and pineal tissue by reversed phase high performance liquid chromatography with electrochemical detection. *J. Chromatogr.*, 181:187–193.
721. Mehta, J., and Mehta, P. (1981): Platelet function in hypertension and effect of therapy. *Am. J. Cardiol.*, 47:331–334.
722. Mehta J., and Mehta, P. (1981): Role of blood platelets and prostaglandins in coronary artery disease. *Am. J. Cardiol.*, 48:366–373.
723. Mehta, J., Mehta, P., Feldman, R. L., and Horalek, C. (1984): Thromboxane release in coronary artery disease: Spontaneous versus pacing-induced angina. *Am. Heart J.*, 107:286–292.
724. Mehta, J., Mehta, P., Zipper, R., and Horalek, C. (1981): Thromboxane/prostacyclin equilibrium at rest and during exercise in man. Its relationship to myocardial ischemia. *Clin. Res.*, 29:223A (abstract).
725. Mellander, S., and Lewis, D. H. (1963): Effect of hemorrhagic shock on the reactivity of resistance and capacitance vessels and on capillary filtration transfer in cat skeletal muscle. *Circ. Res.*, 13:105–118.
726. Mena, M. A., and Vidrio, H. (1979): Reversal of serotonin vasodilatation in the dog external carotid bed by sympathetic denervation. *J. Cardiovasc. Pharmacol.* 1:149–154.
727. Menguy, R. (1969): Gastric mucus and the gastric acid mucous barrier. A review. *Am. J. Surg.*, 117:806–812.
728. Mester, L., Szabados, L., Born, G. V. R., and Michal, F. (1972): Change in aggregation of platelets enriched in sialic acid. *Nature New. Biol.*, 232:213–214.
729. Meuleman, T. R., Hill, D. C., Port, J. D., Stanley, T. H., Pace, N. L., and Mohammad, S. F. (1983): Ketanserin prevents platelet aggregation and endotoxin-induced pulmonary vasoconstriction. *Crit. Care Med.*, 11:606–611.
730. Meyerhoff, J., and Dorsch, C. A. (1981): Platelet aggregation in systemic sclerosis (PSS). *Arthritis Rheum.*, 24:S85.

731. Meyers, K. M., Smith, J. B., Holmsen, H., and Prieur, H. (1979): The dominant role of thromboxane formation in secondary aggregation of platelets. *Nature (Lond.)*, 282:331–333.
732. Michal, F. (1969): D-Receptors for serotonin on blood platelets. *Nature (Lond)*, 221:1253–1254.
733. Michal, F., and Motamed, M. (1975): Time-dependent potentiation and inhibition by 5-hydroxytryptamine of platelet aggregation induced by ADP. *Br. J. Pharmacol.*, 54:221P–222P.
734. Michell, R. H., Kirk, C. J., Jones, L. M., Downes, C. P., and Creba, J. (1981): The stimulation of inositol lipid metabolism that accompanies calcium mobilization in stimulated cells: Defined characteristics and unanswered questions. *Philos. Trans. R. Soc. Lond. (Biol.)*, 296:123–137.
735. Middlemiss, D. N., and Fozard, J. R. (1983): 8-Hydroxy-2-(di-*n*-propyl-amino)-tetralin discriminates between subtypes of the 5-HT_1 recognition site. *Eur. J. Pharmacol.*, 90:151–153.
736. Miller, T. M., and Heuser, J. E. (1984): Endocytosis of synaptic vesicle membrane at the frog neuromuscular junction. *J. Cell Biol.*, 98:685–698.
737. Minson, J. B., Choy, V. J., and Chalmers, J. P. (1984): Bulbospinal serotonin neurons and hypotensive effects of methyldopa in the spontaneously hypertensive rat. *J. Cardiovasc. Pharmacol.*, 6:312–317.
738. Modlin, I. M. (1980): Carcinoid syndrome. *J. Clin. Gastroenterol.*, 2:349–354.
739. Moertel, C. G., Beahrs, O. H., Wollner, L. B., and Tyce G. M. (1965): Malignant carcinoid syndrome associated with noncarcinoid tumors. *N. Engl. J. Med.*, 273:244–248.
740. Moertel, C. G., Dockerty, M. B., and Judd E. S. (1968): Carcinoid tumors of the vermiform appendix. *Cancer*, 21:270–278.
741. Moertel, C. G., Sauer, W.G., Dockerty, M. B., and Baggenstoss, A. H. (1961): Life history of the carcinoid tumor of the small intestine. *Cancer*, 14:901–912.
742. Moir, A. T. B. (1971): Interaction in the cerebral metabolism of the biogenic amines: Effects of intravenous infusion of L-tryptophan on the metabolites of dopamine and 5-hydroxyindoles in brain. *Br. J. Pharmacol.*, 43:715–723.
743. Moncada, S. (1982): Biological importance of prostacyclin. *Br. J. Pharmacol.*, 76:3–31.
744. Moncada, S., and Vane, J. R. (1978): Unstable metabolites of arachidonic acid and their role in haemostasis and thrombosis. *Br. Med. Bull.*, 34:129–135.
745. Moncada, A. S., and Vane, J. R. (1979): Arachidonic acid metabolites and the interaction between platelets and blood vessel wall. *N. Engl. J. Med.*, 300:1142–1147.
746. Moncada, A. S., and Vane, J. R. (1979): Pharmacology and endogenous roles of prostaglandin endoperoxides, thromboxane A_2, and prostacyclin. *Pharmacol. Rev.*, 30:293–331.
747. Moore, F. D. (1968): Pulmonary insufficiency—Discussion. *J. Trauma*, 8:666–675.
748. Moreland, R. S., and Bohr, D. F. (1983): Prior exposure to serotonin attenuates norepinephrine contraction in dog mesenteric artery. *Blood Vessels*, 20:203.
749. Morgan, J. G., Marks, C., and Hearn, D. (1974): Carcinoid tumors of the gastrointestinal tract. *Ann. Surg.*, 180:720–727.
750. Morgan, M., and Mandell, A. J. (1969): Indole (ethyl) amine *N*-methyltransferase in the brain. *Science*, 165:492–493.
751. Morita, K., North, R. A., and Tokimasa, T. (1982): Evidence that substance P is a neurotransmitter in the myenteric plexus. *Nature (Lond.)* 287:151–152.
752. Morris, R. H., Vassalli, P., Beller, F. K., et al. (1964): Immunofluorescent studies of renal biopsies in the diagnosis of toxemia pregnancy. *Obstet. Gynecol.*, 24:32.
753. Morrissey, J. J., Walker, M. N., and Lovenberg, W. (1977): The absence of tryptophan hydroxylase activity in blood platelets. *Proc. Soc. Exp. Biol. Med.*, 154:496–499.
754. Morson, B. C., and Dawson, I. M. P. (1972): Carcinoid tumours. In: *Gastrointestinal Pathology*, edited by B. C. Morson and I. M. P. Dawson, pp. 354–360. Blackwell Scientific Publications, Oxford.
755. Moulds, R. F. W., Iwanov, V., and Medcalf, R. L. (1981): The sensitivity of human vascular smooth muscle to platelet-derived vasoactive agents. Australian Society for Clinical and Experimental Pharmacology, December 14–16, Adelaide (abstract).
756. Moulds, R. F. W., Iwanov, V., and Medcalf, R. L. (1984): The effects of platelet-derived contractile agents on human digital arteries. *Clin. Sci.*, 66:443–451.
757. Muakkassah-Kelly, S. F., Andresen, J. W., Shih, J. C., and Hochstein, P. (1982): Decreased [^{3}H]serotonin and [^{3}H]spiperone binding consequent to lipid peroxidation in rat cortical membranes. *Biochem. Biophys. Res. Commun.*, 104:1003–1010.
758. Muakkassah-Kelly, S. F., Andresen, J. W., Shih, J. C., and Hochstein, P. (1983): Dual effects of

ascorbate on serotonin and spiperone binding in rat cortical membranes. *J. Neurochem.*, 41:1429–1439.

759. Mullane, K. M., Bradley, G., and Moncada, S. (1982): The interactions of platelet-derived mediators on isolated canine coronary arteries. *Eur. J. Pharmacol.*, 84:115–118.
760. Müller-Schweinitzer, E., and Engel, G. (1983): Evidence for mediation by 5-HT_2 receptors of 5-hydroxytryptamine-induced contraction of canine basilar artery. *Naunyn Schmiedebergs Arch. Pharmacol.*, 324:287–292.
761. Mürer, E. H., Davenport, K., and Day, H. J. (1976): Metabolic aspects of the secretion of stored compounds from blood platelets. IV. Effects of ionophore X537A on washed platelets. *Biochim. Biophys. Acta*, 428:369–378.
762. Murphy, E. A., and Mustard, J. F. (1962): Coagulation tests and platelet economy in atherosclerotic and control subjects. *Circulation*, 25:114–125.
763. Murphy, M. B., Scriven, A. J., Brown, M. J., Causon, R., and Dollery, C. T. (1982): The effects of nifedipine and hydralazine induced hypotension on sympathetic activity. *Eur. J. Clin. Pharmacol.*, 23:479–482.
764. Murphy, T. L., Allison, R. C., Weisman, I. M., McCaffree, D. R., and Gray, B. A. (1981): Role of platelet serotonin in the canine pulmonary response to endotoxin. *J. Appl. Physiol.*, 50:178–184.
765. Murray-Lyon, I. M., Rake, M. O., and Marshall, A. K. (1973): Malignant carcinoid tumour with gangrene of the small intestine. *Br. Med. J.*, 4:770–771.
766. Musemici, V., Carloni, L., Zappa-Costa, B., Marra, R., and Elamthuruthil, J. (1980): Influence of blood dilution on the findings of platelet aggregates in native venous blood fixed by EDTA-formalin. *Thromb. Res.*, 20:25–30.
767. Mustard, J. F., Movat, H. Z., Macmorine, D. R. L., and Senyi, A. (1965): Release of permeability factors from the blood platelet. *Proc. Soc. Exp. Biol. Med.*, 119:988–991.
768. Mustard, J. F., and Murphy, E. A. (1963): Effect of smoking on blood coagulation and platelet survival in man. *Br. Med. J.*, 1:846–851.
769. Mutaz, M., Narasimhachari, N., and Friedel, R. O. (1982): Evaluation of fluorometric assay methods for serotonin in platelets, plasma and whole blood samples by comparison with GC-MS-SIM technique. *Res. Commun. Chem. Pathol. Pharmacol.*, 36:45.
770. Myrvold, H. E., and Svalander, C. (1977): Pulmonary microembolism in early experimental septic shock. A morphological study in dogs. *J. Surg. Res.*, 23:65–74.
771. Nachman, R. L., Weksler, B., and Ferris, B. (1970): Increased vascular permeability produced by human platelet granule cationic extract. *J. Clin. Invest.*, 49:274–281.
772. Nachman, R. L., Weksler, B., and Ferris, B. (1972): Characterization of human platelet vascular permeability-enhancing activity. *J. Clin. Invest.*, 51:549–556.
773. Najean, Y., Dassin, E., Renner, C., and Wacquet, M. (1979): Cinétique plaquettaire au cours des maladies artérielles. *Nouv. Presse Med.*, 8:3813–3816.
774. Nelson, D. L., Herbet, A., Enjalbert, A., Bockaert, J., and Hamon, M. (1980): Serotonin-sensitive adenylate cyclase and [^{3}H]serotonin binding sites in the CNS of the rat. I. Kinetic parameters and pharmacological properties. *Biochem. Pharmacol.*, 29:2445–2453.
775. Nelson, D. R., and Huggins, A. K. (1976): The effect of *oxygen* on the enzymes concerned in the metabolism of 5-hydroxytryptamine by rat brain and liver preparations. *Gen. Pharmacol.*, 7:271–276.
776. Neufeld, E. J., and Majerus, P. W. (1983): Arachidonate release and phosphatidic acid turnover in stimulated human platelets. *J. Biol. Chem.*, 258:2461–2467.
777. Neumayr, R. J., Hare, B. D., and Franz, D. D. (1974): Evidence for bulbospinal control of sympathetic preganglionic neurones by monoaminergic pathways. *Life Sci.*, 14:793–806.
778. Nevelsteen, A., De Clerck, F., Loots, W., and De Gryse, A. (1984): Restoration of post-thrombotic peripheral collateral circulation in the cat by ketanserin, a selective 5-HT2 receptor antagonist. *Arch. Int. Pharmacodyn. Ther. (in press).*
779. Niemegeers, C. J. E., Colpaert, F. C., Leysen, J. E., Awouters, F., and Janssen, P. A. J. (1983): Mescaline-induced head-twitches in the rat: An *in vivo* method to evaluate serotonin-S_2 antagonists. *Drug Devel. Res.*, 3:123–135.
780. Niemegeers, C. J. E., Leysen, J. E., Laduron, P. M., and Janssen, P. A. J. (1984): Differential pharmacological and biochemical profiles of serotonin-S_2 antagonists. *Collegium Internationale Neuro-Psychopharmacologicum, 14th CINP Congress, Book of Abstracts*, p. 667.
781. Nilsson, O., Cassuto, J., Larsson, P.-A., Jodal, M., Lidberg, P., Ahlman, H., Dahlström, A.,

and Lundgren, O. (1983): 5-Hydroxytryptamine and cholera secretion: A histochemical and physiological study in cats. *Gut*, 24:542–548.

782. Nishizuka, Y. (1983): Phospholipid degradation and signal translation for protein phosphorylation. *TIBS*, 8:13–16.
783. Niwa, M., Kunisado, K., Himeno, A., Kawaguchi, A., and Ozaki, M. (1984): 5-Hydroxytryptamine content in the rat heart. Quantitation by high performance liquid chromatographic electrochemical detection. *Jpn. J. Pharmacol.*, 24:264–267.
784. Nobin, A., Axelsson, S., Falck, B., Ingemansson, S., Lunderquist, A., Mårtensson, H. and Reichardt, W. (1983): Selective mesenteric vein catheterization and plasma serotonin determination in patients with carcinoid tumors. *World J. Surg.*, 7:223–229.
785. Nolan, P. L., and Brazenor, R. M. (1978): L-Tryptophan induced depression of the pressor response to clonidine in anesthetized rats. *Clin. Exp. Pharmacol. Physiol.*, 5:199–206.
786. Nomura, Y., and Takagi, H. (1972): Studies on plasma factor involved in 5-hydroxytryptamine release from blood platelets by bacterial endotoxin. I. *Jpn. J. Pharmacol.*, 22:37–42.
787. Norheim, I. (1984): *Personal communication.*
788. North, R. A., Henderson, C., Katayama, Y., and Johnson, S. M. (1980): Electrophysiological evidence of presynaptic inhibition of acetylcholine release by 5-hydroxytryptamine in the enteric nervous system. *Neuroscience*, 5:581–586.
789. Northover, A. M., and Northover, B. J. (1969): The effects of histamine, 5-hydroxytryptamine and bradykinin on rat mesenteric blood vessels. *J. Pathol.*, 98:265–276.
790. Norton, W. L., and Nardo, J. M. (1970): Vascular disease in progressive systemic sclerosis (scleroderma). *Ann. Intern. Med.*, 73:317–324.
791. Nossel, H. L., and Vogel, H. J., editors (1982): *Pathobiology of the Endothelial Cell.* Academic Press, New York.
792. Nukiwa, T., Toyama, C., Okita, C., Kataoka, T., and Ichiyama, A. (1974): Purification and some properties of bovine pineal tryptophan 5- monooxygenase. *Biochem. Biophys. Res. Commun.*, 60:1029–1039.
793. Nunez, E. A., and Gershon, M. D. (1972): Synthesis and storage of serotonin by parafollicular (C) cells of the thyroid gland of active, prehibernating and hibernating bats. *Endocrinology*, 90:1008–1024.
794. Nunez, E. A., Gershon, P., and Gershon, M. D. (1980): Serotonin and seasonal variation in the pancreatic structure of bats: Possible presence of serotonergic axons in the gland. *Am. J. Anat.*, 159:347–360.
795. Nunez, E. A., Gershon, M. D., and Silverman, A. J. (1981): Uptake of 5-hydroxytryptamine by gonadotrophs of the bat's pituitary: A combined immunocytochemical and radioautographic analysis. *J. Histochem. Cytochem.*, 29:1336–1346.
796. Oates, J. A., Melmon, K., Sjoerdsma, A., Gillespie, L., and Mason, D. T. (1964): Release of a kinin peptide in the carcinoid syndrome. *Lancet*, 1:514–517.
797. Oates, J. A., and Sjoerdsma, A. (1962): A unique syndrome associated with secretion of 5-hydroxytryptophan in metastatic gastric carcinoids. *Am. J. Med.*, 32:333–342.
798. Öberg, K. (1984): *Personal communication.*
799. Öberg, K., Lundqvist, G., and Boström, H. (1982): The effects of streptozotocin in the treatment of endocrine pancreatic tumours and carcinoids. In: *Streptozotocin*, edited by M. K. Agarwall. Elsevier/North Holland.
800. Oberndorfer, S. (1907): Karzinoide Tumoren des Dünndarms. *Frankfurt Z. Pathol.*, 1:426–432.
801. O'Brien, J. R., Etherington, M. D., Shuttleworth, R. D., and Calwell, W. H. (1981): Platelet function in acute myocardial infarction patients compared with controls. *Thromb. Haemost.*, 44:96–99.
802. Olesen, S. P., and Crone, S. (1984): Serotonin increases microvascular permeability in the brain. *Int. J. Microcirc. Clin. Exp.* (abstract) *(in press).*
803. Oliva, P. B., and Breckinridge, J. C. (1977): Arteriographic evidence of coronary arterial spasm in acute myocardial infarction. *Circulation*, 56:366–374.
804. Olsen, C. R., Colebatch, H. J. H., Mebel, P. E., Nadel, J. A., and Staub, N. C. (1965): Motor control of pulmonary airways studied by nerve stimulation. *J. Appl. Physiol.*, 20:202–208.
805. Olsen, N., and Nielsen, S. L. (1978): Prevalence of primary Raynaud phenomenon in young females. *Scand. J. Clin. Lab. Invest.*, 37:761–764.
806. Omni, C., Folco, G. C., Pasargiklian, R., et al. (1979): Prostacyclin and pregnant human uterus. *Prostaglandins*, 17:113–120.

807. Omvik, P., and Lund-Johansen, P. (1983): Long-term effects on central hemodynamics and body fluid volumes of ketanserin in essential hypertension. Studies at rest and during dynamic exercise. *J. Hypertension*, 1:405–412.
808. Opie, L. H., Nathan, D., and Lubbe, W. F. (1979): Biochemical aspects of arrhythmogenesis and ventricular fibrillation. *Am. J. Cardiol.*, 43:131–148.
809. Orlandi, C., Blackshear, J. L., and Hollenberg, N. K. (1984): Serotonin-induced arterial spasm: Implications for tissue perfusion. *Int. J. Microcirc. Clin. Exp.* (abstract) *(in press)*.
810. Ortmann, R., Bischoff, S., Radeke, E., Buech, O., and Delini-Stula, A. (1982): Correlations between different measures of antiserotonin activity of drugs. Study with neuroleptics and serotonin receptor blockers. *Naunyn Schmiedebergs Arch. Pharmacol.*, 321:265–270.
811. Osaka, M., and Kobayashi, S. (1976): Duodenal basal-granulated cells in the human fetus with specific reference to their relationship to nervous element. In: *Endocrine Gut and Pancreas*, edited by T. Fujita, pp. 145–158. Elsevier, Amsterdam.
812. Oxdemir, I. A., Kusajima, K., Wax, S. D., and Webb, W. (1972): Effects of serotonin on pulmonary vascular resistance and microcirculation. *Circulation*, 46(Suppl. II):56.
813. Packham, M. A. (1978): Methods for detection of hypersensitive platelets. *Thromb. Haemost.*, 40:175–195.
814. Packham, M. A., Nishizawa, E. E., and Mustard, J. F. (1968): Response of platelets to tissue injury. *Biochem. Pharmacol.*, 17(Suppl.):171–184.
815. Page, I. H. (1952): The vascular action of natural serotonin, 5- and 7-hydroxytryptamine and tryptamine. *J. Pharmacol. Exp. Ther.*, 105:58–73.
816. Page, I. H. (1954): Serotonin (5-hydroxytryptamine). *Physiol. Rev.*, 34:563–588.
817. Page, I. H. (1957): Cardiovascular actions of serotonin (5-hydroxytryptamine). In: *5-Hydroxytryptamine*, edited by G. P. Lewis, pp. 93–108. Pergamon Press, London.
818. See ref. 817 for information.
819. Page, I. H. (1958): Serotonin (5-hydroxytryptamine); the last four years. *Physiol. Rev.*, 38:277–235.
820. Page, I. H. (1968): *Serotonin*. Year Book Medical Publishers, Inc., Chicago. 143 pp.
821. Page, I. H., Corcoran, A. C., Udenfriend, S., Sjoerdsma, A., and Weissbach, H. (1955): Argentaffinoma as an endocrine tumor. *Lancet*, 1:198–199.
822. Page, I. H., and McCubbin, J. W. (1953): Modification of vascular responses to serotonin. *Am. J. Physiol.*, 174:436–440.
823. Page, I. H., and McCubbin, J. W. (1953): The variable arterial pressure response to serotonin in laboratory animals. *Circ. Res.*, 1:354.
824. Page, I. H., and McCubbin, J. W. (1956): Arterial pressure response to infused serotonin in normotensive dogs, cats, hypertensive dogs and man. *Am. J. Physiol.*, 184:265–270.
825. Papanas, T., and McDonald, J. C. (1963): The direct removal of portal blood serotonin by the liver. *Surg. Gynecol. Obstet.*, 116:481–485.
826. Pascual, R., and Bevan, J. A. (1980): Evidence that changes in vascular tone may be initiated from the intima of the rabbit aorta. In: *Vascular Neuroeffector Mechanisms*, edited by J. A. Bevan, T. Godfraind, R. A. Maxwell, and P. M. Vanhoutte, pp. 300–304. Raven Press, New York.
827. Pasqui, A. L., Laghi Pasini, F., Guerrini, M., Pieragalli, D., Acciavatti, A., Del Bigo, C., Galigani, C., Ceccatelli, L., Forconi, S., Di Perri, T. (1983): Whole blood filterability versus red blood filterability. III. *In vitro* influence of FMLP-induced polymorphonuclear leucocyte activation. A possible direct role of FMLP. *Clin. Haemorheol.*, 3 (abstract).
828. Paul, S. M., Hsu, L. L., and Mandell, A. J. (1974): Extrapineal *N*-acetyltransferase activity in rat brain. *Life Sci.*, 15:2135–2143.
829. Pawloski, A. (1963): The nerve network of the skin in diffuse scleroderma and clinically similar conditions. *Arch. Dermatol.*, 83:868–873.
830. Payette, R. F., Gershon, M. D., and Nunez, E. A. (1984): Serotonergic elements of the mammalian pituitary. *Neurosci. Abst.*, 10.
831. Pearce, L. B., and Roth, J. A. (1983): Human brain monoamine oxidase: Solubilization and kinetics of inhibition by octyglucoside. *Arch. Biochem. Biophys.*, 224:464–472.
832. Pearse, A. G. E. (1968): Common cytochemical and ultrastructural characteristics of cells producing hormones (the APUD series) and their relevance to thyroid and ultimo-brachial C-cells and calcitonin. *Proc. R. Soc. Lond. (Biol.)*, 170:71–80.

833. Pearse, A. G. E. (1980): The APUD concept and hormone production. *Clin. Endocrinol. Metab.*, 17:211–222.
834. Pearse, A. G. E., and Polak, J. M. (1971): Cytochemical evidence for neuronal crest origin of mammalian ultimo-branchial C-cells. *Histochemie*, 27:96–102.
835. Pearson, C. M., and Fitzgerald, P. J. (1949): Carcinoid tumors—A reemphasis of their malignant nature. Review of 140 cases. *Cancer*, 2:1005–1026.
836. Pedigo, N. W., Yamamura, H. I., and Nelson, D. L. (1981): Discrimination of multiple [^{3}H]5-hydroxytryptamine binding sites by the neuroleptic spiperone in rat brain. *J. Neurochem.*, 36:220–226.
837. Perez, J. E., Saffitz, J. E., Gutierrez, F. A., and Henry, P. D. (1983): Coronary artery spasm in intact dogs induced by potassium and serotonin. *Circ. Res.*, 52:423–431.
838. Peroutka, S. J., Lebovitz, R. M., and Snyder, S. H. (1979): Serotonin receptor binding sites affected differentially by guanine nucleotides. *Mol. Pharmacol.*, 16:700–708.
839. Peroutka, S. J., Lebovitz, R. M., and Snyder, S. H. (1981): Two distinct central serotonin receptors with different physiological functions. *Science*, 212:827–829.
840. Peroutka, S. J., Noguchi, M., Tolner, D. J., and Allen, G. S. (1983): Serotonin-induced contraction of canine basilar artery: Mediation by 5-HT_1 receptors. *Brain Res.*, 259:327–330.
841. Peroutka, S. J., and Snyder, S. H. (1979): Multiple serotonin receptors: Differential binding of ^{3}H-5-hydroxytryptamine, ^{3}H-lysergic acid diethylamide and ^{3}H-spiroperidol. *Mol. Pharmacol.*, 16:687–699.
842. Peroutka, S. J., and Snyder, S. H. (1980): Long term antidepressant treatment decreases spiroperidol-labelled serotonin receptor binding. *Science*, 210:88–90.
843. Peroutka, S. J., and Snyder, S. H. (1980): Regulation of serotonin (5-HT_2) receptors labeled with [^{3}H]spiroperidol by chronic treatment with the antidepressant amitriptyline. *J. Pharmacol. Exp. Ther.*, 215:582–587.
844. Peroutka, S. J., and Snyder, S. H. (1981): [^{3}H]Mianserin: Differential labeling of $serotonin_2$ and $histamine_1$ receptors in rat brain. *J. Pharmacol. Exp. Ther.*, 216:142–148.
845. Persson, B., Hedner, T., and Berglund, G. (1983): Ketanserin in essential hypertension: Effects during rest and exercise. *Eur. J. Clin. Pharmacol.*, 25:307–312.
846. Persson, B., Hedner, T., and Henning, M. (1982): Cardiovascular effects in the rat of ketanserin, a novel 5-hydroxytryptamine receptor blocking agent. *J. Pharm. Pharmacol.*, 34:442–445.
847. Peskor, B., and Spector, S. (1973): Serotonin: Radioimmunoassay. *Science*, 179:1340–1341.
848. Peters, J. R., and Grahame-Smith, D. G. (1980): Human platelet 5-HT receptors: Characterization and functional association. *Eur. J. Pharmacol.*, 68:243–256.
849. Pettersson, A., Hedner, T., Henning, M., and Persson, B. (1984): Short and long term antihypertensive effects of ketanserin in the SHR. IUPHAR 9th International Congress of Pharmacology.
850. Phillips, D. R., and Baughan, A. K. (1983): Fibrinogen binding to human platelet plasma membranes. Identification of two steps requiring divalent cations. *J. Biol. Chem.*, 258:10240–10246.
851. Pierce, E. T., Foote, W. E., and Hobson, J. A. (1976): The efferent connection of the nucleus raphe dorsalis. *Brain Res.*, 107:137–144.
852. Pietra, G. G., Szidon, J. P., Leventhal, M. M., and Fishman, A. P. (1971): Histamine and interstitial pulmonary edema in the dog. *Circ. Res.*, 29:323–337.
853. Pignatti, P. F., and Cavalli-Sforza, L. L. (1975): Serotonin binding proteins from human blood platelets. *Neurobiology*, 5:65–74.
854. Pimparker, B. D., Senesky, D., and Kalser, M. H. (1961): Blood serotonin in nontropical sprue. *Gastroenterology*, 40:504–506.
855. Pitt, B. R., Hammond, G. L., and Gillis, C. N. (1982): Comparison of pulmonary and extrapulmonary extraction of biogenic amines. *J. Appl. Physiol.*, 52:1545–1551.
856. Pletscher, A. (1968): Metabolism, transfer and storage of 5-hydroxytryptamine in blood platelets. *Br. J. Pharmacol. Chemother.*, 32:1–16.
857. Pletscher, A., and Affolter, H. (1983): The 5-HT receptor of blood platelets. *J. Neural. Transm.*, 57:233–242.
858. Pletscher, A., and Da Prada, M. (1975): The organelles storing 5-hydroxytryptamine in blood platelets. In: *Biochemistry and Pharmacology of Platelets. Ciba Foundation Symposium 35*, pp. 261–279. Elsevier-Excerpta Medica, North Holland (Amsterdam).
859. Poggi, A., Niewiarowski, S., Steward, G. J., Sobel, E., and Smith, J. B. (1983): Human platelet

secreted proteins and prostacyclin production by bovine aortic endothelial cells. *Proc. Soc. Exp. Biol. Med.*, 172:543–550.

860. Polak, J. M., Heitz, P., and Pearse, A. G. E. (1976): Differential localization of substance P and motilin. *Scand. J. Gastroenterol. (Suppl.)*, 39:39–42.
861. Poplawski, A., Skorulska, M., and Micwiarowski, S. (1968): Increased platelet adhesiveness in hypertensive cardiovascular disease. *J. Atheroscl. Res.*, 8:721–723.
862. Pott, J. W., Moreland, R. S., and Bohr, D. F. (1984): Interaction of serotonin (5-HT) and norepinephrine (NE) in femoral arteries from renal hypertensive rabbits. *Fed. Proc.*, 43:557.
863. Poulson, E., Botros, M., Robson, J. M. (1960): Effect of 5-hydroxytryptamine and iproniazid on pregnancy. *Science*, 131:1101.
864. Poulson, E., and Robson, J. M. (1963): Prevention by antagonists of the toxic action of 5-hydroxytryptamine on pregnancy. *Br. J. Pharmacol.*, 21:150.
865. Preston, F. E., Ward, J. D., Marcola, B. H., Porter, N. R., and Timperley, W. R. (1978): Elevated β-thromboglobulin levels and circulating platelet aggregates in diabetic microangiopathy. *Lancet*, 1:238–240.
866. Putney, J. W. (1981): Recent hypotheses regarding the phosphatidyl inositol effect. *Life Sci.*, 29:1183–1194.
867. Pyles, J. D., Silberman, H. R., and Feldman, J. M. (1980): Histamine antagonists and carcinoid flush. *N. Engl. J. Med.*, 302:234–235.
868. Rådegran, K., Bergentz, S. E., Lewis, D. H., Ljungqvist, U., and Olsson, P. (1971): Pulmonary effects of induced platelet aggregation. Intravascular obstruction or vasoconstriction? *Scand. Lab. Invest.*, 28:423–427.
869. Rakoczi, I., Tallian, F., Bagdany, S., and Gati, I. (1979): Platelet lifespan in normal pregnancy and preeclampsia as determined by a nonradioisotope technique. *Thromb. Res.*, 15:553.
870. Rambaud, J. C., and Matuchansky, D. (1974): Diarrhea and digestive endocrine tumors. *Clin. Gastroenterol.*, 3:657–670.
871. Ramirez, A. J., Chemerinski, E., and Enero, M. A. (1983): Central serotonergic participation on blood pressure regulation. *Gen. Pharmacol.*, 13:427–432.
872. Rane, A., Wilkinson, G. R., and Shand, D. G. (1977): Prediction of hepatic extraction ratio from *in vitro* measurement of intrinsic clearance. *J. Pharmacol. Exp. Ther.*, 200:420–424.
873. Rapoport, R. M., and Murad, F. (1983): Agonist-induced endothelium dependent relaxation in rat thoracic aorta may be mediated through cGMP. *Circ. Res.*, 52:352–357.
874. Rapport, M. M., Green, A. A., and Page, I. H. (1948): Serum vasoconstrictor (serotonin). IV. Isolation and characterization. *J. Biol. Chem.*, 176:1243.
875. Ratz, P. H., and Flaim, S. F. (1984): Mechanism of 5-HT contraction in isolated bovine ventricular coronary arteries. Evidence for transient receptor-operated calcium influx channels. *Circ. Res.*, 54:135–143.
876. Raynaud, M. (1888): On local asphyxia and symmetrical gangrene of the extremities and new researches on the nature and treatment of local asphyxia of the extremities. In: *Selected Monographs of the New Sydenham Society*, translated by T. Barlow, *Vol. 121*, pp. 1–199. London.
877. Read, N. W., Miles, C. A., Fisher, D., Holgate, A. M., Kime, N. D., Mitchell, M. A., Reeve, A. M., Roche, T. B., and Walker, M. (1980): Transit of a meal through the stomach and small intestine and its role in the pathogenesis of diarrhea. *Gastroenterology*, 79:1276–1282.
878. Redman, C. W. G., Allington, M. J., Bolton, F. G., and Stirrat, G. M. (1977): Plasma beta-thromboglobulin in preeclampsia. *Lancet*, 2:248.
879. Redman, C. W. G., Bonnar, J., and Beilin, L. (1978): Early platelet consumption in preeclampsia. *Br. Med. J.*, 1:467.
880. Redman, C. W. G., Denson, K. W. E., Beilin, L. J., et al. (1977): Factor VIII consumption in preeclampsia. *Lancet*, 2:1249.
881. Reichardt, W., Ingemansson, S., Lunderquist, A., and Nobin, A. (1979): Selective mesenteric phlebography in patients with carcinoid tumors. *Gastrointest. Radiol.*, 4:179–189.
882. Reid, J. V. O. (1970): Trial of tryptophan prophylaxis in patients liable to "African" cardiomyopathy. *S. Afr. Med. J.*, 44:732–735.
883. Reid, J. V. O., and Berjak, P. (1966): Dietary production of myocardial fibrosis in the rat. *Am. Heart J.*, 71:240–250.
884. Reid, J. V. O., and Berjak, P. (1967): Tryptophan and serotonin levels in patients with, or susceptible to, African cardiomyopathy. *Am. Heart J.*, 74:337–339.

885. Reimann, I. W., and Frölich, J. C. (1983): Mechanism of antihypertensive action of ketanserin in man. *Br. Med. J.*, 287:381–383.
886. Reimers, H. J., Packham, M. A., Kinlough-Rathbone, R. I., and Mustard, J. F. (1973): Effect of repeated treatment of rabbit platelets with low concentrations of thrombin on their function, metabolism and survival. *Br. J. Haematol.*, 25:675–689.
887. Reinhard, J. F., Jr., Moskowitz, M. A., Sved, A. F., and Fernstrom, J. D. (1980): A simple, sensitive and reliable assay for serotonin and 5HIAA in brain tissue using liquid chromatography with electrochemical detection. *Life Sci.*, 27:905–911.
888. Remuzzi, G., Marchesi, D., Zoja, C., et al. (1980): Reduced umbilical and placental vascular prostacyclin in severe preeclamptics. *Prostaglandins*, 20:105.
889. Reneman, R. S. (1975): The anterior and the lateral compartmental syndrome of the leg due to intensive use of muscles. *Clin. Orthop.*, 113:69–80.
890. Reneman, R. S., and Jageneau, A. H. M. (1973): The influence of weighted exercise on tissue (intramuscular) pressure in normal subjects and patients with intermittent claudication. *J. Clin. Lab. Invest.*, 31(*Suppl.* 128):37–42.
891. Reneman, R. S., Slaaf, D. W., Lindbom, L., Tangelder, G. J., and Arfors, K. E. (1980): Muscle blood flow disturbances produced by simultaneously elevated venous and total tissue pressure. *Microvasc. Res.*, 20:307–318.
892. Renskers, K. J., Feor, K. D., and Roth, J. A. (1980): Sulfation of dopamine and other biogenic amines by human brain phenol sulfotransferase. *J. Neurochem.*, 34:1362–1368.
893. Resnick, R. H., and Gray, S. J. (1962): Chemical and histological demonstration of hydrochloric acid-induced release of serotonin from intestinal mucosa. *Gastroenterology*, 42:48–55.
894. Reyntjens, A., Waelkens, J., Gelders, Y., Ceulemans, D., and Janssen, P. A. J. (1984): A novel approach for psychosomatic normalization: A placebo-controlled pilot trial of R 55 667, a potent indoleamine antagonist. *Collegium Internationale Neuro-Psychopharmacologicum, 14th CINP Congress, Book of Abstracts*, p. 251.
895. Richardson, P. D. I., and Withrington, P. G. (1978): Responses of the simultaneously-perfused hepatic arterial and portal venous vascular beds of the dog in histamine and 5-hydroxy-tryptamine. *Br. J. Pharmacol.*, 64:581–588.
896. Rickaby, D. A., Dawson, C. A., and Maron, M. B. (1980): Pulmonary inactivation of serotonin and site of serotonin pulmonary vasoconstriction. *J. Appl. Physiol.*, 48:606–612.
897. Rickaby, D. A., Linehan, J. H., Bronikowski, T. A., and Dawson, C. A. (1981): Kinetics of serotonin uptake in the dog lung. *J. Appl. Physiol.*, 51:405–414.
898. Riddle, J. M., Bluhm, G. B., and Pitchford, W. C. (1981): A comparative study of platelet reactivity in arthritis. *Ann. N.Y. Acad. Sci.*, 370:22–29.
899. Rink, T. J., Smith, S. W., and Tsien, R. Y. (1982): Cytoplasmic free Ca^{2+} in human platelets: Ca^{2+} thresholds and Ca-independent activation for shape change and secretion. *FEBS Lett.*, 148:21–26.
900. Robb, H. J., Margulis, R. R., and Jabs, C. M. (1972): Role of pulmonary microembolism in the hemodynamics of endotoxin shock. *Surg. Gynecol. Obstet.*, 135:777–783.
901. Robertson, J. I. S., Peart, W. S., and Andrews, T. M. (1962): The mechanism of facial flushes in the carcinoid syndrome. *Q. J. Med.*, 31:103–123.
902. Robertson, R. M., Robertson, D., Roberts, L. J., Maas, R. L., Fitzgerald, G. A., Friesinger, G. C., and Oates, J. A. (1981): Thromboxane A_2 in vasotonic angina pectoris. *N. Engl. J. Med.*, 304:998–1003.
903. Robinson, R., and Gershon, M. D. (1971): Synthesis and uptake of 5-hydroxytryptamine by the myenteric plexus of the small intestine of the guinea pig. *J. Pharmacol. Exp. Ther.*, 178:311–324.
904. Robinson, S. E. (1982): Interaction of the median raphe nucleus and hypothalamic serotonin with cholinergic agents and pressor responses in the rat. *J. Pharmacol. Exp. Ther.*, 223:662–668.
905. Robinson-White, A., Peterson, S., Hechtman, H. B., and Shepro, D. (1981): Serotonin uptake by isolated adipose capillary endothelium. *J. Pharmacol. Exp. Ther.*, 216:125–128.
906. Rodbell, M. (1980): The role of hormone receptors and GTP-regulatory proteins in membrane transduction. *Nature (Lond.)*, 284:17–22.
907. Rodnan, G. P., Myerowitz, R. L., and Justh, G. O. (1980): Morphologic changes in the digital arteries of patients with progressive systemic sclerosis (scleroderma) and Raynaud phenomenon. *Medicine*, 59:393–408.

908. Rodriguez, D. L., Arnaiz, G., and De Robertis, E. (1964): 5-Hydroxytryptophan decarboxylase activity in nerve endings of the rat brain. *J. Neurochem.*, 11:213–219.
909. Rogawski, M. A., and Aghajanian, G. K. (1981): Serotonin autoreceptors on dorsal raphe neurons: Structure-activity relationships of tryptamine analogs. *J. Neurosci.*, 1:1148–1154.
910. Rogers, R. C., Kita, H., Butcher, L. L., and Novin, D. (1980): Afferent projections to the dorsal motor nucleus of the vagus. *Brain Res. Bull.*, 5:365–373.
911. Romero, R., Snider, E., Rickles, F., et al. (1982): The clinical significance and mechanism of thrombocytopenia in pregnancy-induced hypertension. *Proc. Soc. Gynecol. Invest.*, Dallas, Texas, abstract 69.
912. Rosenberg, J. C. (1966): Carcinoid and other amine producing tumors. *Prog. Clin. Cancer*, 2:297–308.
913. Rosenberg, J. C., Lillehei, R. C., Moran, W. H., and Zimmerman, B. (1959): A comparison of the effects of irreversible hemorrhagic and endotoxin shock on serotonin concentration in blood. *Surg. Forum*, 10:397–400.
914. Rosoff, C. B., Salzman, E. W., and Gurewich (1971): Reduction of platelet serotonin and the response to pulmonary emboli. *Surgery*, 70:12–19.
915. Ross, R. (1979): The arterial wall and atherosclerosis. *Annu. Rev. Med.*, 30:1–15.
916. Ross, R. (1981): Atherosclerosis: A problem of the biology of arterial wall cells and their interactions with blood components. *Arteriosclerosis*, 1:293–311.
917. Ross, R., and Glomset, J. (1973): Atherosclerosis and the arterial smooth muscle cell. *Science*, 180:1332–1339.
918. Ross, R., and Glomset, J. (1976): The pathogenesis of atherosclerosis. *N. Engl. J. Med.*, 295:369–377;420–425.
919. Ross, R., and Vogel, A. (1978): The platelet derived growth factor. *Cell*, 14:203–210.
920. Ross, R. L., and Wilkerson, R. D. (1979): The role of serotonin in arrhythmias produced by myocardial infarction in dogs. *Alabama J. Med. Sci.*, 16:38.
921. Roswel, H. C., Hegardt, B., Downie, H. G., Mustard, J. F., and Murphy, E. A. (1966): Adrenaline and experimental thrombosis. *Br. J. Haematol.*, 12:66–73.
922. Roth, R. A., and Gillis, C. N. (1975): Multiple forms of amine oxidase in perfused rabbit lung. *J. Pharmacol. Exp. Ther.*, 194:537–544.
923. Roth, R. A., and Wallace, K. B. (1980): Disposition of biogenic amines and angiotensin I by lungs of spontaneously hypertensive rats. *Am. J. Physiol.*, 239:H736–H761.
924. Rothman, T. P., and Gershon, M. D. (1982): Phenotypic expression in the developing murine enteric nervous system. *J. Neurosci.*, 2:381–393.
925. Rubenstein, M. D., Wall, R. T., Baim, D. S., and Harrisen, D. C. (1981): Platelet activation in clinical coronary artery disease and spasm. *Am. Heart J.*, 102:363–367.
926. Rucinski, B., Niewiarowski, S., James, P., Walz, D. A., and Budzynski, A. Z. (1979): Antiheparin proteins secreted by human platelets. Purification, characterization and radio-immunoassay. *Blood*, 153:47–62.
927. Rudnick, G., Bencuyer, R., Nelson, P. J., and Zito, R. A. (1981): Inhibition of platelet serotonin transport by propanolol. *Mol. Pharmacol.*, 20:118–123.
928. Rulter, J. W., Pictet, R. L., and Harding, J. D. (1978): An analysis of pancreatic development; role of mesenchymal factors and other extracellular factors. In: *Molecular Control of Proliferation and Differentiation*, edited by E. Papaconstantinou, pp. 205–227. Academic Press, New York.
929. Saavedra, J. M., Brownstein, M., and Axelrod, J. (1973): A specific and sensitive enzymatic-isotopic microassay for serotonin in tissues. *J. Pharmacol. Exp. Ther.*, 186:508–515.
930. Saavedra, J. M., Palkovits, M., Brownstein, M. J., and Axelrod, J. (1974): Serotonin distribution in the nuclei of the rat hypothalamus and preoptic region. *Brain Res.*, 77:157–165.
931. Sadavongvivad, C. (1970): Pharmacological significance of biogenic amines in the lungs: 5-Hydroxytryptamine. *Br. J. Pharmacol.*, 38:353–365.
932. Sako, M., Lunderquist, A., Owman, T., Mårtensson, H., and Nobin, A. (1982): Angiographic and computed tomographic appearance of secondary carcinoid of the liver. *Cardiovasc. Intervent. Radiol.*, 5:90–96.
933. Saman, S., Thandroyen, F., and Opie, L. H. (1984): Serotonin and the heart. Effects of ketanserin on myocardial function, heart rate, and arrhythmias. *J. Cardiovasc. Pharmacol.* (Suppl.) *(submitted)*.
934. Samuelsson, B., Paoletti, R., and Ramwell, P. W. (1983): Cardiovascular system. *Advances in*

Prostaglandin, Thromboxane, and Leukotriene Research, Vol. 11, pp. 359–498. Raven Press, New York.

935. Sanders-Bush, E., Bushing, J. A., and Sulser, F. (1972): *p*-Chloroamphetamine inhibition of cerebral tryptophan hydroxylation. *Biochem. Pharmacol.*, 21:1501–1510.
936. Sandler, M., and Coveney, J. (1962): Placental monoamine-oxidase activity in toxemia of pregnancy. *Lancet*, 1:1096.
937. Sandler, M., and Snow, P. J. D. (1958): An atypical carcinoid tumor secreting 5-hydroxytryptophan. *Lancet*, 1:137–139.
938. Sawchenko, P. E., and Swanson, L. W. (1981): A method for tracing biochemically defined pathways in the central nervous system using combined fluorescence retrograde transport and immunohistochemical techniques. *Brain Res.*, 210:31–51.
939. Saxena, P. R., and Verdouw, P. D. (1982): Redistribution by 5-hydroxytryptamine of carotid arterial blood at the expense of arteriovenous anastomotic blood flow. *J. Physiol. (Lond.)*, 332:501–520.
940. Scatchard, G. (1949): The attractions of proteins for small molecules and ions. *Ann. N.Y. Acad. Sci.*, 51:660–672.
941. Schalekamp, M. A. D. H., and Wentling, G. J. (1983): Mechanism of antihypertensive action of ketanserin in man. *Br. Med. J.*, 287:1300–1301.
942. Schaub, R. G., Meyers, K. M., Sande, R., and Hamilton, G. (1976): Inhibition of feline collateral vessel development following thrombotic occlusion. *Circ. Res.*, 39:736–743.
943. Schaub, R. G., Meyers, K. M., and Sande, R. D. (1977): Serotonin as a factor in the depression of collateral blood flow following experimental arterial thrombosis. *J. Lab. Clin. Med.*, 90:645–653.
944. Scherbel, A. L., and Harrison, J. W. (1959): Response to serotonin and its antagonists in patients with rheumatoid arthritis and related diseases. *Angiology*, 10:29–38.
945. Schick, P. K., and McKean, M. L. (1979): Serotonin binding in human platelets. *Biochem. Pharmacol.*, 28:2667–2670.
946. Schmid-Schönbein, H., Born, G. V. R., Richardson, P. P., Cusack, N., Rieger, H., Forst, R., Rohling-Winbel, I., Blasberg, P., and Wehmeyer, A. (1981): Rheology of thrombotic processes in flow: The interaction of erythrocytes and thrombocytes subjected to high flow forces. *Biorheology*, 18:415–444.
947. Schmid-Schönbein, H., Rieger, H., and Fischer, T. (1980): Blood fluidity as a consequence of red cell fluidity: Flow properties of blood and flow behaviour of blood in vascular diseases. *Angiology*, 31:301–319.
948. Schneider, J. A., and Yonkman, F. F. (1953): Action of serotonin (5-hydroxytryptamine) on vagal afferent impulses in the cat. *Am. J. Physiol.*, 174:127–134.
949. Schofield, S. P. M., and Everitt, B. J. (1981): The organization of indoleamine neurons in the brain of the rhesus monkey *(Macaca mulatta)*. *J. Comp. Neurol.*, 197:369–383.
950. Scholz, P. M., Karis, J. H., Gump, F. E., and Chien, S. (1975): Correlation of blood rheology with vascular resistance in critically ill patients. *J. Appl. Physiol.*, 39:1008–1011.
951. Schönbaum, E., Vargaftig, B. B., Lefort, J., Lamar, J. C., and Hasenack, T. (1975): An unexpected effect of serotonin antagonists on the canine nasal circulation. *Headache*, 15:180–187.
952. Schotte, A., Maloteaux, J. M., and Laduron, P. M. (1983): Characterization and regional distribution of serotonin S_2-receptors in human brain. *Brain Res.*, 276:231–235.
953. Scroop, G. C., and Walsh, J. A. (1968): Interactions between angiotensin, noradrenaline and serotonin on the peripheral blood vessels in man. *Aust. J. Exp. Biol. Med. Sci.*, 46:573–580.
954. Segal, I. H., editor (1975): *Enzyme Kinetics. Behavior and Analysis of Rapid Equilibrium and Steady-State Enzyme Systems*. J. Wiley & Sons, New York, London, Sydney, Toronto.
955. Seibold, J. R., and Harris, J. N. (1984): Plasma β-thromboglobulin in the differential diagnosis of Raynaud's phenomenon. *J. Rheumatol. (in press)*.
956. Seibold, J. R., and Harris, J. N. (1984): Platelet aggregation in systemic sclerosis and primary Raynaud's phenomenon *(manuscript in preparation)*.
957. Seibold, J. R., and Jageneau, A. H. M. (1984): Treatment of Raynaud's phenomenon with ketanserin, a selective antagonist of the $serotonin_2$ $(5\text{-}HT_2)$ receptor. *Arthritis Rheum.*, 27:139–146.
958. Seibold, J. R., Medsger, T. A., Winkelstein, A., Kelly, R. H., and Rodnan, G. P. (1982): Immune complexes in progressive systemic sclerosis (scleroderma). *Arthritis Rheum.*, 25:1167–1173.
959. Seibold, J. R., and Terregino, C. A. (1984): Ketanserin relieves but does not prevent cold-induced vasoconstriction in primary Raynaud's phenomenon. *Clin. Res. (in press)*.

960. Seiler, K., Tarum, G., and Wasserman, O. (1974): On the role of serotonin in the pathogenesis of pulmonary hypertension induced by anorectic drugs, an experimental study in the isolated perfused rat lung. *Clin. Exp. Pharmacol. Physiol.*, 1:463–471.

961. Senior, J. B., Fahim, I., Sullivan, F. M., et al. (1963): Possible role of 5-hydroxytryptamine in toxemia of pregnancy. *Lancet*, 2:553.

962. Shano, A., Davi, G., Avellone, G., Novo, S., and Pinto, A. (1983): Haemostatic alterations in peripheral arteriopathies. In: *Advances in Haemostasis and Thrombosis*, edited by A. Cajozzo, P. Perricone, and P. Di Marco, pp. 217–226. Ettore Majoranci Center for Scientific Culture.

963. Shapoval, L. N., and Pobegailo, L. S. (1981): Changes in renal nerve efferent activity in response to serotonin stimulation of bulbar nuclei. *Bull. Eksp. Biol. Med.*, 92(7):3–5.

964. Shepherd, J. T., and Vanhoutte, P. M. (1979): *The Human Cardiovascular System. Facts and Concepts*. Raven Press, New York. 363 pp.

965. Shepro, D., and D'Amore, P. (1980): Endothelial cell metabolism. In: *Vascular Endothelium and Basement Membranes*, edited by B. M. Altura, pp. 161–201. S. Karger, Basel.

966. Sherry, S., and Scriabine, A. (1974): *Platelets and Thombosis*. University Park Press, Baltimore.

967. Shields, P. J., and Eccleston, D. (1972): Effects of electrical stimulation of rat midbrain on 5-hydroxytryptamine synthesis as determined by a sensitive radioenzymatic method. *J. Neurochem.*, 19:265–272.

968. Shuttleworth, R. D., and O'Brien, J. R. (1981): Intraplatelet serotonin and plasma hydroxyindoles in health and disease. *Blood*, 57:505–509.

969. Sibbald, W., Peters, S., and Lindsay, R. M. (1980): Serotonin and pulmonary hypertension in human septic ARDS. *Critical Care Med.*, 8:490–494.

970. Refer to ref. 957 for information.

971. Siegel, M. I., McConnel, R. T., and Cuatrecasas, P. (1979): Aspirin-like drugs interfere with arachidonate metabolism by inhibition of the 12-hydroperoxy-5,8,10-14,-eicosatetraenoic acid peroxidase activity of the lipoxygenase pathway. *Proc. Natl. Acad. Sci. U.S.A.*, 76:3774–3778.

972. Siegel, R. C., and Fries, J. F. (1974): Intraarterially administered reserpine and saline in scleroderma. *Arch. Intern. Med.*, 134:515–518.

973. Sih, C. J., and Takeguchi, C. (1970): Mechanism of prostaglandin biosynthesis. III. Catecholamines and serotonin as coenzymes. *J. Am. Cancer Soc.*, 92:6670.

974. Singer, H. A., and Peach, M. J. (1982): Calcium- and endothelial-mediated vascular smooth muscle relaxation in rabbit aorta. *Hypertension*, 4(Suppl. II):19–25.

975. Singleton, W., Saxton, C. A. P. D., Hernandez, J., and Prichard, B. N. C. (1982): Postjunctional selectivity of α-blockade with prazosin, trimazosin, and UK-33,274 in man. *J. Cardiovasc. Pharmacol.*, 4:S145–S151.

976. Sirek, A., and Sirek, O. V. (1970): Serotonin: A review. *C.M.A. Journal*, 102:846–849.

977. Sjoerdsma, A., and Melmon, K. (1964): Carcinoid spectrum. *Gastroenterology*, 47:104–107.

978. Sjoerdsma, A., Oates, J. A., Zaltzman, P., and Udenfriend, S. (1960): Serotonin synthesis in carcinoid patients. Its inhibition by alphamethyl DOPA with measurement of associated increases in urinary 5-hydroxytryptophan. *N. Engl. J. Med.*, 263:585–588.

979. Skolasinska, K., Kihara, M., Horie, R., and Yamori, Y. (1982): CBF response to *l*-dopa influenced by peripheral decarboxylase inhibitor and serotonin in conscious SHR-SP, SHR-SR and WK rats. *Clin. Exp. Pharmacol. Physiol.*, 9:631–637.

980. Small, R., Macarak, E., and Fisher, A. B. (1976): Production of 5-hydroxyindoleacetic acid from serotonin by cultured endothelial cells. *J. Cell Physiol.*, 90:225–232.

981. Smith, J. E., Lane, J. D., Shea, P. A., McBride, W. J., and Aprison, M. H. (1975): A method for the concurrent measurement of picomole quantities of acetylcholine, choline, dopamine, norepinephrine, serotonin, 5-hydroxytryptophan, 5-hydroxyindoleacetic acid, tryptophan, tyrosine, glycine, aspartate, glutamate, alanine, and gamma-aminobutyric acid in single tissue samples from different areas of rat central nervous system. *Anal. Biochem.*, 64:149–169.

982. Smith, M. E., Gunther, R., Gee, M., Flynn, J., and Demling, R. H. (1981): Leukocytes, platelets and thromboxane A_2 in endotoxin-induced lung injury. *Surgery*, 90:102–107.

983. Smits, J. F., and Struyker-Boudier, H. A. J. (1976): Intrahypothalamic serotonin and cardiovascular control in rats. *Brain Res.*, 11:422–427.

984. Smits, J. F. M., van Essen, H., and Struyker-Boudier, H. A. J. (1978): Serotonin-mediated cardiovascular responses to electrical stimulation of raphe nuclei in the rat. *Life Sci.*, 23:173–178.

985. Sneddon, J. M. (1969): Sodium dependent accumulation of 5-hydroxytryptamine by rat platelets. *Br. J. Pharmacol.*, 37:680–688.

986. Snyder, S. H., Axelrod, J., Wurtman, R. J., and Fischer, J. E. (1965): Control of 5-hydroxytryptamine decarboxylase activity in the rat pineal gland by sympathetic nerve. *J. Pharmacol. Exp. Ther.*, 147:371.

987. Socol, M., Weiner, C., Louis, G., et al. (1984): Platelet activation in preeclampsia. *Am. J. Obst. Gynecol. (in press).*

988. Soga, J., and Tazawa, K. (1971): Pathologic analysis of carcinoids. Histologic reevaluation of 62 cases. *Cancer*, 28:990–998.

989. Sole, M. J., Shum, A., and Van Loon, G. R. (1979): Serotonin: Metabolism in the normal and failing heart. *Circ. Res.*, 45:629–634.

990. Somlyo, A. P., and Somlyo, A. V. (1970): Vascular smooth muscle. II. Pharmacology of normal and hypertensive vessels. *Pharmacol. Rev.*, 22:249–353.

991. Spannhake, E. W., Colombo, J. L., Craigo, P. A., McNamara, D. B., Hyman, A. L., and Kadowitz, P. J. (1983): Evidence for modification of pulmonary cyclo-oxygenase activity by endotoxin in the dog. *J. Appl. Physiol.*, 54:191–198.

992. Spannhake, E. W., Levin, J. L., Mellion, B. T., Gruetter, C. A., Hyman, A. L., and Kadowitz, P. J. (1980): Reversal of 5HT-induced bronchoconstriction by PGI_2: Distribution of central and peripheral action. *J. Appl. Physiol.*, 49:521–527.

993. Spatz, M. (1964): Pathogenetic studies of experimentally induced heart lesions and their relation to the carcinoid syndrome. *Lab. Invest.*, 13:288–300.

994. Spatz, M. (1969): Tryptophan metabolism and cardiac disease. *Ann. N.Y. Acad. Sci.*, 156:152–163.

995. Stallones, R. A. (1984): The epidemiology of systemic sclerosis. In: *Epidemiology of the Rheumatic Diseases*, edited by R. C. Lawrence and L. E. Shulman, pp. 169–175. Gower Medical Publishing, New York.

996. Starling, E. H., and Verney, E. G. (1925): The secretion of urine as studied on the isolated kidney. *Proc. R. Soc. Lond.*, 97:321–363.

997. Stein, M., Hirose, T., Yasutake, T., and Khan, M. A. (1970): The effects of platelet amines on airway function. In: *Airway Dynamics*, edited by Arend Bouhuys, pp. 283–295. Charles C. Thomas, Springfield, IL.

998. Stein, M., and Thomas, D. P. (1967): Role of platelets in the acute pulmonary response to endotoxin. *J. Appl. Physiol.*, 23:47–52.

999. Steinberg, H., and Fisher, A. (1977): Serotonin in the lung. In: *Serotonin in Health and Disease, Vol. 5*, edited by W. B. Essman, pp. 69–109. Spectrum Publications, New York.

1000. Steinbusch, H. W. M. (1981): Distribution of serotonin-immunoreactivity in the central nervous system of the rat—Cell bodies and terminals. *Neuroscience*, 6:557–618.

1001. Steinbusch, H. W. M., Verhofstad, A. A. J., and Joosten, H. W. J. (1978): Localization of serotonin in the central nervous system by immunocytochemistry: Description of a specific and sensitive technique and some applications. *Neuroscience*, 3:811–819.

1002. Stemerman, M. B., Baumgartner, H., and Spaet, T. H. (1971): The subendothelial microfibril and platelet adhesion. *Lab. Invest.*, 24:179–187.

1003. Stoltz, J. F. (1981): Ionized or ionizable groups of blood cells. In: *Cells: Rheology of Blood, Blood Vessels and Associated Tissues*, edited by D. R. Gross and N. M. C. Hwang. NATO Advanced Study Institutes Series. Applied Sciences Serie E, 41:184–213. Sijhoff and Noordhoff, Netherlands.

1004. Stoltz, J. F. (1983): Role of platelets and biophysical factors in thrombosis and atherosclerosis. In: *Heart Perfusion, Energetics and Ischemia*, edited by L. Dintenfass, D. G. Julian, and G. V. F. Seaman. NATO Advanced Study Institutes Series. Life Sciences Serie A, 62:295–313. Plenum Press, New York, London.

1005. Stoltz, J. F. (1983): Serotonin and platelets. Abstracts of the Satellite Symposium of the First European Meeting on Hypertension, "Serotonin and Cardiovascular Diseases," Milan 29th May.

1006. Stranden, E., Roald, O. K., and Krohg, K. (1982): Treatment of Raynaud's phenomenon with the 5-HT_2-receptor antagonist ketanserin. *Br. Med. J.*, 285:1069–1071.

1007. Strandness, D. E., and Sumner, D. S. (1975): Raynaud's disease and Raynaud's phenomenon. In: *Hemodynamics for Surgeons*, edited by D. E. Strandness and D. S. Sumner, pp. 543–581. Grune & Stratton, New York.

1008. Strodel, W. E., Vinik, A. I., Thompson, N. W., Eckhauser, F. E., and Talpos, G. B. (1983):

Small bowel carcinoid tumors and the carcinoid syndrome. In: *Endocrine Surgery Update*, edited by N. W. Thompson and A. I. Vinik, pp. 277–291. Grune & Stratton, New York.
1009. Sullivan, J. M. (1981): Prostaglandins and regulation of blood pressure: Clinical implications. *Pharmacol. Ther.*, 15:447–465.
1010. Sved, A. F., Van Itallie, C. M., and Fernstrom, J. D. (1982): Studies on the antihypertensive action of L-tryptophan. *J. Pharmacol. Exp. Ther.*, 221:329–333.
1011. Swank, R. L., Hissen, W., and Fellman, J. H. (1964): 5-Hydroxytryptamine (serotonin) in acute hypotensive shock. *Am. J. Physiol.*, 207:215–222.
1012. Sweetman, H. E., Shepro, D., and Hechtman, H. B. (1981): Inhibition of thrombocytopenic petechiae by exogenous serotonin administration. *Haemostasis*, 10:65–78.
1013. Sworn, M. J., Reasbeck, P., and Buchanan, R. (1978): Intestinal ischemia associated with ileal carcinoid tumors. *Br. J. Surg.*, 65:313–315.
1014. Symoens, J. (1982): Are the therapeutic effects of ketanserin compatible with a serotonergic mechanism? In: *5-Hydroxytryptamine in Peripheral Reactions*, edited by F. De Clerck and P. M. Vanhoutte, pp. 199–212. Raven Press, New York.
1015. Symoens, J., and Vanhoutte, P. M. (1984): Role of serotonin in blood pressure regulation. In: *Care of the Postoperative Surgical Patient*, edited by J. A. R. Smith and J. Watkins. Butterworths and Co., London *(in press)*.
1016. Tabakoff, B., and Erwin, V. G. (1970): Purification and characterization of a reduced nicotinamine adenine dinucleotide phosphate-linked aldehyde reductase from brain. *J. Biol. Chem.*, 245:3263–3268.
1017. Tachiki, K. H., and Aprison, M. H. (1975): Fluorometric assay for 5-hydroxytryptophan with sensitivity in the picomole range. *Anal. Chem.*, 47:7–11.
1018. Tada, M., Kuzuya, T., and Inove, M. (1981): Elevation of thromboxane B_2 levels in patients with classic and variant angina pectoris. *Circulation*, 64:1107–1115.
1019. Tadepalli, A. S. (1980): Inhibition of reflex vagal bradycardia by a central action of 5-hydroxytryptophan. *Br. J. Pharmacol.*, 69:647–650.
1020. Takagi, H., Shiosaka, S., Tohyama, M., Senba, E., and Sakanaka, M. (1980): Ascending components of the medial forebrain bundle from the lower brainstem in the rat, with special reference to raphe and catecholamine cell groups. A study by the HRP method. *Brain Res.*, 193:315–337.
1021. Takai, Y., Kaibuchi, K., Sano, K., and Nishizuka, Y. (1982): Counteraction of calcium-activated, phospholipid-dependent protein kinase activation by adenosine-3′,5′-monophosphate and guanosine-3′,5′-monophosphate in platelets. *J. Biochem.*, 91:403–406.
1022. Tamir, H. (1983): Serotonin binding protein: Function in synaptic vesicles. *Trans. N.Y. Acad. Sci.*, 41:237–242.
1023. Tamir, H., Barasch, J., Muller, F. S., and Liu, K. P. (1982): Serotonin storage: Identification of essential sulfhydryl group and bicarbonate binding site in SBP. *Trans. Am. Soc. Neurochem.*, 13:245.
1024. Tamir, H., Bebirian, R. J., Casper, D., and Muller, F. (1980): Differences between intracellular platelet and brain proteins that bind serotonin. *J. Neurochem.*, 35:1033–1044.
1025. Tamir, H., Brunner, W., Casper, D., and Rapport, M. M. (1980): Enhancement of the binding of serotonin to serotonin binding protein (SBP) by gangliosides. *J. Neurochem.*, 34:1719–1724.
1026. Tamir, H., and Gershon, M. D. (1979): Storage of serotonin and serotonin binding protein in synaptic vesicles. *J. Neurochem.*, 33:35–44.
1027. Tamir, H., and Gershon, M. D. (1984): Intracellular proteins that bind serotonin in neurons, paraneurons, and platelets. *J. Physiol. (Paris)*, 77 *(in press)*.
1028. Tamir, H., and Huang, Y. L. (1974): Binding of serotonin to soluble protein from synaptosomes. *Life Sci.*, 14:83–93.
1029. Tamir, H., Klein, A., and Rapport, M. M. (1976): Serotonin binding protein: Enhancement of binding by Fe^{+2} and inhibition of binding by drugs. *J. Neurochem.*, 26:871–878.
1030. Tamir, H., and Kuhar, M. J. (1975): Association of serotonin binding protein with projections of the midbrain raphe nuclei. *Brain Res.*, 83:169–172.
1031. Tamir, H., Kupsky, W. J., Huang, Y. L., and Gershon, M. D. (1983): Serotonin-binding glycoprotein of rat platelets. *J. Cell Sci.*, 62:439–459.
1032. Tamir, H., Payette, R. F., Huang, Y. L., Liu, K. P., and Gershon, M. D. (1984): Human serotonectin: A blood glycoprotein that binds serotonin and is associated with platelets and white blood cells. *J. Cell Sci. (in press)*.

1033. Tamir, H., Theoharides, T. C., Gershon, M. D., and Askenase, P. W. (1982): Serotonin storage pools in basophil leukemia and mast cells: Characterization of two types of serotonin binding protein and radioautographic analysis of the intracellular distribution of [^{3}H]serotonin. *J. Cell. Biol.*, 93:638–647.

1034. Tapia, F. J., Polak, J. M., Barbosa, A. J. A., Bloom, S. R., Marangos, P. J., Dermody, C., and Pearse, A. G. E. (1981): Neuron-specific enolase is produced by neuroendocrine tumors. *Lancet* i:808–811.

1035. Tate, S. S., Sweet, R., McDowell, F. H., and Meister, A. (1971): Decrease of the 3,4-dihydroxyphenylalanine (DOPA) decarboxylase activities in human erythrocytes and mouse tissues after administration of DOPA. *Proc. Natl. Acad. Sci. U.S.A.*, 68:2121–2123.

1036. Thiele, B. L., Young, I. V., Chikos, P. M., Hirsch, J. H., and Strandness, D. E. (1980): Correlation of arteriographic findings with symptoms in patients with cerebrovascular disease. *Neurology*, 30:1041–1046.

1037. Thomas, D. P., Gurewich, F., and Ashford, T. (1966): Platelet adherence to thromboemboli in relation to the pathogenesis and treatment of pulmonary embolism. *N. Engl. J. Med.*, 274:935–955.

1038. Thomas, D. P., and Vane, J. R. (1967): 5-Hydroxytryptamine in the circulation of the dog. *Nature (Lond.)*, 216:335–338.

1039. Thompson, J. H. (1971): Serotonin and the alimentary tract. *Res. Commun. Chem. Pathol. Pharmacol.*, 2:687–781.

1040. Thorén, L. (1979): Fluid therapy in shock. A historical survey. *Ann. Acad. Reg. Sci. Ups.*, 22:73–114.

1041. Thulin, L., and Holm, J. (1977): Effect of substance P on the flow of hepatic bile and pancreatic juice. In: *Substance P*, edited by U. S. von Euler and B. Pernow, pp. 247–251. Raven Press, New York.

1042. Tilney, N. L., and Lazarus, J. M. (1983): Acute renal failure in surgical patients. Causes, clinical patterns, and care. *Surg. Clin. North. Am.*, 63:357–377.

1043. Tilson, M. D. (1974): Carcinoid syndrome. *Surg. Clin. North. Am.*, 54:409–423.

1044. Toda, N., and Fujita, Y. (1973): Responsiveness of isolated cerebral and peripheral arteries to serotonin, norepinephrine, and transmural electrical stimulation. *Circ. Res.*, 33:98–104.

1045. Todd, M. M., Forrest, J. B., and Cragg, D. B. J. (1983): The effects of aspirin and methysergide on responses to clot-induced pulmonary embolism. *Am. Heart J.*, 105:769–776.

1046. Tollenaere, J. P., Moereels, H., and Raymaekers, L. A. (1979): *Atlas of the Three-Dimensional Structure of Drugs*. Elsevier/North Holland, Amsterdam. 321 pp.

1047. Tong, J. H., and Kaufman, S. (1975): Tryptophan hydroxylase. Purification and some properties of the enzyme from rabbit hindbrain. *J. Biol. Chem.*, 250:4152–4158.

1048. Tranzer, J. P., Da Prada, M., and Pletcher, A. (1966): Ultra-structural localization of 5-hydroxytryptamine in blood platelets. *Nature (Lond.)*, 212:1574–1575.

1049. Traub, Y. M., Shapiro, A. P., Rodnan, G. P., Medsger, T. A., McDonald, R. H., Steen, V. D., Osial, T. A., and Tolchin, S. F. (1983): Hypertension and renal failure (scleroderma renal crisis) in progressive systemic sclerosis. Review of a 25-year experience with 68 cases. *Medicine*, 62:335–352.

1050. Trubatch, J. R., and Van Harreveld, A. (1981): Anatomical correlates of synaptic transmission in chemical neurotransmission, edited by L. Stjarne, P. Hedqvist, H. Lagercrantz, and A. Wennmalm. Academic Press, New York.

1051. Turcani, P., Nossal, R., and Turi-Nagy, L. (1982): On the serotonin liberation from rat platelets due to beta-adrenoceptor blocking drugs. *Thromb. Res.*, 28:213–221.

1052. Turitto, V. T., and Weiss, H. J. (1982): Platelet and red cell involvement in mural thrombogenesis. Presentation at New York Academy of Sciences Meeting.

1053. Turner, A. J., and Tipton, K. F. (1972): The characterization of two nicotinamide-adenine dinucleotide phosphate-linked aldehyde reductases from pig brain. *Biochem. J.*, 130:765–772.

1054. Turpie, A. G. G., de Boer, A. C., and Genton, E. (1982): Platelet consumption in cardiovascular disease. *Semin. Thromb. Hemost.*, 8:161–185.

1055. Tyce, G. M., Flock, E. V., and Owen, C. A., Jr. (1968): 5-Hydroxytryptamine metabolism in brains of ethanol-intoxicated rats. *Mayo Clin. Proc.*, 43:668–673.

1056. Tyce, G. M., Flock, E. V., and Owen, C. A., Jr. (1968): Metabolism of serotonin by the isolated perfused rat liver—Effect of glucuronyl transferase deficiency or monoamine oxidase inhibition. *Biochem. Pharmacol.*, 17:1543–1552.

1057. Tyce, G. M., Flock, E. V., and Owen, C. A., Jr. (1968): Uptake and metabolism of 5-hydroxytryptamine by the isolated perfused rat liver. *Am. J. Physiol.*, 215:611–619.
1058. Tyce, G. M., Stockard, J., Sharpless, N. S., and Muenter, M. D. (1983): Excretion of amines and their metabolites by two patients in hepatic coma treated with L-dopa. *Clin. Pharmacol. Ther.*, 34:390–398.
1059. Udaka, K., Tacheuchi, Y., and Mouht, H. Z. (1970): Simple methods for quantification of enhanced vascular permeability. *Proc. Soc. Exp. Biol. Med.*, 133:1384–1387.
1060. Udenfriend, S. (1964): *Fluorescence Assay in Biology and Medicine*. Academic Press, New York.
1061. Udenfriend, S., and Weissbach, H. (1958): Turnover of 5-hydroxytryptamine (serotonin) in tissues. *Proc. Soc. Exp. Biol. (NY)*, 97:748–751.
1062. Ugurbil, K., Fukami, M. H., and Holmsen, H. (1984): Proton NMR studies of nucleotide and amine storage in the dense granules of pig platelets. *Biochemistry*, 23:416–428.
1063. Utsunomiya, T., Krausz, M. M., Dunaham, B., Valeri, C. R., Levine, L., Shepro, D., and Hechtman, H. B. (1982): Modification of inflammatory response to aspiration with ibuprofen. *Am. J. Physiol.*, 243:H903–H910.
1064. Utsunomiya, T., Krausz, M. M., Shepro, D., and Hechtman, H. B. (1981): Prostaglandin control of plasma and platelet 5-hydroxytryptamine in normal and embolized animals. *Am. J. Physiol.*, 141:H766–H771.
1065. Uzbekov, M. G., Murphy, S., and Rose, S. P. R. (1979): Ontogenesis of serotonin 'receptors' in different regions of rat brain. *Brain Res.*, 168:195–199.
1066. Vanable, J. W. (1963): A ninhydrin reaction giving a sensitive quantitative fluorescence assay for 5-hydroxytryptamine. *Anal. Biochem.*, 6:393–403.
1067. Van Beek, J., Janssen, P. A. J., and Van Nueten, J. M. (1983): Inhibitory effect of ketanserin on ergonovine-induced contractions of isolated canine coronary arteries. *Arch. Int. Pharmacodyn. Ther.*, 265:167–171.
1068. Van de Voorde, J., and Leusen, I. (1982): Vascular endothelium and the relaxation effect of histamine on aorta of the rat. *Arch. Int. Pharmacodyn. Ther.*, 256:329–330.
1069. Vane J. R. (1957): A sensitive method for the assay of 5-hydroxytryptamine. *Br. J. Pharmacol.*, 12:344–349.
1070. Vanhoutte, P. M. (1978): Heterogeneity in vascular smooth muscle. In: *Microcirculation*, edited by G. Kaley and B. M. Altura, *Vol. II*, pp. 181–309. University Park Press, Baltimore.
1071. Vanhoutte, P. M. (1980): Physical factors of regulation. In: *Handbook of Physiology*, edited by D. F. Bohr, A. P. Somlyo, and H. V. Sparks, pp. 443–474. American Physiological Society, Bethesda, MD.
1072. Vanhoutte, P. M. (1982): Does 5-hydroxytryptamine play a role in hypertension? *Trends Pharmacol. Sci.*, 3:370–373.
1073. Vanhoutte, P. M. (1982): 5-Hydroxytryptamine, vasospasm and hypertension. In: *5-Hydroxytryptamine and Peripheral Reactions*, edited by F. DeClerck and P. M. Vanhoutte, pp. 163–174. Raven Press, New York.
1074. Vanhoutte, P. M. (1982): Hyperreactivity of the blood vessel wall in hypertension. In: *Cardiology, Vol. 1: Hypertension*, edited by P. Sleight and E. Freis, pp. 216–228. Butterworth, England.
1075. Vanhoutte, P. M. (1983): 5-Hydroxytryptamine and vascular disease. *Fed. Proc.*, 42:233–237.
1076. Vanhoutte, P. M. (1983): Introductory remarks: Why 5-hydroxytryptamine? *Fed. Proc.*, 42:211–212.
1077. Vanhoutte, P. M. (1984): Can peripheral serotonergic blockade explain the hypotensive effects of ketanserin? *J. Cardiovasc. Pharmacol. (in press).*
1078. Vanhoutte, P. M., and Cohen, R. A. (1983): The elusory role of serotonin in vascular function and disease. *Biochem. Pharmacol.*, 32:3671–3674.
1079. Vanhoutte, P. M., Cohen, R. A., and Van Nueten, J. M. (1984): Serotonin and arterial vessels. *J. Cardiovasc. Pharmacol. (in press).*
1080. Vanhoutte, P. M., and Janssens, W. J. (1980): Thermosensitivity of cutaneous vessels and Raynaud's disease. *Am. Heart J.*, 100:263–265.
1081. Vanhoutte, P. M., and Rimele, T. J. (1983): Role of the endothelium in the control of vascular smooth muscle function. *J. Physiol. (Paris)*:.
1082. Vanhoutte, P. M., and Shepherd, J. T. (1970): Effect of temperature on reactivity of isolated cutaneous veins of the dog. *Am. J. Physiol.*, 218:187–190.
1083. Vanhoutte, P. M., and Van Nueten, J. M. (1983): Ketanserin, a novel antihypertensive drug. *J. Pharm. Pharmacol.*, 35:339–340.

1084. Vanhoutte, P. M., Van Nueten, J. M., Symoens, J., and Janssen, P. A. J. (1983): Antihypertensive properties of ketanserin (R 41 468). *Fed. Proc.*, 42:182–185.

1085. Vanhoutte, P. M., Verbeuren, T. J., and Webb, R. C. (1981): Local modulation of the adrenergic neuroeffector interaction in the blood vessel wall. *Physiol. Rev.*, 61:151–247.

1086. Vanhoutte, P. M., Webb, R. C., and Collis, M. G. (1980): Pre- and postjunctional adrenergic mechanisms and hypertension. *Clin. Sci.*, 59:211s–223s.

1087. Van Nueten, J. M. (1982): Serotonergic amplification mechanisms in vascular tissues. In: *5-Hydroxytryptamine in Peripheral Reactions*, edited by F. De Clerck and P. M. Vanhoutte, pp. 77–82. Raven Press, New York.

1088. Van Nueten, J. M. (1983): 5-Hydroxytryptamine and precapillary vessels. *Fed. Proc.*, 42:223–227.

1089. Van Nueten, J. M. (1984): Antivasoconstrictor effects of drugs used in migraine therapy. In: *The Pharmacological Basis of Migraine Therapy*, edited by W. K. Amery, J. M. Van Nueten, and A. Wauquier, pp. 19–35. Pitmann, London.

1090. Van Nueten, J. M., De Ridder, W., and Van Beek, J. (1982): Hypoxia and spasms in the cerebral vasculature. *J. Cereb. Blood Flow Metab.*, 2(*Suppl.* 1):S29–S31.

1091. Van Nueten, J. M., De Ridder, W., and Vanhoutte, P. M. (1984): Ketanserin and vascular contractions in response to cooling. *Eur. J. Pharmacol.*, 99:329–332.

1092. Van Nueten, J. M., Janssen, P. A. J., De Ridder, W., and Vanhoutte, P. M. (1982): Interaction between 5-hydroxytryptamine and other vasoconstrictor substances in the isolated femoral artery of the rabbit: Effect of ketanserin (R 41 468). *Eur. J. Pharmacol.*, 77:281–287.

1093. Van Nueten, J. M., Janssen, P. A. J., Van Beek, J., Xhonneux, R., Verbeuren, T. J., and Vanhoutte, P. M. (1981): Vascular effects of ketanserin (R 41 468), a novel antagonist of 5-HT_2 serotonergic receptors. *J. Pharmacol. Exp. Ther.*, 218:217–230.

1094. Van Nueten, J. M., Janssen, P. A. J., and Vanhoutte, P. M. (1983): Pharmacological properties of serotonergic responses in vascular bronchial and gastrointestinal smooth muscle. In: *Vascular Neuroeffector Mechanisms. 4th International Symposium*, edited by J. A. Bevan et al., pp. 243–250. Raven Press, New York.

1095. Van Nueten, J. M., Leysen, J. E., De Clerck, F., and Vanhoutte, P. M. (1984): Serotonergic receptor subtypes and vascular reactivity. *J. Cardiovasc. Pharmacol.*, 6:5564–5574.

1096. Van Nueten, J. M., Leysen, J. E., Schuurkes, J. A. J., and Vanhoutte, P. M. (1983): Ketanserin: A selective antagonist of 5-HT_2 serotonergic receptors. *Lancet*, 1:297–298.

1097. Van Nueten, J. M., Leysen, J. E., Vanhoutte, P. M., and Janssen, P. A. J. (1982): Serotonergic responses in vascular and non-vascular tissues. *Arch. Int. Pharmacodyn. Ther.*, 256:331–334.

1098. Van Nueten, J. M., Van Beek, J., and Vanhoutte, P. M. (1980): Inhibitory effects of lidoflazine on contractions of isolated canine coronary arteries caused by norepinephrine, 5-hydroxytryptamine, high potassium, anoxia and ergonovine maleate. *J. Pharmacol. Exp. Ther.*, 213:179–187.

1099. Van Nueten, J. M., Van Beek, J., and Vanhoutte, P. M. (1983): Effect of Ca^{2+}-entry blockade and serotonergic antagonism in contractions of hypoxic isolated canine coronary arteries. *Arch. Int. Pharmacodyn. Ther.*, 265:172–176.

1100. Van Nueten, J. M., and Vanhoutte, P. M. (1980): Effect of the Ca^{2+} antagonist lidoflazine on normoxic and anoxic contractions of canine coronary arterial smooth muscle. *Eur. J. Pharmacol.*, 64:173–176.

1101. Van Nueten, J. M., and Vanhoutte, P. M. (1981): Calcium entry blockers and vasospasm. In: *Vasodilatation*, edited by P. M. Vanhoutte and I. Leusen, pp. 459–468. Raven Press, New York.

1102. Van Nueten, J. M., and Vanhoutte, P. M. (1981): Selectivity of calcium antagonism and serotonin antagonism with respect to venous and arterial tissues. *Angiology*, 32:476–484.

1103. Van Nueten, J. M., and Vanhoutte, P. M. (1984): Serotonin and vascular function. *Clin. Anaesthesiol.*, 1:363–381.

1104. VanOost, B., Smith, J. B., Vladutiu, G., and Holmsen, H. (1984): Lysosomotropic agents selectively potentiate thrombin-induced acid hydrolase secretion from platelets. *Proc. Natl. Acad. Sci. U.S.A. (submitted)*.

1105. Verbeuren, T. J., Jordaens, F. H., and Herman, A. G. (1983): Accumulation and release of [^{3}H]5-hydroxytryptamine in saphenous veins and cerebral arteries of the dog. *J. Pharmacol. Exp. Ther.*, 226:579–588.

1106. Verhaeghe, R. (1982): Platelets, thrombosis and peripheral arterial disease. In: *Drugs: Studies in Cerebrovascular Disease and Peripheral Vascular Disease*, edited by Cahn and Witzmann, pp. 131–134. Pergamon Press, Paris.

1107. Verhaeghe, R., and Janssens, W. (1981): Blood coagulation factors causing changes in vascular resistance. In: *Vasodilatation*, edited by P. M. Vanhoutte and I. Leusen, pp. 141–149. Raven Press, New York.
1108. Verhaeghe, R., Piessens, J., Suy, R., and Verstraete, M. (1984): Platelet survival in patients with angiographically detectable diseases and normal coronary arteries. *Thromb. Haemost. (in press)*.
1109. Verheyen, A., Vlaminckx, E., Lauwers, F., Van den Broeck, C., and Wouters, L. (1984): Serotonin-induced blood flow changes in the rat hindlegs after unilateral ligation of the femoral artery. Inhibition by the S_2 receptor antagonist ketanserin. *Arch. Int. Pharmacodyn. Ther. (submitted)*.
1110. Verheyen, A., Vlaminickx, E., Van Den Broeck, C., and Lauwers, F. (1983): The effect of serotonin on the peripheral vasculature of the rat hind-legs after unilateral ligation of the femoral artery. Influence of the S_2-receptor antagonist ketanserin. *Ric. Scient. Educ. Permanenti* (*Suppl.* 33):462 (abstract).
1111. Verhofstad, A. J. J., and Jonsson, G. (1978): Immunohistochemical and neurochemical evidence for the presence of serotonin in adrenal medulla of the rat. *Neuroscience*, 10:1443–1453.
1112. Verhofstad, A. A. J., Steinbusch, H. W. M., Penke, B., Varga, J., and Joosten, H. W. J. (1981): Serotonin-immunoreactive cells in superior cervical ganglion of the rat: Evidence for the existence of separate serotonin and catecholamine containing small ganglionic cells. *Brain Res.*, 212:39–49.
1113. Vermylen, J., Badenhorst, P. N., Deckmyn, H., and Arnout, J. (1983): Normal mechanisms of platelet function. *Clin. Haematol.*, 12:107–151.
1114. Vidrio, H., and Hong, E. (1976): Vascular tone and reactivity to serotonin in the internal and external carotid vascular beds of the dog. *J. Pharmacol. Exp. Ther.*, 197:49–56.
1115. Vincent, J. L., Degaute, J. P., Domb, M., Simon, P., Berre, J., and Vandesteen, A. (1984): Administration of ketanserin, a serotonin antagonist, in patients with acute respiratory failure. *Chest (in press)*.
1116. Vlachakis, N. D., and Aldedort, L. (1980): Hypertension and propranolol therapy: Effect on blood pressure, plasma catecholamines, and platelet aggregation. *Am. J. Cardiol.*, 45:321–324.
1117. Vlikorkala, O., Makila, U.-M., and Viinikka, L. (1981): Amniotic fluid prostacyclin and thromboxane in normal, preeclamptic and some other complicated pregnancies. *Am. J. Obstet. Gynecol.*, 141:487.
1118. Votavova, M., Boullin, D. J., and Costa, E. (1971): Specificity of action of 6-hydroxydopamine in peripheral cat tissue: Depletion of noradrenaline with depletion of 5-hydroxytryptamine. *Life Sci.*, 10:87–91.
1119. Vyden, J. K., Lent, D., Nagasawa, K., Carvalho, M., Serraya, A., and Corday, E. (1974): The effects of serotonin on regional hemodynamics in the vascular system. *J. Clin. Pharmacol.*, Aug-Sept:434–441.
1120. Waldenström, J. (1958): Clinical picture of carcinoidosis. *Gastroenterology*, 35:565–569.
1121. Wallenburg, H. C. S., and Kessel, P. H. (1979): Platelet life span in pregnancies resulting in small for gestational age infants. *Am. J. Obstet. Gynecol.*, 134:739.
1122. Walling, M. W., Brasitus, T. A., and Kimberg, D. V. (1977): Effects of calcitonin and substance P on the transport of Ca, Na and Cl across rat ileum *in vitro*. *Gastroenterology*, 73:89–94.
1123. Wallis, D., and Nash, H. (1981): Relative activities of substances related to 5-hydroxytryptamine as depolarizing agents of superior cervical ganglion cells. *Eur. J. Pharmacol.*, 70:381–392.
1124. Warsh, J. J., Chiu, A., Li, P. P., and Godse, D. D. (1980): Comparison of liquid chromatography—electrochemical and gas chromatography—mass spectrometry methods for brain dopamine and serotonin. *J. Chromatogr.*, 183:483–486.
1125. Watts, A. D., Feniuk, W., and Humphrey, P. P. A. (1981): A pre-junctional action of 5-hydroxytryptamine and methysergide on noradrenergic nerves in dog isolated saphenous vein. *J. Pharm. Pharmacol.*, 33:515–520.
1126. Waugh, D., and Pearl, M. J. (1960): Serotonin induced acute nephrosis and renal corticonephrosis in rats—A morphologic study with pregnancy correlations. *Am. J. Pathol.*, 36:431.
1127. Webb, R. C. (1982): Increased vascular sensitivity to serotonin and methysergide in hypertension in rats. *Clin. Sci.*, 63:73s–75s.
1128. Webb, R. C. (1984): Vascular changes in hypertension. In: *Cardiovascular Pharmacology*, edited by M. J. Antonaccio, pp. 215–255. Raven Press, New York.
1129. Webb, R. C., and Bohr, D. F. (1984): The membrane of the vascular smooth muscle cell in

experimental hypertension and its response to serotonin. In: *Smooth Muscle Contractions*, edited by N. Stephens. Marcel Dekker, Inc., New York *(in press)*.

1130. Weenink, G. H., Born, J. J. J., TenCate, J. W., and Treffers, P. E. (1983): Antithrombin III levels in normotensive and hypertensive pregnancy. *Gynecol. Obstet. Invest.*, 16:230.

1131. Weiner, C. P., and Brandt, J. (1982): Plasma antithrombin III activity: An aid in the diagnosis of preeclampsia-eclampsia. *Am. J. Obstet. Gynecol.*, 142:275–281.

1132. Weiner, C. P., Gelfan, R., and Socol, M. L. (1984): Intrapartum treatment of preeclamptic hypertension by ketanserin—A serotonin receptor antagonist. *Am. J. Obstet. Gynecol. (in press)*.

1133. Weiner, C. P., Kwaan, H., Hauck, W. W., et al. (1984): Fibrin generation in normal pregnancy. *Obstet. Gynecol.*, 64:46.

1134. Weiner, C. P., Kwaan, H. C., Xu, C., et al. (1984): Antithrombin III activity in women with hypertension during pregnancy. *Obstet. Gynecol. (in press)*.

1135. Weiner, C. P., Socol, M. L., and Vaisrub, N. (1984): Control of preeclamptic hypertension by ketanserin—A serotonin receptor antagonist. *Am. J. Obstet. Gynecol. (in press)*.

1136. Weiner, C. P., Xu, C., Kwaan, H., et al. (1983): Distinguishing preeclampsia from chronic hypertension using antithrombin III. *Proceedings of the Society for Gynecologic Investigation*, Washington, D.C.

1137. Weisel, R. D., Vito, L., Dennis, R. D., Valeri, C. R., and Hechtman, H. B. (1977): Myocardial depression during sepsis. *Am. J. Surg.*, 133:512–521.

1138. Weiss, H. J., Tschopp, T. B., Rogers, J., and Brand, H. (1974): Studies of platelet 5-hydroxytryptamine (serotonin) in storage pool disease and albinism. *J. Clin. Invest.*, 54:421–429.

1139. Weisul, J. P., O'Donnell, T. F., Stone, M. A., and Clowes, G. H. A., Jr. (1975): Myocardial performance in clinical septic shock. *J. Surg. Res.*, 18:357–363.

1140. Weksler, B., Ley, C. W., and Jaffe, E. A. (1978): Stimulation of endothelial cell prostacyclin production by thrombin, trypsin, and the ionophore A23187. *J. Clin. Invest.*, 62:923–930.

1141. Welbourn, R. B. (1983): Islet cell tumors. In: *Endocrine Surgery Update*, edited by N. W. Thompson and A. I. Vinik, pp. 173–193. Grune & Stratton, London.

1142. Wenting, G. J., Man in 't Veld, A. J., Woittiez, A. J., Boomsma, F., and Schalekamp, M. A. D. H. (1982): Hemodynamic effects of ketanserin, a selective 5-hydroxytryptamine (serotonin) receptor antagonist, in essential hypertension. *Clin. Sci.*, 63:435S–438S.

1143. Wenting, G. J., Man in 't Veld, A. J., Woittiez, A. J., Boomsma, F., and Schalekamp, M. A. D. H. (1982): Treatment of hypertension with ketanserin, a new selective 5-HT_2 receptor antagonist. *Br. Med. J.*, 284:537–539.

1144. Wenting, G. J., Woittiez, A. J., Man in 't Veld, A. J., and Schalekamp, M. A. D. H. (1984): 5-HT, alpha-adrenoceptors, and blood pressure. Effects of ketanserin in essential hypertension and autonomic insufficiency. *Hypertension*, 6:100–109.

1145. Wessendorf, M. W., and Elde, R. P. (1983): Nucleus raphe magnus and its serotonergic neurons: Quantitative anatomical studies in the rat. *Neurosci. Abst.*, 9:1149.

1146. Whigham, K. A., Howie, P. W., Drummond, A. H., and Prentile, C. R. M. (1978): Abnormal platelet function in pre-eclampsia. *Br. J. Obstet. Gynecol.*, 85:28–32.

1147. Whigham, K. A. E., Howie, P. W., Shaw, M. M., et al. (1980): Factor VIII related antigen/coagulant activity ratio as a predictor of fetal growth retardation: A comparison with hormone and uric acid measurements. *Br. J. Obstet. Gynecol.*, 87:797–803.

1148. White, J. G. (1970): A biphasic response of platelets to serotonin. *Scand. J. Haematol.*, 7:145–151.

1149. White, M. K., Hechtman, H. B., and Shepro, D. (1975): Canine lung uptake of plasma and platelet serotonin. *Microvascular Res.*, 9:230–241.

1150. Wiersma, D. A., and Roth, R. A. (1980): Clearance of 5-hydroxytryptamine by rat lung and liver: The importance of relative perfusion and intrinsic clearance. *J. Pharmacol. Exp. Ther.*, 212:97–102.

1151. Wiklund, L., Leger, L., and Persson, M. (1981): Monoamine cell distribution in the cat brainstem. A fluorescence histochemical study with quantification of indoleaminergic and locus coeruleus cell groups. *J. Comp. Neurol.*, 203:613–647.

1152. Williams, E. D., and Sandler, M. (1963): The classification of carcinoid tumors. *Lancet*, 1:238–239.

1153. Williams, P. (1980): Response of adrenergic agonists and antagonists by collateral arteries from the hind limb of the dog. *J. Pharmacol. Exp. Ther.*, 214:239–245.

1154. Willis, W. D. (1980): Role of serotonin in processing nociceptive information. In: *Serotonin:*

Current Aspects of Neurochemistry and Function, edited by B. Haber, S. Gabay, M. R. Issidores, and S. G. A. Alivisatos, pp. 101–123. Plenum Press, New York.
1155. Wilson, H., Cheek, R. C., Sherman, R. T., and Storer, E. H. (1970): Carcinoid tumors. *Curr. Probl. Surg.*, Nov.:1–51.
1156. Wing, L. M. H., and Chalmers, J. P. (1974): Participation of central serotonergic neurons in the control of the circulation in the rabbit; a study using 5,6-dihydroxytryptamine in experimental neurogenic and renal hypertension. *Circ. Res.*, 35:504–518.
1157. Wingert, T. D., Morley, J. E., and Hershman, J. M. (1980): Histamine antagonists and carcinoid flush. *N. Engl. J. Med.*, 302:234.
1158. Wingren, U., Enerbäck, L., Ahlman, H., Allenmark, S., and Dahlström, A. (1983): Amines of the mucosal mast cell of the gut in normal and nematode-infected rats. *Histochemistry*, 77:145–158.
1159. Winkelmann, R. K., Goldyne, M. E., and Linscheid, R. L. (1976): Hypersensitivity of scleroderma cutaneous vascular smooth muscle to 5-hydroxytryptamine. *Br. J. Dermatol.*, 95:51–56.
1160. Winkelmann, R. K., Goldyne, M. E., and Linscheid, R. L. (1977): Influence of cold on catecholamine response of vascular smooth muscle strips from resistance vessels of scleroderma skin. *Angiology*, 28:330–339.
1161. Winquist, R. J., and Bohr, D. F. (1983): Structural and functional changes in cerebral arteries from spontaneously hypertensive rats. *Hypertension*, 5:292–297.
1162. Wintenitz, M. C., Thomas, R. M., and LeCompete, P. M. (1938): *The Biology of Atherosclerosis*. Charles C Thomas, Springfield, IL.
1163. Wise, R. A., Wigley, F., Newball, H. H., and Stevens, M. B. (1982): The effect of cold exposure on diffusing capacity in patients with Raynaud's phenomenon. *Chest*, 81:695–698.
1164. Wolf, W. A., and Kuhn, D. M. (1984): Antihypertensive effects of L-tryptophan are not mediated by brain serotonin. *Brain Res.*, 295:356–359.
1165. Wolf, W. A., Kuhn, D. M., and Lovenberg, W. (1981): Blood pressure responses to local application of serotonergic agents in the nucleus tractus solitarii. *Eur. J. Pharmacol.*, 69:291–299.
1166. Wolf, W. A., Kuhn, D. M., and Lovenberg, W. (1981): Pressor effects of dorsal raphe stimulation and intrahypothalamic application of serotonin in the spontaneously hypertensive rat. *Brain Res.*, 208:192–197.
1167. Wood, J. D., and Mayer, C. J. (1979): Serotonergic activation of tonic-type enteric neurons in guinea pig small bowel. *J. Neurophysiol.*, 42:582–593.
1168. Wright, C. E., and Angus, J. A. (1983): Haemodynamic response to ketanserin in rabbits with Page hypertension: Comparison with prazosin. *J. Hypertens.*, 1:183–190.
1169. Wu, K. K., and Hoak, J. C. (1974): A new method for the quantitative detection of platelet aggregates in patients with peripheral arterial insufficiency. *Lancet*, 2:924–926.
1170. Yen, C. T., Blum, P. S., and Spath, J. A., Jr. (1983): Control of cardiovascular function by electrical stimulation within the medullary raphe region of the cat. *Exp. Neurol.*, 79:666–679.
1171. Yokoyama, M., Akita, H., Mizutani, T., Fukuzaki, H., and Watanabe, Y. (1983): Hyperreactivity of coronary arterial smooth muscles in response to ergonovine from rabbits with hereditary hyperlipidemia. *Circ. Res.*, 53:63–71.
1172. Youdim, M. B. H., and Ashkenazi, R. (1982): Regulation of 5HT catabolism. *Adv. Biochem. Psychopharmacol.*, 34:35–60.
1173. Yuwiler, A. (1973): Conversion of D- and L-tryptophan to brain serotonin and 5-hydroxyindoleacetic acid and to blood serotonin. *J. Neurochem.*, 20:1099–1109.
1174. Zabludowski, J. R., Zoccali, C., Isles, C. G., Murray, G. D., Robertson, J. I. S., Inglis, G. C., Fraser, R., and Ball, S. G. (1984): Effect of the 5-hydroxy-tryptamine type 2-receptor antagonist, ketanserin, on blood pressure, the renin-angiotensin system and sympatho-adrenal function in patients with essential hypertension. *Br. J. Clin. Pharmacol.*, 17:309–316.
1175. Zahavi, J., Hamilton, W. A. P., O'Reilly, M. J. G., Leyton, J., Cotton, L. T., and Kakkar, V. V. (1980): Plasma exchange and platelet function in Raynaud's phenomenon. *Thromb. Res.*, 19:85–93.
1176. Zahavi, J., Jones, N. A. G., Leyton, J., Dubiel, M., and Kakkar, V. V. (1980): Enhanced *in vivo* platelet release reaction in old healthy individuals. *Thromb. Res.*, 17:329–336.
1177. Zahavi, J., and Kakkar, V. V. (1980): β-Thromboglobulin—A specific marker of *in vivo* platelet release reaction. *Thromb. Haemost.*, 44:23–29.
1178. Zannad, R., Pointel, J. P., Stoltz, J. R., and Gilgenkrantz, J. M. (1984): Effects of ketanserin on

hemorheology and platelet function: *In vivo* studies. Abstracts of the IXth European Congress of Cardiology, Dusseldorf, Germany, July 8–12, 1984.

1179. Zarafonetis, C. J. D., Lorber, S. H., and Hansen, S. M. (1958): Association of functioning carcinoid syndrome and scleroderma. *Am. J. Med. Sci.*, 236:1–14.
1180. Zeegen, R., Rothwell-Jackson, R., and Sandler, M. (1969): Massive hepatic resection for the carcinoid syndrome. *Gut*, 10:617–621.
1181. Zimmerman, T. W., and Binder, A. J. (1984): Serotonin-induced alteration of colonic electrolyte transport in the rat. *Gastroenterology*, 86:310–317.
1182. Zoccali, C., Zabludowski, J. R., Isles, C. G., Murray, G. D., Inglis, G. C., Robertson, J. I. S., Fraser, R., and Ball, S. G. (1983): The effect of a 5-HT antagonist, ketanserin, on blood pressure, the renin-angiotensin system and sympatho-adrenal function in normal man. *Br. J. Clin. Pharmacol.*, 16:305–311.
1183. Zucker, M. B., and Borrelli, J. (1956): Absorption of serotonin (5-hydroxytryptamine) by canine and human platelets. *Am. J. Physiol.*, 186:105–110.
1184. Zwaal, R. F. A., Comfurius, P., and Van Deenen, L. L. M. (1977): Membrane asymmetry and blood coagulation. *Nature (Lond.)*, 268:360–362.

Subject Index